AF762993

Paul Rempp and Edward W. Merrill
Polymer Synthesis

Paul Rempp and Edward W. Merrill

Polymer Synthesis

2nd, revised Edition

Hüthig & Wepf Verlag Basel · Heidelberg · New York

Addresses of Authors
Paul Rempp Institut Ch. Sadron (CNRS/ULP)
6 Rue Boussingault
67083 Strasbourg, France

Edward W. Merrill C. P. Dubbs Professor of Chemical Engineering
Massachusetts Institute of Technology
Cambridge MA, 02139, USA

CIP-Titelaufnahme der Deutschen Bibliothek

Rempp, Paul:
Polymer synthesis / Paul Rempp u. Edward W. Merrill. – 2. rev. ed. – Basel : Heidelberg : New York : Hüthig u. Wepf, 1991
ISBN 3-85739114-6

NE: Merrill, Edward:

9809461

Das Werk ist urheberrechtlich geschützt. Die dadurch begründeten Rechte, insbesondere die der Übersetzung, des Nachdruckes, der Entnahme von Abbildungen, der Funksendung, der Wiedergabe auf photomechanischem oder ähnlichem Wege und der Speicherung in Datenverarbeitungsanlagen bleiben, auch bei nur auszugsweiser Verwertung, vorbehalten.

Bei Vervielfältigung für gewerbliche Zwecke ist gemäß § 54 UrhG eine Vergütung an den Verlag zu zahlen, deren Höhe mit dem Verlag zu vereinbaren ist.

© 1991, Hüthig & Wepf Verlag Basel, Heidelberg, New York
Printed in Germany
Satz: Lichtsatz Michael Glaese GmbH, 6944 Hemsbach

THE UNIVERSITY OF WARWICK LIBRARY
TP 156 .P6 R35

Foreword

Polymer Science is a rather young domain of science, but it has developed very rapidly over the past 50 years. Synthesized polymers are utilized increasingly in our daily life, and manifold industrial applications have contributed to their expansion. They are employed as substitutes for metals, glass, ceramics, wood, cardboard, and paper. Chemical industry devotes a growing part of its activity to the synthesis of monomers and polymers. Following this gigantic industrial development which requires skilled specialists, many universities throughout the world have set up teaching programs in Polymer Science. Their students are drawn from Chemistry, Materials Sciences, Solid State Physics, and the Engineering Disciplines, especially Chemical Engineering.

This textbook on "Polymer Synthesis" describes the major methods of polymerization, with emphasis on the special characteristics of macromolecules that influence the synthesis.

The first part of the book is concerned with the reaction mechanisms involved in polymer synthesis, the kinetic aspects of these reactions, and their implications with respect to molecular weight distribution and molecular structure (Molecular structure refers to such issues as chain defects, branches, stereoregularity, and so on). Many different polymerizations and polymers are described, and the scale of synthesis is that of the laboratory. The second part concerns engineering aspects of polymerization reactors and processes, as used in industry. In contrast to laboratory scale synthesis, which is usually carried out homogeneously and in batches, industrial synthesis seeks heterogeneous processes and continuous processes where possible. In contrast to the myriads of polymers which can be synthesized in the laboratory, very few achieve industrial production, and even at the industrial level the scale of manufacture varies by orders of magnitude according to the value of the product. These are some of the issues considered in the second part of the book.

This book originates from a course in Polymer Synthesis that has been given jointly by the two authors, at the Department of Chemical Engineering of the Massachusetts Institute of Technology, for the past twelve years. Students in this course have come mostly from Chemical Engineering and from Materials Science and Engineering, while a few have come from Chemistry and from Mechanical Engineering. The book has been tested and improved year after year to its present state, and it should prove useful for any student having some background in organic chemistry and chemical kinetics.

The focus of the book is *synthesis*. Physical chemistry, characterization, structure and properties, and other important aspects of polymer science are not addressed except as relevant to synthesis.

Our object has been to provide students with the central concepts and features of polymer synthesis that are commonly accepted, in as compact a form as possible. We have therefore intentionally omitted the historical account of developments in the field, as well as all references. In no way does this imply disrespect for the publications of the founders of macromolecular science, many of them still living, or of the many active workers of today. It is in polymer synthesis that industrial research has often preceded research in the university, or has advanced research to a very high level within itself, before significant activity developed in the university. If we sought to be comprehensive in references, a large number of pioneering patents would also have to be cited.

We would like to express our deep appreciation to Prof. Wei, Head of the Department of Chemical Engineering at M. I. T., for having invited one of us, P. R., to M. I. T. as visiting Professor, and for having encouraged us to write this book as a contribution to teaching Polymer Chemistry.

E. W. MERRILL dedicates his part of this work to ELSIE A. WILSON, superb editor and beloved aunt

Preface to the Second Edition

As it has been conceived initially, the Textbook addresses students in Universities and Engineering Schools to help them acquire a good basic knowledge in Polymer Chemistry. The chief methods of polymer synthesis — polycondensations, free radical and anionic polymerizations, copolymerizations, chemical transformations of polymers — are reviewed, as well as some of the main industrial processes of manufacturing polymers.

As the opportunity was granted to them to revise the book, the authors felt it was of interest to include into the textbook some recent and promising developments in the field of Polymer Synthesis. It is necessary that teaching keeps up with the progress of Science.

The authors have also done their best to excise the mistakes and misprints which had been left in the first edition, and to improve the style whenever necessary.

The introductory chapter has been shortened, as the discussion of the most appropriate industrial processes is now located in chapters 11 to 13.

No major changes have been performed in the chapters dealing with polycondensations and with free radical polymerizations, except a short paragraph on the recently developed "Iniferter" polymerization methods.

The scope of chapter 7 has been widened: It is now devoted to a general review of all polymerization processes involving ions or dipoles as the active sites, including the so-called "activated monomer" polymerizations. Major changes have also been introduced into chapter 8 which now contains a paragraph on the newly developed metathesis polymerizations. Chapters 9 and 10 have also been complemented.

On the other hand, some lengthy demonstrations of the initial edition have been removed: The authors felt that they were not of utmost interest to students. Examples of such cuts are the detailed calculation of the molecular weight distribution in AA + BB polycondensates, and the discussion of all possible reactions in cationic vinylic polymerizations.

In the second part of the textbook reactors and processes for synthesis of industrially important polymers are considered in a revised format (in chapter 11), emphasizing broadly applicable concepts, starting with the inter-relationship of polymer structure, chemistry of synthesis, and process engineering. These are illustrated by the exercise of following, and answering the questions sequentially posed in the form of a *decision tree*.

Following the chapter dealing with these basic concepts, chapter 12 presents an account of emulsion and suspension polymerizations, two processes both having water as the continuous phase, both producing polymer in a dispersed phase by radical propagation, but differing completely in most other respects, with comments on the reasons for choosing one or the other process.

The final chapter (13) is almost completely new and under the title: "Heterogeneous polymerizations and interfacial reactions", it takes up, in turn, industrially important examples illustrating:

— polymerizations without solvent leading to precipitated polymer
— polymerizations in a solvent which causes precipitation of the polymer
— polymerizations with particulate stereoregulating catalysts
— polymerizations at the interface between immiscible phases
— reactions on polymers within interfacial zones.

Two appendices complete the text. Appendix A shows six cases in which the shape of the reactor becomes a central issue in that it defines the shape of the polymer product. Appendix B outlines the high pressure free radical synthesis of polyethylene, still useful because it is the sole route to the valuable copolymers of ethylene with vinylacetate, ethylene with an acrylate, ethylene with carbon monoxide, the latter being biodegradable.

A number of new problems pertinent to the respective chapters have been added with solutions. General references as a short bibliography are given at the end.

Table of Contents

Chapter 3: Free Radical Chain Polymerization

Chapter 4: Free Radical Copolymerization

Chapter 5: Anionic Polymerizations

Chapter 6: Cationic Polymerizations

Chapter 7: Special Features of Ionic and Pseudo-Ionic Polymerizations

Chapter 8: Polymerizations Involving Stereoregulation

Chapter 9: Chemical Reactions on Polymers

Chapter 10: Functional Polymers, Block and Graft Copolymers, Model Networks

Part II: Reactors and Processes for Manufacturing Polymers

Chapter 11: Interrelationships of Polymer Structure, Chemistry of Synthesis, and Process Engineering

Chapter 12: Emulsion and Suspension Polymerization

List of Symbols

$\underline{a}$, $\underline{b}$	Chemical functions which upon reaction yield a -$\underline{c}$-linkage with or without release of a $\underline{d}$ molecule (chapter 2)
$\underline{aa}$, $\underline{bb}$, $\underline{ab}$	Bifunctional monomers in a polycondensation process
a	Factor characterizing the mode of termination in a free radical polymerization (chapter 3)
a_S	Area of packed monolayer of soap molecules (chapter 12)
A, B	Refer to monomers in a copolymerization process (chapter 4)
c	Concentration (by weight)
c_i	Concentration of species exhibiting a degree of polymerization i in a polydisperse polymer sample
C, AH	Refer to catalyst-cocatalyst systems in cationic polymerizations
C_p	Heat capacity
C_T	Transfer constant (chapter 3)
C_S, C_I, C_M, C_{Pol}	Transfer constants to solvent, initiator, monomer, polymer
D_{mp}	Diffusion constant of a monomer molecule in a polymer matrix
D_{pp}	Diffusion constant of a polymer molecule in a polymer matrix
E_p, E_t	Activation energies characterizing the propagation and the termination reactions
E_d	Activation energy characterizing the homolytic decomposition of the free radical initiator
E_R	Apparent activation energy for the overall rate of polymerization
E_{DP}	Apparent activation energy for the DP of the polymer formed
f	Functionality of a crosslinking agent
f	Initiator efficiency factor (chapter 3)
f_a, f_b	Instantaneous A (or B) content of the monomer mixture in a copolymerization process (chapter 4)
f_n	Fraction of units left over after a cyclization reaction involving neighbouring monomer units in a sequence of $\underline{n}$ such units (chapter 9)
F_a, F_b	Instantaneous A (or B) content of the copolymer formed
$\langle F_a \rangle$	Integrated (experimental) A content of a copolymer formed to finite conversion
F, S	Fast and slow reacting functions in the chemical modification of a polymer (chapter 9)
$\dot{G}$	Mass flow rate
GPC	Gel permeation chromatography, also called size exclusion chromatography (SEC)
$\langle h^2 \rangle$	Mean square end-to-end distance of a polymer chain

h_i	GPC refractometric response to a polymer species *i* $\left(\sim c_i \left(\frac{dn}{dc}\right)_{i,\lambda,T}\right)$; $\left(\frac{dn}{dc}\right)_{i,\lambda,T}$ is the refractive index increment ν at temperature *T* and at wavelength λ of the incident radiation
h, i, s	Proportions of heterotactic, isotactic and syndiotactic triads in a polymer sample
i (as subscript)	Refers to a macromolecule constituted of *i* monomer units
i	Degree of polymerization of the above macromolecule
I	Refers to the initiating species (initiator)
k	Rate constant characterizing the reaction of functions -$\underline{a}$ and -$\underline{b}$ to yield a -$\underline{c}$ link (chapter 2)
k_d	Rate constant of homolytic dissociation of a free radical initiator (chapter 3)
$k_{d,r}$	Rate constant of dissociation of a redox pair (chapter 12)
k_i, k_p, k_t	Rate constants of initiation, of propagation and of termination
k_0, k_1, k_2	Rate constants characterizing a chemical reaction on a polymer (chapter 9)
k_{tr}	Rate constant of transfer
k_{aa}, k_{ab}	Rate constants of reaction of an $A^{\bullet}$ radical with an A or a B monomer molecule (chapter 4)
$k_{p\pm}$, k_{p-}	Rate constants of propagation of ion-pairs and of free anion (chapter 5)
$k_{p//}$, k_{pc}	Rate constants of propagation of solvent separated ion-pairs and of contact ion-pairs
k_r, k_s	Rate constants of propagation involving the enantiomeric monomers R and S (chapter 5)
k_u	Rate constant of depropagation (unzipping) (chapter 7)
k_f, k_s	Rate constants involving "fast" and "slow" units with a reagent in the chemical transformation of a polymer
K	Equilibrium constant in a reversible reaction
K_D, K_ϕ	Electrolytic dissociation constants
K_e	Equilibrium constant of aggregation
K_{Al}, K_M	Equilibrium constants of adsorption onto the catalyst surface (chapter 8)
l	Distance between anion and cation
l, d	Optical isomers: *d* (dextro), *l* (laevo), *dl* (racemic mixture)
m, r	Meso and racemic diads
m_0	Molecular weight of a monomer unit (or of a base unit)
M	Refers to the monomer
M^*, $M^{\bullet}$, $M^{\oplus}$, $M^{\ominus}$	Refer to monomer units carrying an active site
M_i^*	Refers to a polymer molecule with *i* monomer units and carrying an active site at chain end
M_i	Molecular weight of a polymer species with *i* monomer units
$\overline{M}_n$	Number average molecular weight (chapter 1)

$\overline{M}_w$	Weight average molecular weight (chapter 1)
n	Refractive index, chiefly used in the refractive index increment $dn/dc = v$
n	Number of particles per unit volume in emulsion polymerization (chapter 12)
n_1	Refractive index of pure solvent
n^*	Number of micelles per unit volume in emulsion polymerization (chapter 12)
n_i	Number of molecules of DP $= i$ in a polymer sample (chapter 1)
N_a, N_b	Number of $\underline{a}$ and $\underline{b}$ functions available at a given time in a polycondensation process
N_A	Avogadro's number
$\dot{p}$	Rate of production of polymer (chapter 14)
p_a, p_b	Probability for a given function $\underline{a}$ or $\underline{b}$ to have undergone reaction at time t in a polycondensation (chapter 2)
P	Probability for a given polymeric radical to add a monomer molecule rather than to transfer or terminate (chapter 3)
p_{aa}, p_{ab}	Probability for an $A^\bullet$ radical to react with an A or with a B monomer molecule. It is also the probability to find AA or AB diads in the macromolecule formed (chapter 4)
p_i, p_s	Probability of forming isotactic or syndiotactic diads
P_{ab}	Probability to find an AB diad at a given location on the copolymer chain
P_{aab}	Probability to find an AAB triad at a given location. It is also the proportion of such triads in the copolymer
P_{alt}	Probability of alternation
P_i	Probability of formation of an i-mer (chapter 2)
$\overline{P}_n$	Number average degree of polymerization (chapter 1)
$\overline{P}_w$	Weight average degree of polymerization (chapter 1)
Q, e	Parameters of the Q-e-scheme (chapter 4)
Q_+, Q_-	Electric charge of cation and anion
$\dot{Q}$	Rate of heat removal or addition per unit volume. Dots refer to derivatives with respect to time ($\dot{Q} = dQ/dt$)
r	See *m, r* (chapter 1)
r	Particle radius in emulsion polymerization (chapter 12)
r	Stoichiometric ratio in a polycondensation (chapter 2)
R, S	Refer to monomeric enantiomers (chapter 5)
r_a, r_b	Reactivity ratios of radicals involved in a binary A-B copolymerization (chapter 4)
R	Refers to reaction rates
R_p	Rate of propagation (i. e. of polymerization)
R_i, R_t	Rates of initiation and termination
Re	Reynolds number
S	See *R, S*

s	See *i, h, s*
S	See *F, S*
S_n	Average number of units left over upon cyclization involving neighbouring units in a sequence of *n* units
SEC	See GPC
t	Refers always to the time
T	Refers always to the temperature (whether in *K* or in °C)
T_G	Glass transition temperature
T_M	Melting temperature of a crystalline polymer
T_c	Critical temperature, especially "ceiling" temperature in reversible polymerizations (chapter 7)
υ	Specific volume of the monomer (chapter 12)
$\bar{\upsilon}$	Average flow velocity
V	Molar volume [ml/gmol]
V_t	Volume of a stirred tank reactor
V_e	Elution volume in GPC
$\dot{V}$	Volumetric flow rate
$\overline{V}$	Average Velocity
w_i	Weight fraction of species exhibiting a polymerization degree *i* in a polymer sample
x_i	Mole fraction of molecules exhibiting a polymerization degree *i* in a polymer samples
x	Fraction of fast reacting functions (chapter 9)
y	Fraction of ion-pairs that are solvent separated in anionic polymerization
α	Optical activity of an enantiomeric monomer mixture
$\bar{\alpha}$	Probability for a branching unit to be connected to another one, regardless of the length of the connecting chain, in a polycondensation process involving multifunctional units (chapter 2)
α_c	Critical value of α at which gelation occurs (chapter 2)
α, α_ϕ	Fraction of active sites present as free ions in anionic polymerization (degree of dissociation)
ΔM	Monomer consumed (difference between $[M]_0$ and $[M]$)
ΔG, ΔH, ΔS	Free enthalpy, enthalpy and entropy of polymerization
ε	Dielectric constant
η	Viscosity (of the reaction medium)
θ_{AL}, θ_m	Fractions of the active surface of the transition metal compound coated with AIR_3 and with monomer (chapter 8)
ν	Kinetic chain length (chapter 3)
$\bar{\nu}$	Average number of growth steps per "living" chain (chapter 5)
ν	Refractive index increment (chapter 1), at given temperature *T*, wavelength λ, fixed solvent and polymer type *i*, defined by $\nu = \left(\frac{dn}{dc}\right)_{i, \lambda, T}$

π	Extent of reaction, degree of conversion of a polymerization reaction
π_a, π_b	Extent of reaction of functions $\underline{a}$ and $\underline{b}$ at time t in a polycondensation reaction
π_c	Critical degree of conversion at which gelation occurs in a polycondensation involving multifunctional base molecules
ρ	Proportion of $\underline{a}$ functions belonging to multifunctional units (chapter 2)
ρ	Ratio of the propagation constants k_r and k_s for the enantiomers R and S (chapter 5)
ρ	Ratio of the reaction rate constants k_f and k_s (chapter 9)
ρ	Density
ρ	Rate of production of radicals per unit volume in an emulsion polymerization (chapter 12)
σ	Interfacial tension between water and a polymer particle
τ	Mean life time of a polymeric radical (chapter 3)
τ	Mean residence time ($= V_t/\dot{V}$) in a stirred tank reactor (chapter 11)
Φ_2	Volume fraction of polymer in a reaction medium
Φ	Volume fraction of organic phase in an emulsion polymerization (chapter 12)
χ	Flory-Huggins interaction parameter (polymer-solvent)
[ ]	Brackets are used systematically for molar concentrations of reagents (monomers, initiators, growing radicals, transfer agents), whenever the reaction medium is homogeneous. Dimensions are also included in brackets
$[\ ]_0$	Refers to initial molar concentrations
$[\ ]_e$	Refers to equilibrium concentrations

Part I
Polymerization Reactions

Chapter 1
General Considerations about Polymers and Polymer Synthesis

1.1 Introduction

Polymer chemistry is not just a branch of organic chemistry. It has developed its own methods of synthesis, and its own methods of characterization. Polymer Science and Technology have gained major economic importance as the various uses of plastics, fibers, rubbers, composites, coatings, and all kinds of synthetic materials, in our daily life as well as in industry, have become one of the major characteristics of the second half of this century.

Polymeric compounds exhibit a number of specific features:

- Their molecular weight is high, ranging from a few thousands to several millions (or even infinity in crosslinked polymeric materials).
- They are constituted of repeat units (or *mers*) generally named after the low molecular weight constituents (*monomers*) from which they originate. For instance polystyrene is a chain polymer in which the repeat units originate from styrene.
- Polymer synthesis always involves a large number of steps, each of them resulting in the formation of a bond, and implying extension of the chain.
- Polymers are usually polydisperse, which means that macromolecules exhibiting the same chemical composition but different chain lengths coexist within a sample.

The aim of this book is to present the various methods of polymer synthesis, and to show how polymerization kinetics and mechanisms can determine the structure, molecular weight, polydispersity, and consequently the properties of the polymers to be formed.

1.2 Polymerization Reactions

The chemical processes which result in linking together small molecules (monomers), to build a polymer molecule (macromolecule), are called *polymerization reactions*. It is self evident that monomer molecules must be bifunctional*) to allow linkages to be formed with the neighboring mers in the polymer chain. A molecule exhibiting only one reactive site cannot be incorporated in the chain. If a monomer molecule possesses more than two potential sites to form linkages, it may become a branch point, and eventually lead to crosslinking between macromolecules: A polymeric network would result.

Two major types of polymerization reactions have to be distinguished: *step growth reactions* and *chain growth reactions.*

1.2.1 Step Growth Polymerizations

A step growth polymerization, also referred to as *polycondensation* is a straightforward application of a well known reaction of organic chemistry to the synthesis of a polymer. The *monomers* are small molecules carrying antagonist functional sites **a** and **b**, capable of reacting together to build a linkage **c**. Any **a** function can react with any **b** function, regardless of whether they belong to a monomer molecule or whether they are located at chain end of an already existing polymer molecule. For instance, a molecule constituted of i base units (mers) and fitted with an **a** function at chain end can react with a molecule exhibiting j mers and a **b** function, to form a polymer with (i + j) units, still fitted at both chain ends with **a** or **b** functions. Each individual reaction of this type results in a decrease of the total number of molecules by one. The formation of a **c** link from **a** and **b** functions does not depend on similar events having taken place previously.

Typical examples of polymers arising from polycondensation reactions include:

— polyesters, from a diacid and a dialcohol
— polyurethanes, from a diisocyanate and a dialcohol
— polyamides, from a diamine and a diacid.

Step polymerizations will be considered in detail in chapter 2.

1.2.2 Chain Growth Polymerizations

Chain polymerizations are specific to polymer science. Monomers susceptible to such reactions include unsaturated compounds of various kinds (vinylic, acrylic, dienic compounds) or strained heterocyclics which may undergo ring opening.

The basic feature of a chain polymerization is the existence of an active site which is capable of attacking a monomer molecule to have it linked to the chain, whereby the active site is restored at chain end.

*) An unsaturation is obviously bifunctional, and vinylic and acrylic monomers yield linear polymer molecules. In conjugated dienes (butadiene, isoprene) the reactivity of both double bonds is interrelated, and in most cases such monomers would be considered bifunctional. Heterocyclic compounds are potentially bifunctional, as well, upon ring opening reactions. Unconjugated dienes, or compounds such as divinylbenzene in which the two unsaturations react independently, are tetrafunctional molecules, yielding branching and crosslinking.

Several events have to be considered in a chain polymerization process:

— Initiation, which results in formation of an active site on a monomer molecule.
— Chain growth (or propagation) comprises a large number of steps. Each of them implies the addition of a monomer molecule to an active site. This can occur by opening of the double bond or by ring opening. As a result, the monomer is incorporated into the chain and the site is restored at chain end. (n − 1) growth steps are required to build a n-mer (polymer constituted of n base units).
— Termination involves the destruction of the site whereby the growth of the chain is stopped.
— Transfer reactions imply removal of the active site from a macromolecule, which stops growing. The site being transferred to another molecule can start a new polymer molecule. In other words, the "kinetic chain" is not interrupted: the site still exists. The consequence is that one active site can give rise to several chain molecules, instead of a single one if there were no transfer.

A chain molecule involves formation of high molecular weight polymer at early stages of the reaction, i.e. for low degrees of conversion. The situation is totally different in step growth polymerizations.

The most frequently encountered active sites are:

— Unpaired electrons in free radical polymerizations (described in chapter 3)
— Anions (carbon-metal or alkoxide bonds) in anionic polymerizations (chapter 5)
— Cations as carbenium or oxonium ions, (with their counterions) in cationic polymerizations (chapter 6)
— Coordination bonds with a transition metal in Ziegler-Natta type polymerizations (chapter 8)

1.2.3 Thermodynamics

Polymerizations involve thermodynamic considerations in two major ways:

(1) The limiting conversion of monomer corresponding to *equilibrium*, under process conditions (temperature, pressure, concentrations of reaction species).

(2) The total heat to be exchanged and, in combination with kinetic expressions, *the rate of heat exchange.*

Many, if not most, polymerizations are accomplished under conditions far from equilibrium. In any case polymerization is thermodynamically possible, if the free energy ΔG of the process Monomer → Polymer is negative:

$$\Delta G = \Delta H - \mathrm{T}\Delta S < 0$$

Since the entropy change ΔS is usually negative (monomer units having less degrees of freedom after being tied up in polymer chains than single monomer molecules), it follows that ΔH, the enthalpy change upon polymerization, must be negative.

Therefore heat must be removed from the process of polymerization. The rate of heat release will be of major consideration in "fast" polymerizations like those of vinyl chloride, of ethylene under high pressure, and of isobutene, and it is of little consequence in slow processes like the melt polymerization of bis(hydroxyethyl)terephthalate.

In general, the rate of heat release $\dot{Q}$ per unit volume will be given by

$$\dot{Q} = R_p \Delta H$$

where R_p is the rate of conversion of monomer, given in $[mol \cdot l^{-1} \cdot s^{-1}]$, and ΔH is the enthalpy of polymerization, given in $[kJ/mol]$. Thus, kinetic expressions by which R_p can be estimated will be important.

The melt polymerization leading to nylon-6,6 or the cationic living polymerization of heterocycles, allowed to proceed over many hours, are examples where the final composition is near or at equilibrium. Relevant expressions will be presented in connection with these polymers.

1.3 Structural Characteristics of Polymers: Shape and Size

The most obvious feature of a polymer species is the shape of its molecules. They can be *linear**), *branched, ladder-like, crosslinked,* as shown on figure 1.1.

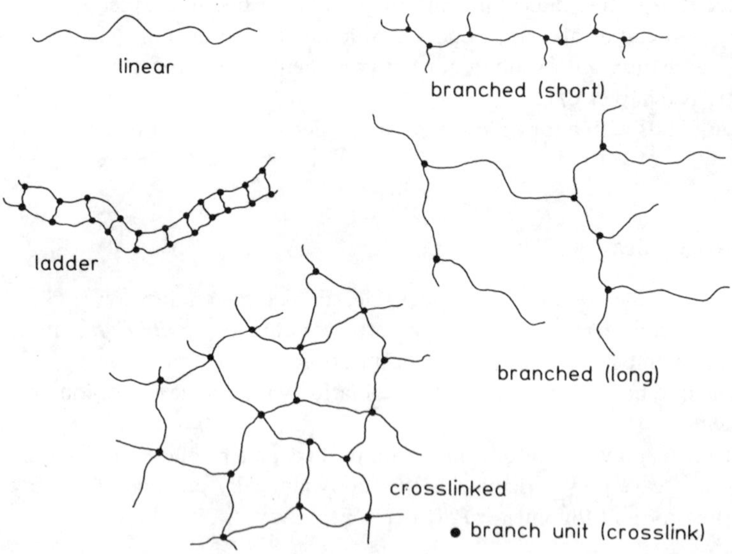

Fig. 1.1. Various types of polymer molecules.

Linear macromolecules are usually randomly coiled, unless they exhibit a strong tendency to crystallize. They then form crystallites of various sizes, implying the chains to fold several times. The crystallites are connected by parts of chains that have not

*) Though "linear" is the adjective most commonly used, almost any other word could be less confusing: stringlike, chain of beads, neither of which implies a straight line. Very rarely do linear macromolecules assume a straight line conformation.

undergone crystallization: Even in the most favorable cases, a polymer will never be crystallized to 100%. In amorphous bulk polymers the macromolecules are largely interpenetrated, as a single random coiled macromolecule occupy a volume of space much larger than the sum of the van der Waals volumes of their constituent monomer units.

Crosslinked macromolecules cannot dissolve in a solvent though they may swell to some extent, depending upon the degree of crosslinking. *Thermoset polymers* are so tightly crosslinked that they cannot be swollen, nor melted, and they retain the shape of the molds in which they were manufactured. Upon building bridges between initially linear macromolecules crosslinking results, as soon as each of the initial macromolecules is connected chemically to more than two others. This bridge-forming reaction (also referred to as vulcanization) leads to species that are swellable, deformable, but which exhibit a marked tendency to retain their shape. These are the chief characteristics of *elastomers.*

1.3.1 Placement of the Monomer Units and Stereoregularity

In the case of linear macromolecules made of one type of repeat units (*mers*), generally referred to as *homopolymers,* the placement of these monomer units is generally *head-to-tail* (if the monomer is vinylic or acrylic), which implies that every other carbon atom of the chain carries a substituent.

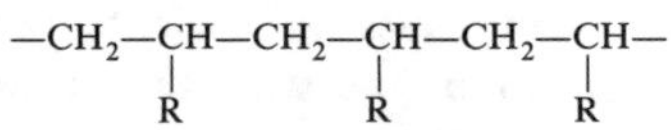
head-to-tail

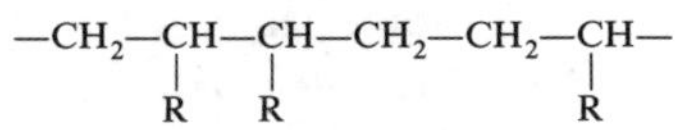
head-to-head

Head-to-head placement can however occur in some instances, especially with monomers such as vinylidene chloride, vinylidene fluoride, and vinylacetate, even though the probability for such "wrong" placement is always low.

1.3.1.1 Tacticity of Vinyl Polymers

Along a polyvinylic chain each substituted carbon atom is a potential center of asymmetry and there are two possible placements of the substituent with respect to the chain. Obviously the site of steric isomerism does not pre-exist in the vinylic monomer, but it is formed upon its insertion into the polymer chain. In most polymerizations there is no steric control of the way monomers add to the chain and the polymer is called *atactic.* Each of the two possible placements of the substituent R — let us call them *l* and *d* — are present in equal amounts, and distributed at random along the chain.

Polymerizations are called *stereospecific* when they provide control of the way the double bond of the incoming monomer is opened, and the corresponding polymers are referred to as *stereoregular.* If all asymmetric carbon atoms along the chain have the same steric structure, the polymer is called *isotactic.* The substituents of such a polyvinylic chain — as looked at in the extended planar zigzag form — are all located on the same side of the plane defined by the chain itself. If there is a regular alternation in the placement of the substituents at every second carbon atom (in front or behind the plane of the extended chain) the polymer is called *syndiotactic.* This can be schematized as shown below:

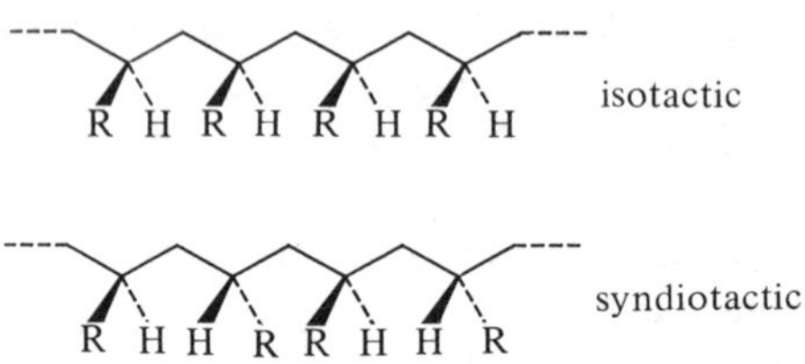

Natta's projection
(Planar zig–zag)

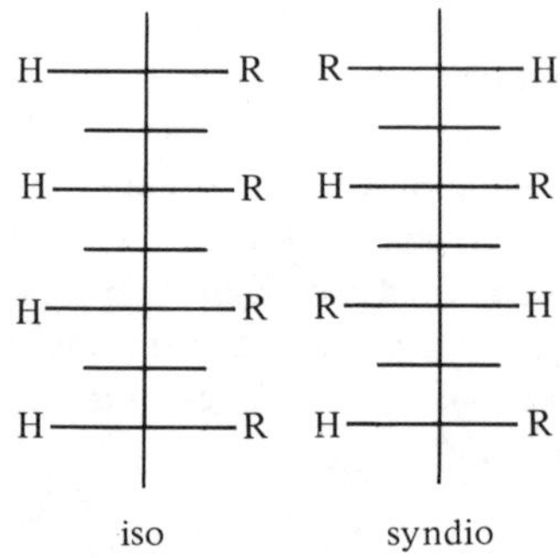

Fischer's projection

Some important remarks should be made at this point:

1. Isotactic vinyl polymers do not exhibit any optical activity, because the neighboring methylenes of a substituted carbon atom are identical. However, the isotactic polymers differ appreciably from atactic homologues in their tendency to crystallize. The steric regularity of the placement of the substituents makes it possible for the chain to fit into a crystal lattice. The ability of a polymer to crystallize generally improves noticeably its thermomechanical behavior.
2. It is not possible to decide whether a given substituted carbon atom along the chain is *l* or *d* in its steric structure. However, one can establish whether two successive substituted carbon atoms along the chain exhibit the same steric structure or whether they are optically inverse. In the first case the diad is called *meso (m)*, and in the second case it is referred to as *racemic (r)*. In an isotactic polymer the fraction of meso diads is close to unity; in a syndiotactic polymer it is the fraction of racemic diads that is close to unity, and in an atactic polymer both types of diads are found in equal amounts.
3. Even more interesting is the distribution of triads along the chain, experimentally accessible by *nuclear magnetic resonance* (NMR) and in some cases by *infrared spectroscopy* (IR). An *isotactic triad i* consists of two successive meso diads, a *syndiotactic triad s* consists of two successive racemic diads and a triad involving a meso and a racemic diad (or vice versa) is referred to as *heterotactic h.* Thus, remembering that the fraction of a given type of diad in a polymer chain is also its probability of occurrence, one can write the following equations:

$$i = m^2$$
$$s = r^2$$
$$h = 2\,mr$$

provided the placement of the substituents obeys Bernouillian statistics (i. e. is determined by one single parameter).

1.3.1.2 Stereoregularity of Diene Polymers

In the case of polydienes several possibilities of addition of the monomer onto the growing chain have to be considered. In the case of isoprene, for instance, the polymerization can involve only one double bond (1,2-addition, or 3,4-addition), or it can involve both together (1,4-addition). In the former cases the stereo placement of the substituent in 1,2- or 3,4-units can be controlled in some instances. In the latter case, the double bonds in the chain also involve stereoisomerism: They can be 1,4-cis or 1,4-trans, the properties of the corresponding polymers being widely different.

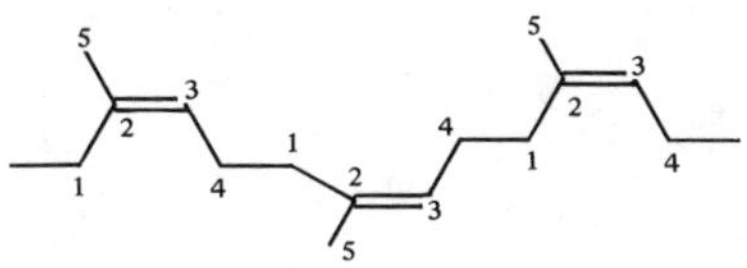

cis–1,4–polyisoprene (natural rubber)

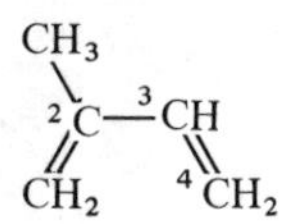

monomer isoprene

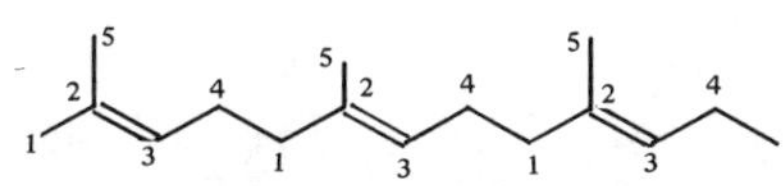

trans–1,4–poly(isoprene) (Gutta Percha)

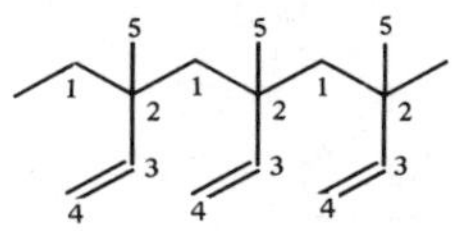

1,2–poly(isoprene)

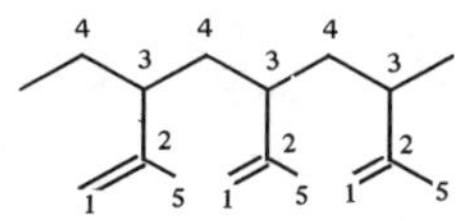

3,4–poly(isoprene)

For each polymerization system investigated the probabilities of each type of addition along the chain have to be determined. *Stereo-regulating systems* will tend to privilege one type of addition with respect to the others. This is of major importance for the applications of diene polymers; for instance, high 1,4-unit contents are required for polybutadiene or polyisoprene to exhibit rubber-like properties, which must be preferably cis-1,4, or at least mixed cis and trans, but not all 1,4-trans which leads to high melting points.

1.3.1.3 Stereoregularity of Polymers Resulting from Ring Opening Polymerization

In ring opening polymerization of *substituted heterocycles*, such as propylene oxide, the asymmetric carbon atom pre-exists in the monomer, and the process usually does not involve racemization: The steric structure of the substituted carbon atom is not affected: The polymerization of a pure enantiomer (optically active *l* or *d* monomer) therefore obviously results in an isotactic polymer.

If a racemic mixture of *l* and *d* monomers is treated with an adequate polymerization initiator, three cases have to be distinguished*).

(A) Both enantiomers take part in the process. They copolymerize randomly, and an *atactic polymer* is formed.
(B) The growing sites exhibit a preference for one of the enantiomers (say *d*) and the polymer contains more than 50% of such units; consequently the remaining unpolymerized monomer is enriched in the other enantiomer, and it exhibits rotatory power. This type of polymerization is called *stereoelective;* however, the degree of stereoelection is usually low (see chapter 5).
(C) Both optical isomers polymerize at the same rate, but *d* molecules are incorporated preferentially into polymer chains already containing *d* units, and *l* molecules into chains containing *l* units. The polymer is optically inactive though it contains isotactic chains (poly *l*-chains and poly *d*-chains being present in equal amounts). This type of polymerization is called *stereoselective.*

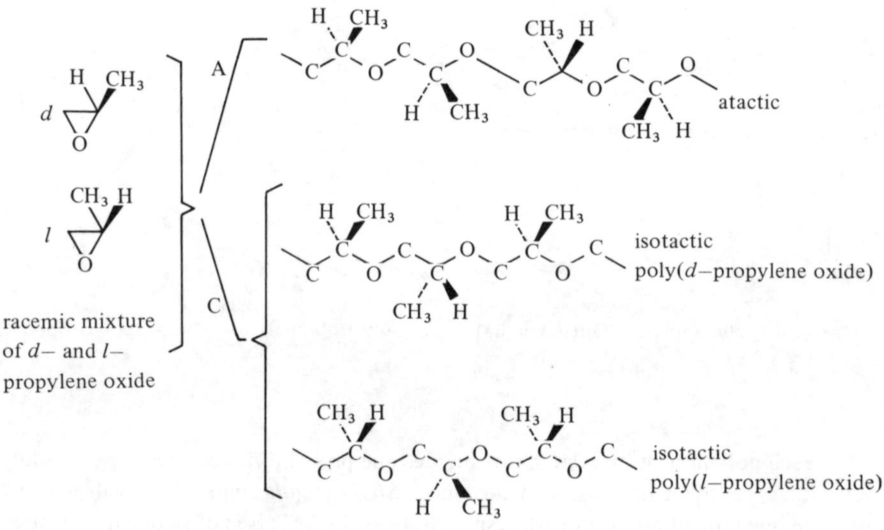

1.3.2 Homopolymers and Copolymers

Homopolymeric macromolecules contain only one single type of structural units in their chains. Polyethylene, polystyrene, poly(vinyl chloride) are homopolymers. Polycondensates originating from one single diacid-dialcohol system are also usually quoted as homopolymers. In this case, however, the repeat unit is constituted of two mers (base units), one arising from the diacid, the other from the dialcohol.

---

*) Similar cases have been encountered in vinyl monomers exhibiting an asymmetric carbon atom denoted by the asterisk in their substituent, such as 3-methyl pentene

$CH_2{=}CH{-}{}^*CH\langle^{CH_3}_{C_2H_5}$, the polymerization of which can involve some stereoselection.

$$[\text{—O—(CH}_2)_m\text{—O—} \mid \text{—CO—(CH}_2)_p\text{—CO]}_n$$

Dialcohol | Diacid

Numerous other polymer species contain two or more types of units. They are referred to as *copolymers*. Several types of copolymers have to be distinguished, to take account of the placement of the units along the chain, and of their molecular structure (linear or branched), as shown on the scheme below:

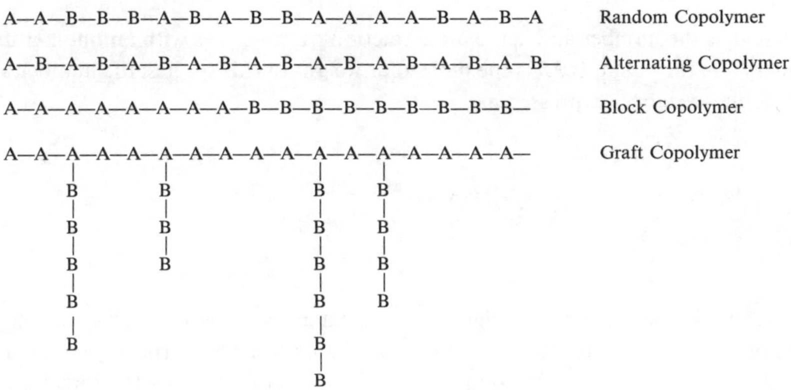

In random copolymers the two types of mers, A and B, are distributed statistically along the chain.

In alternating copolymers the A and B units alternate along the chain. This type of copolymer can be considered in fact as a homopolymer with (AB) repeat units.

Block copolymer molecules are made of a small number (usually 2 or 3) of blocks (or sequences), each of which being homopolymeric — poly A and poly B — which are linked together. The length of the individual blocks can range from a few units to several thousands. Block copolymers have gained much interest in recent years because of the properties they exhibit. Their synthesis will be discussed in chapter 10.

In graft copolymers, homopolymeric sequences of poly B are attached as side chains to a homopolymeric backbone of poly A. Grafting processes will be reviewed in chapter 10.

1.4 Molecular Weight Distribution (MWD) and Molecular Weight Averages

In classical organic chemistry a pure compound is defined by its chemical composition, its molecular structure, and its molecular weight. In a polymer sample, on the contrary, the molecular weights of the individual macromolecules can vary widely, though the composition remains the same for all molecules, at least within a homopolymer sample. It is merely the number of *"mers"* (structural units) that varies from one molecule to another. Therefore one should not speak of molecular weight (or degree of polymerization)*), but rather of *average molecular weight*, and it should be specified what average is being considered.

The *number average degree of polymerization* $\overline{P}_n$ is defined as:

$$\overline{P}_n = \frac{\sum_{i=1}^{\infty} i\, n_i}{\sum_{i=1}^{\infty} n_i} = \sum_{i=1}^{\infty} i x_i$$

where n_i is the number and x_i the mole fraction of molecules with i monomer units**) ("mers") in the chain. If M_i is the molecular weight of this species, the number average molecular weight is expressed as:

$$\overline{M}_n = \frac{\sum_{i=1}^{i=\infty} n_i M_i}{\sum_{i=1}^{i=\infty} n_i} = \sum_{i=1}^{i=\infty} x_i M_i = m_0 \sum_{i=1}^{i=\infty} i x_i$$

where m_0 is the molecular weight of a repeat unit, here assumed constant (for a homopolymer) or as the average of two constants (for an alternating copolymer).

Thus $\overline{M}_n$ is defined as the total weight of polymer divided by the total number of polymer molecules in the sample. $\overline{M}_n$ can be determined by *osmotic measurements*, or by *end-group analysis* if the polymer is fitted with a well defined function at chain end, and provided the $\overline{M}_n$ values are not too high.

The *weight average degree of polymerization* $\overline{P}_w$ is defined as:

$$\overline{P}_w = \frac{\sum_{i=1}^{i=\infty} W_i\, i}{\sum_{i=1}^{i=\infty} W_i} = \sum_{i=1}^{i=\infty} w_i\, i = \frac{\sum_{i=1}^{i=\infty} n_i\, i^2}{\sum_{i=1}^{i=\infty} n_i\, i}$$

where W_i is the weight and w_i is the weight fraction of macromolecules exhibiting a polymerization degree i. The weight fraction w_i is given by:

$$w_i = \frac{W_i}{\sum_{i=1}^{i=\infty} W_i} = \frac{n_i M_i}{\sum_{i=1}^{i=\infty} n_i M_i} = \frac{i n_i}{\sum_{i=1}^{\infty} i n_i}$$

Consequently, the weight average molecular weight $\overline{M}_w$ is expressed as:

*) The degree of polymerization (DP) is defined as the number of repeat units per molecule.
**) Also referred to as structural units, or base units.

$$\overline{M}_w = \overline{P}_w \cdot m_0 = \sum_{i=1}^{i=\infty} w_i M_i = \frac{\sum_{i=1}^{i=\infty} n_i M_i^2}{\sum_{i=1}^{i=\infty} n_i M_i}$$

Weight average molecular weights can be measured by light scattering.

The width of the molecular weight distribution within a sample can be characterized roughly by the ratio of its two moments (molecular weight averages) M_w and M_n: For example, the light scattering value divided by the osmotic value. This ratio, called polydispersity index*), is equal to 1 for a "monodisperse" polymer, in which all molecules would be strictly identical in number of units. No synthetic polymer is monodisperse, though**). M_w is found to be higher than M_n as soon as there are fluctuations in molecular weight within a sample, consequently their ratio is found higher than unity. The larger the value of the polydispersity index, the broader the molecular weight distribution.

Useful as it is, the polydispersity index does not carry information on the shape of the molecular weight distribution. For example, it will not reveal the bimodal distribution of the sample shown in Fig. 1.2c, nor would it indicate the tail of the distribution shown in Fig. 1.2d.

Access to the actual molecular weight distribution is provided by various fractionation techniques, or by sedimentation experiments. A very versatile technique is now commonly used to characterize, upon calibration, the molecular weight distribution within a polymer sample: Size exclusion chromatography, also called gel permeation chromatography (GPC). Fig. 1.2 displays four examples of such chromatograms.

GPC is based upon the fact that chromatographic retention in the column is determined by the hydrodynamic volume of the solute molecules. The higher the molecular weight, the smaller the elution volume. By adequate calibration with well defined polymer samples the elution volume V_e can be related to the molecular weight of the polymer eluted (Fig. 1.3).

Detection of the outcoming polymer can be achieved in various ways (spectroscopic, light scattering), but the method most commonly used is differential refractometry. The

---

*) Note that $\frac{\overline{P}_w}{\overline{P}_n} = \frac{\overline{M}_w}{\overline{M}_n}$ if the molecular weight of all units is identical, or if a clear average can be obtained, as in alternating copolymers. The definition has to be carefully reconsidered for block copolymers, graft copolymers, and even random copolymers, when the weights of the A and B units are not equal, as is usually the case. The *osmotic experiment* will still yield $\overline{M}_n$ absolutely. *Light scattering* and SEC depend on the knowledge of dn/dc. If dn/dc varies with the mole ratio of A to B, characterization of the whole polymer by GPC may be affected by changes in composition with size; in that case the refractive index difference is no longer a measure of polymer concentration, unless the composition can be independently tracked for each aliquot. By the same line of reasoning, light scattering cannot yield $\overline{M}_w$ unless at the same time the correct average dn/dc is known.

**) In a step-by-step synthesis of oligopeptides (for instance by the Merrifield method), all chains, though short, have the same number of, and the same sequence of aminoacid residues. However this synthesis is not a polymerization as defined above.

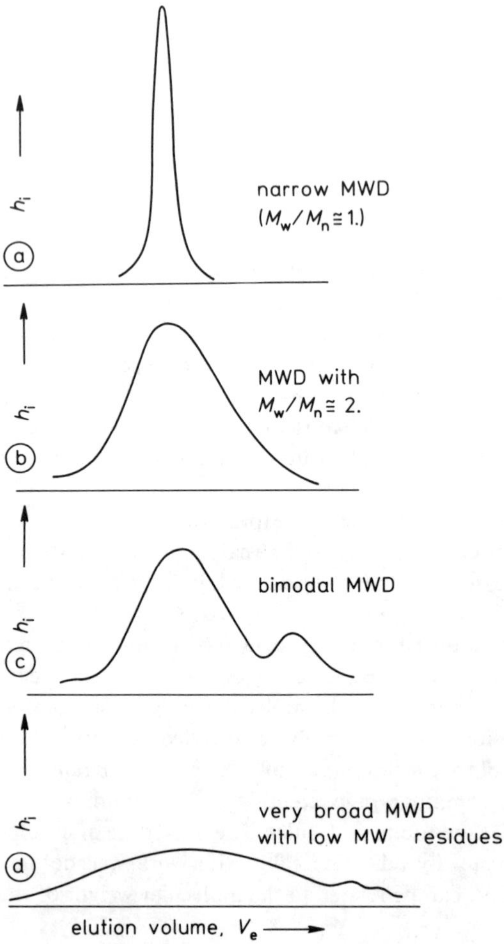

Fig. 1.2. Examples of SEC chromatograms with different molecular weight distributions (MWD).

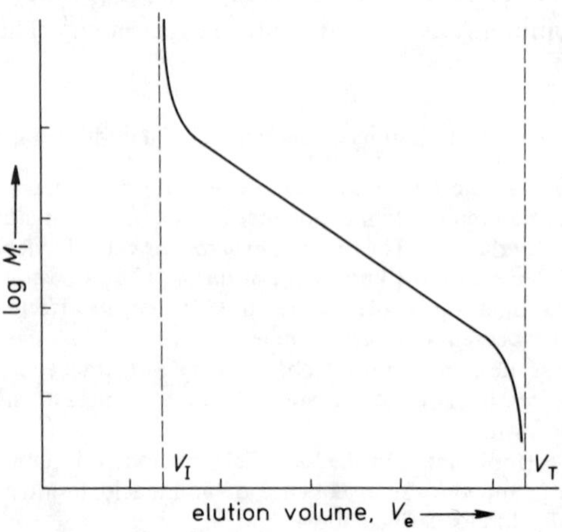

Fig. 1.3. Size exclusion chromatography (SEC) calibration curve. V_I = Intersticial volume of the column; V_T = Total volume of solvent in the column.

detector response n is then proportional to the instantaneous concentration of the eluate, the proportionality factor being the refractive index increment dn/dc characterizing the polymer-elution solvent system:

$$n = (\mathrm{d}n/\mathrm{d}c)\, c$$

For each elution volume $V_{e,i}$ the corresponding molecular weight M_i is known from the calibration curve. The chromatogram is a plot of Δn_i, versus $V_{e,i}$. Equal elution volume increments ΔV_i are considered, as shown on Fig. 1.4. The area of each such element, under the curve, $\Delta V_i \cdot n_i$ is a measure of the actual concentration c_i of the constituent i in the sample. Thus access to the molecular weight distribution is provided (Fig. 1.4).

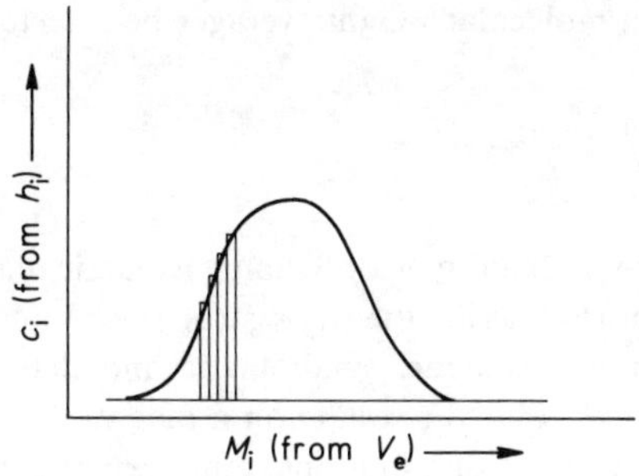

Fig. 1.4. Interpretation of a SEC diagram to calculate the molecular weight distribution of a polymer sample.

The molecular weight averages can be calculated from the following expressions, c_i being the concentration (i.e. the mass) and N_i the number of mole of species i in the polymer sample:

$$\overline{M}_w = \frac{\sum_{i=1}^{\infty} c_i M_i}{\sum_{i=1}^{\infty} c_i}$$

$$\overline{M}_n = \frac{\sum_{i=1}^{\infty} N_i M_i}{\sum_{i=1}^{\infty} N_i} = \frac{\sum_{i=1}^{\infty} c_i}{\sum_{i=1}^{\infty} c_i/M_i}$$

The molecular weight distribution within a sample largely depends upon the mechanism and kinetics of the polymerization process. This will be studied in detail in later chapters of this book.

Problem 1.1

A poly(methyl methacrylate) sample is submitted to a fractional precipitation. Five fractions are obtained and characterized by light scattering. Assuming that they are monodisperse — which is of course not the case — calculate the weight and number average molecular weights of the polymer.

Fraction	weight-%	M_i [g/mole]
1	11	530 000
2	22	290 000
3	31	180 000
4	24	75 000
5	12	30 000

If some low molecular weight constituent had been lost during the fractionation, would both molecular weight averages be affected equally or not?

Problem 1.2

Given the probability p_i of forming isotactic diads and assuming this sole parameter is sufficient to describe the steric structure of any diad along the chain, calculate the proportions of isotactic, syndiotactic and heterotactic triads in the polymer as a function of p_i. Plot the results on a diagram. As an application decide whether the following two triad distributions found experimentally are bernouillian (i. e. governed by a single parameter):

a) $i = 0.49$ $s = 0.09$ $h = 0.42$
b) $i = 0.80$ $s = 0.01$ $h = 0.19$

Problem 1.3

A pool of red and black beads, equal in number, each weighing one gram, are strung in various ways to make chains, until all beads are strung. The only rule is that black alternate with red: If the first bead is black, the next is red, etc.. . . , but the colour of the first is chosen at random. The result is that 100 chains are made of 5 beads each and 50 chains are constituted of 10 beads each.

a) What is the number average of beads per chain?
b) What is the weight average of beads per chain?
c) These chains are scrambled together in a bowl. What is the probability of picking blind-folded the end bead of a chain, and later finding out that not only it is red, but the bead at the other end of the chain is red, as well?

Chapter 2
Step-Growth Polymerization (Polycondensation)

2.1 Introduction

Step growth polymerization reactions, also referred to as polycondensations, are of major economic importance. The first synthetic polymers to be manufactured industrially were polycondensates: the production of phenol-formaldehyde resins (bakelites), which are thermosets, i. e. crosslinked polymeric materials, has expanded widely during the first half of this century. Urea-formaldehyde resins also originate from this period. After world war II, however, the major industrial expansion of polycondensates arose from the development of polyamides and of polyesters as textile fibers but also as moldable thermoplastic materials of good mechanical strength. More recently polycarbonates have found a rapidly expanding market, and polyurethanes have been applied increasingly as elastomeric materials, as foams and as thermoplastics. High temperature resistant polymers such as polybenzimidazoles have found a large number of applications. They also originate from polycondensation reactions. It should be mentioned finally that step growth reactions are also used for purpose of increasing the length of already existing polymer chains, provided these chains are fitted at both chain ends with reactive functions. Such processes are known as chain extension reactions.

As already stated step growth polymerizations are based on classical organic reactions, that are used to produce linear macromolecules, starting from bifunctional base molecules, or "monomers". To produce polymer networks the presence of at least one multifunctional constituent is required.

The basic feature of these reactions is the attack of function *a* (for example, hydroxyl) onto a function *b* (for instance carboxylic acid) to form a linkage -*c*- (ester from hydroxyl and carboxylic acid).

Thus, if the initial monomers are designated as *a-a* and *b-b**), many possibilities are to be considered:

$$\left.\begin{array}{l} \mathrm{a{-}a + b{-}b \longrightarrow a{-}c{-}b} \\ \mathrm{a{-}c{-}b + a{-}a \longrightarrow a{-}c{-}c{-}a} \\ \mathrm{a{-}c{-}c{-}a + b{-}b \longrightarrow a{-}c{-}c{-}c{-}b} \end{array}\right\} \text{addition of monomer}$$

and

$$\left.\begin{array}{l} \mathrm{a{-}c{-}c{-}a + b{-}c{-}c{-}c{-}a} \\ \quad \longrightarrow \mathrm{a{-}c{-}c{-}c{-}c{-}c{-}c{-}a} \end{array}\right\} \text{combination of oligomers}$$

---

*) In short *aa* and *bb*.

and, provided the stoichiometric ratio of a to b is unity, the chains can grow extremely large.

Similarly, *networks* can be formed, for example, from bifunctional *a-a* molecules and trifunctional molecules.

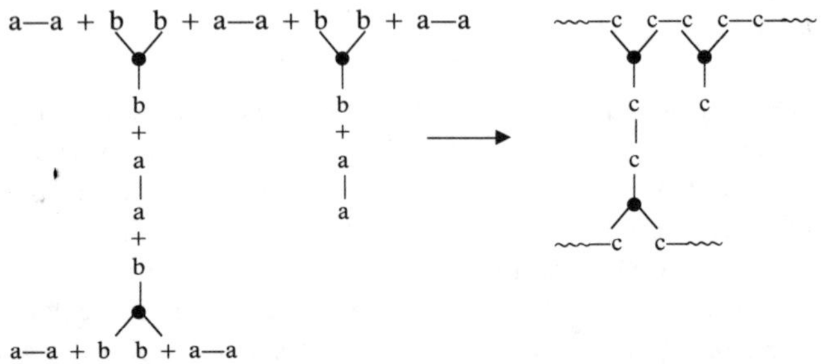

Regardless of whether linear molecules or networks are formed, it is useful to distinguish between two broad, qualitatively defined categories of step-polymerizations, because of differences in reaction kinetics and molecular weight distributions.

I. "Near Equilibrium" Reversible Step Polymerizations

Group a reacts with function b to form a c link with by-product d, but d can react with c to reform a and b. An example is the polyesterification reaction, with a = —OH, b = —COOH, c = —COO—, d = H_2O. This equilibrium can be schematized as:

$$\text{-}a + \text{-b} \underset{}{\overset{K}{\rightleftharpoons}} \text{-}c\text{-} + d$$

where the equilibrium constant K is defined as:

$$K \equiv \frac{[\text{-}c\text{-}][d]}{[\text{-}a][\text{-}b]} \tag{2.1}$$

This kind of step-polymerization is also usually characterized by *exchange reactions,* by which an a or a b group reacts with a -c- link to reshuffle molecules without a net change in concentrations of a, b, or c.

Thus, for example, the molecule X shown below can react with the indicated link c in the molecule Y - Y′ to form two new molecules X - Y and Y′ without changing the total number of molecules.

II. Irreversible "Far From Equilibrium" Step Polymerizations

The general scheme may be the same as above, i. e. $-\underline{a} + -\underline{b} \rightarrow -\underline{c} + \underline{d}$, but there is no counterpart: $-\underline{c} + \underline{d} \rightarrow -\underline{a} + \underline{b}$, i. e., $K \rightarrow \infty$. There may be no by-product $\underline{d}$ formed, so that the equilibrium constant K would then be defined as

$$K \equiv \frac{[\text{-}c\text{-}]}{[\text{-}a][\text{-}b]} \tag{2.2}$$

but again it would be several powers of ten in order of magnitude.

Another hallmark of irreversible reactions is the absence of *exchange reactions* (such as schematically indicated above).

As a general rule, the "reversible" polycondensation reactions have to be coaxed to produce polymers of high degrees of polymerization, even when exact stoichiometric balance of $\underline{a}$ to $\underline{b}$ is achieved. They are characterized by low equilibrium constants and slow rates of reaction.

In contrast, the irreversible polycondensation reactions are characterized by very large equilibrium constants, orders of magnitude greater than for the first category. In addition, some of these are extremely rapid so that *uniform* mixing of $\underline{a}$ with $\underline{b}$ is difficult or impossible to achieve before significant reaction occurs. In this category, very high degrees of polymerization can be achieved if stoichiometric equivalence is maintained. Usually, stoichiometric imbalance constitutes the practical limitation.

The implications of "irreversible" reactions will be made more clear by comparison with the features of reversible polycondensations.

Step Growth Polymerization, Reversible

Two types of *reversible* polycondensations have to be distinguished.

- Those involving one single "monomer" species, carrying one $\underline{a}$ and one $\underline{b}$ function. This type of molecule is irrelevant to *irreversible polycondensation,* since it could not exist ($K \rightarrow \infty$ for $-\underline{a} + -\underline{b} \rightarrow -\underline{c}-$). The self-condensation of $\underline{ab}$ type molecules involves by nature exact stoichiometry between the antagonist functions. At any stage of the reaction, each polymer molecule carries one $\underline{a}$ function and one $\underline{b}$ type function at the chain ends.
- The more common reversible polycondensations involving two different bifunctional monomers, say $\underline{aa}$ and $\underline{bb}$, are not necessarily in exact stoichiometric balance. Even if this condition is fulfilled, the $\underline{a}$ and $\underline{b}$ functions carried at chain end are distributed at random, which means that one can find 3 types of polymer molecules in the reaction mixture:

a—▭—*a*, *b*—▭—*b*, *a*—▭—*b*

If exact stoichiometric equivalence of $\underline{a}$ with $\underline{b}$ is achieved, the relative amounts of these 3 species would be 0.25, 0.25 and 0.50, respectively, regardless of their degrees of polymerization. Furthermore, *structural units* must now be distinguished from *repeat units*, since the macromolecules are constituted of alternating segments originating from $\underline{aa}$ and from $\underline{bb}$ type base molecules.

Tab. 2.1.

I. Reversible (near-equilibrium) polycondensation

Name of reaction	a	b	c	d
Hydroxyl-carboxylic acid condensation	—OH	—COOH	—C(O)O—	HOH
Hydroxyl-ester interchange	—OH	—C(O)OR	—C(O)O—	HOR
Amine-carboxylic acid condensation	$—NH_2$	—C(O)NHR	—NHC(O)—	HOH
Amine-amide interchange	$—NH_2$	—C(O)NHR	—C(O)NH—	H_2NR

II. Irreversible (far from equilibrium) polycondensations

Name of reaction	a	b	c	d
Amide-acid chloride condensation	$—NH_2$	—C(O)Cl	—C(O)NH—	HCl
Hydroxyl-acid chloride condensation	—OH	—C(O)Cl	—C(O)O—	HCl
Isocyanate-hydroxyl condensation*)	—NCO	—OH	—NHC(O)O—	—
Isocyanate-amine condensation*)	—NCO	$—NH_2$	—NHC(O)NH—	—
Oxirane-amine condsation*) ("Epoxy" reaction)	$—CH—CH_2$ (epoxide ring, —O—)	—NHR	$—C(OH)—CH_2—NR—$	—

Phenol-formaldehyde condensation

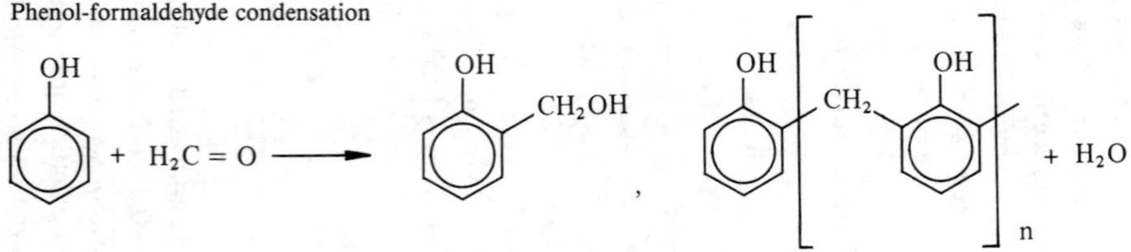

	a	b	c	d
Silane hydrogen-vinyl coupling	—SiH	$CH_2=CH—Si—$	$—Si—CH_2—CH_2—Si—$	—
Silanol-silanol conensation	—Si—OH	HO—Si—	—Si—O—Si—	H_2
Acetoxysilane condensation	$—Si—OC(O)—CH_3$	HO—Si—	—Si—O—Si—	$CH_3—C(O)—OH$

*) ... formation of low molecular weight by-products

Example: poly(hexamethylene adipamide)

$$+NH{-}(CH_2)_6{-}NH{-}CO{-}(CH_2)_4{-}CO+_n$$

Structural (or base) unit originating from hexamethylene diamine	Structural (or base) unit originating from adipic acid

Repeat Unit of the Polymer Molecules

The degree of polymerization is defined as the number of structural units that constitute the polymer molecule, and it is therefore twice the number of *repeat units* in polymers such as nylon-6,6 or other "alternating" polycondensates.

No limit is placed on the number of different bifunctional monomers introduced into the reaction, provided that stoichiometric equivalence of $\underline{a}$ with $\underline{b}$ is achieved.

For example, 9-amino nonanoic acid, $H_2N(CH_2)_8COOH$ and 11-amino undecanoic acid, $H_2N(CH_2)_{10}COOH$ could be reacted together ($\underline{a}$ = —NH_2, $\underline{b}$ = —COOH, $\underline{c}$ = —NHCO—).

Likewise, adipic acid, shown above, and sebacic acid $HOOC(CH_2)_8COOH$ could be mixed with hexamethylene diamine (shown above) and ethylene diamine ($H_2NCH_2CH_2NH_2$), provided that equivalence of —NH_2 and —COOH is achieved.

In principle, no limit is placed on the size of the bifunctional monomer, and in the frame of reference of polycondensation, the molecule obtained by anionic polymerization of styrene using a bifunctional initiator, and deactivated with CO_2, i. e.:

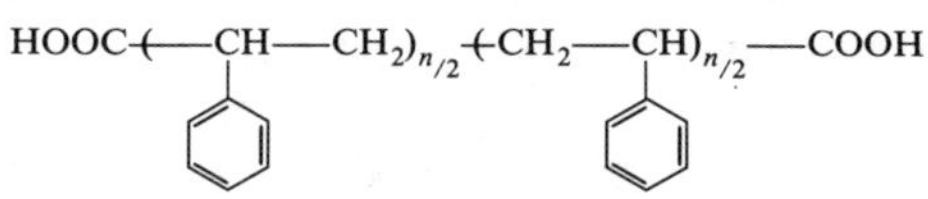

whatever the value of n, is a bifunctional monomer like adipic acid.

A few remarks should be added at this point.

- Contrary to free radical or ionic polymerization, there is neither initiation nor growth. Any function $\underline{a}$ can react with any function $\underline{b}$. In particular, a polymer molecule with i base molecules and fitted with an a function at one chain end can react with a polymer molecule with j base molecules fitted with an end standing $\underline{b}$ function, to yield a polymer molecule containing $i + j$ structural units, whereby the number of molecules in the reaction mixture decreases by one unit. In the sense used here, "molecule" includes every species present including monomers ($i = 1, j = 1$).
- In a batch polycondensation process high conversion degrees π must be reached whenever elevated molecular weights are wanted. If the fraction of functions ($\underline{a}$ or $\underline{b}$) of either kind that have reacted is 0.5 the total number of molecules present has decreased by a factor of 2, which means that *on the average* two base molecules are linked together (with rather large fluctuations, of course); in other words, the number average polymerization degree is only 2. If a chemical reaction does not allow high conversion to be reached, it cannot be used for step growth polymerization.

● Modelling of homogeneous polycondensation reactions can be performed by the *Monte Carlo method,* implying equal probability of reaction for any a (or b) function, regardless of the length of the chain to which it is attached.

2.2 Calculation of Molecular Weights and Molecular Weight Distributions of Polycondensates

2.2.1 Molecular Weight Averages

The number average degree of polymerization can be expressed as a function of the extent of reaction, i. e. of the fraction of a or of b functions that have been consumed to form links between the base molecules initially present. If π_a is the degree of conversion, i. e. the fraction of groups a initially present which have undergone reaction after a given time*), their initial number being $(N_a)_0$, one can write:

$$\pi_a = 1 - (N_a)/(N_a)_0 \tag{2.3}$$

Likewise for b groups:

$$\pi_b = 1 - (N_b)/(N_b)_0 \tag{2.4}$$

If, as here assumed, the only possible reaction is of a with b, and we start with $(N_a)_0$ and $(N_b)_0$, *with b in excess of a,* there will be some unreacted b groups after all a have reacted; thus when $\pi_a = 1$, $\pi_b < 1$. If extent of reaction π is written *without* subscript, it always refers to the group in *lesser* amount, and if $(N_a)_0 = (N_b)_0$, then necessarily: $\pi = \pi_a = \pi_b$.

For sake of illustration, let us consider the reaction between aa and bb molecules. Assume we have initially $(N_a)_0$ a functions or $\frac{(N_a)_0}{2}$ aa-molecules, and $(N_b)_0$ b functions, or $\frac{(N_b)_0}{2}$ bb-molecules. Let us call r the initial mole ratio of a to b. As assumed above,

$$r \equiv \frac{(N_a)_0}{(N_b)_0} \leqslant 1 \tag{2.5}$$

The total number of functions present *initially* $(N)_0$ can then be written as

$$(N)_0 = (N_a)_0 + (N_b)_0 = (N_a)_0 \left[1 + \frac{1}{r}\right] \tag{2.6}$$

After a given time of reaction, a fraction π_a of the a functions has been consumed, and the number of functions remaining at that time is given by

---

*) In later chapters on chain polymerization π is the fraction of monomer that has been converted to polymer.

$$(N)_t = (N_a)_0 (1 - \pi_a) + (N_b)_0 - (N_a)_0 \pi_a = (N_a)_0 \left(1 - 2\pi_a + \frac{1}{r}\right) \tag{2.7}$$

Since the average degree of polymerization is defined as the average number of structural units per chain, it is given by the ratio of the initial number of molecules present, consisting of *aa* and *bb* types, $(N)_0/2$, to the number of molecules remaining, $(N)_t/2$, when the extent of reaction is π*). This implies that all molecules, regardless of their size, have *two* functional ends, and that these *two* ends as pairs are *a* + *a*, *a* + *b* or *b* + *b*, and no other kind. Further, by counting molecules as pairs of ends, trifunctional and other branched molecules, are excluded.

Division of eq. 2.6 by 2.7 gives the *number average degree of polymerization* $\bar{P}_n$ as:

$$\bar{P}_n = \frac{1 + 1/r}{1 - 2\pi + 1/r} \tag{2.8}$$

or

$$\bar{P}_n = \frac{1 + r}{1 - r + 2r(1 - \pi)} \tag{2.9}$$

- If the polycondensation is stoichiometric, $r = 1$ and the expression (2.9) reduces to

$$\bar{P}_n = \frac{1}{1 - \pi} \tag{2.10}$$

Thus when 99% of the original *a* groups (and thus 99% of the original *b* groups) have reacted, $\pi = 0.99$ and $\bar{P}_n = 100$.

- If the polymerization is not stoichiometric, it is possible, by setting $\pi = 1$, to calculate the *maximum* possible degree of conversion that can be obtained for a given ratio *r* between *a* and *b* functions.

$$\bar{P}_n = \frac{1 + r}{1 - r} \tag{2.11}$$

Example: An excess of 1% *bb* molecules at start of the reaction implies that the number average degree of polymerization cannot exceed 199.

Maintaining *r* slightly less than unity is a procedure currently employed to limit the molecular weight of polyamides. All the molecules formed by the end of the process (when $\pi = 1$) carry *b* groups at their ends, and these *b* functions cannot find any *a* partners to react with, as the *a* functions have already undergone reaction quantitatively.

An alternate method to limit the molecular weights of the polycondensates is to insert a small proportion of a monofunctional compound *Rb* together with the *bb* base molecules, *a* and *b* functions being in stoichiometric ratio. Whenever a monofunctional *Rb* molecule has reacted, any further reaction is prevented because the corresponding chain end *R* does not possess a reactive group anymore. This means

*) π refers to *a* functions, and it can take any value from 0 to 1, contrary to *b* functions that are supposedly in excess.

that if the initial mixture contains $(N_{aa})_0$ a groups as $\underline{aa}$ molecules, $(N_{bb})_0$ $\underline{b}$ functions as $\underline{bb}$ molecules and $(N_{Rb})_0$ $\underline{b}$ functions as $\underline{Rb}$ molecules, r should be redefined as:

$$r = \frac{(N_{aa})_0}{(N_{bb})_0 + 2(N_{Rb})_0} \tag{2.12}$$

The factor 2 arises from the fact that in the stoichiometric mixture each missing $\underline{bb}$ molecule must be replaced by 2 monofunctional R$\underline{b}$ units.

2.2.2 Molecular Weight Distributions (MWD)

2.2.2.1 MWD of Polycondensates Obtained from $\underline{ab}$ Type Base Molecules

The molecular weight distribution in a polycondensate obtained by selfcondensation of $\underline{ab}$ type base molecules (for example, 11-amino undecanoic acid, to yield nylon-11) can be calculated easily since the formation of an i-mer requires (i – 1) linkages to be formed, each of them with a probability factor p_i.

The concept of equal reactivity has to be introduced, greatly simplifying the calculation. If the probability of reaction of a given function decreased as the chain length increases, the growth of the chains would be limited and the formation of long chains unlikely. This kind of argument was made initially for polyester formation, though erroneously. In fact the reaction of a function can be assumed to be independant of the length of the chain it is linked to. This assumption is not only reasonable, it has never been contradicted by experimental results. Thus, the formation of an i-mer by self condensation of $\underline{ab}$ base molecules implies the formation of (i – 1) links and it is given by:

$$P_i = p^{(i-1)}(1 - p) \tag{2.14}$$

If all end groups are equally reactive the value of p becomes identical to the experimentally measurable value of the extent of reaction π. If, for example, the end group analysis tells us that 95% of the initial $\underline{b}$ functions have reacted, only 5% remaining untouched, it implies that the probability for any $\underline{b}$ function to have reacted with an $\underline{a}$ group to form a $\underline{c}$ group is also 0.95.

Thus in the following it is possible to replace all symbols p with π. We will retain p with its conotation of probability (unlike π) because of a generalization to be made concerning molecular weight distributions in chain polymerization.

P_i given in equation (2.14) is also the fraction x_i of molecules that have reached a degree of polymerization i. The number average degree of polymerization $\bar{P}_n$, defined as indicated above can thus be written as*):

$$\bar{P}_n = \sum_{i=1}^{\infty} i x_i = (1 - p) \sum_{i=1}^{\infty} i p^{(i-1)} = \frac{1}{1 - p} \tag{2.15}$$

*) Summation formulae are given in Appendix 2A.

This confirms the result of equation (2.10), as self condensation of $\underline{ab}$ base molecules implies stoichiometric equivalence: $r = 1$.

The weight average degree of polymerization $\bar{P}_w$ can be calculated in the same manner, from the weight fraction w_i of the molecules that have reached the degree of polymerization i in the sample:

$$\bar{P}_w = \sum_{i=1}^{\infty} i w_i = \frac{\sum_{i=1}^{\infty} i^2 x_i}{\sum_{i=1}^{\infty} i\, x_i} = (1 - p)^2 \sum_{i=1}^{\infty} i^2\, p^{(i-1)} \tag{2.16}$$

Thus*)

$$\bar{P}_w = \frac{1 + p}{1 - p} \tag{2.17}$$

and the polydispersity index is given by

$$\frac{\bar{P}_w}{\bar{P}_n} = 1 + p = 1 + \pi \tag{2.18}$$

It is concluded from this set of equations that the molecular weight distribution within an $(\underline{ab})_n$ type polycondensate is the so-called *most probable* or Schulz-Flory distribution. The polydispersity index tends toward 2 as the degree of conversion π goes to completion.

An example of Schulz-Flory distribution is given in Fig. 2.1

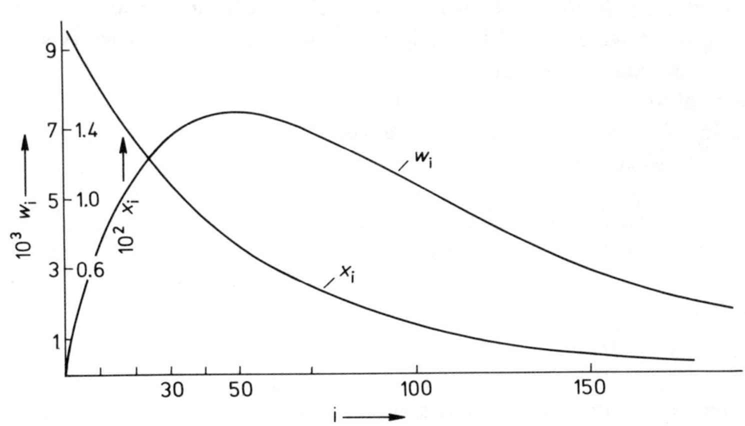

Fig. 2.1. Most probable (Schulz-Flory) distributions (x_i by number, or mole fraction; w_i by weight fraction) of polymerization degrees, calculated for an average degree of polymerization of $\bar{P}_n = 50$.

*) Summation formulae are given in Appendix 2A.

2.2.2.2 MWD of *aa* + *bb* Type Polycondensates

The case of *aa* + *bb* step growth polycondensation is far more complicated as stoichiometric equivalence is not implied. Extents of reaction must be considered separately for the *a* functions, π_a, and for the *b* functions, π_0. However, these parameters are algebraically related, provided the only possible reaction is between *a* and *b* functions, to yield *c* links*):

$$(N_a)_0 \, \pi_a = (N_b)_0 \, \pi_b \tag{2.19}$$

To calculate the distribution of molecular weights in a polycondensate originating from *aa* and *bb* base molecules**) one has to evaluate separately the proportions of macromolecules with an even number of structural units, carrying an *a* and a *b* function at their chain ends, and the proportions of macromolecules with odd numbers of base units that have either two *a* functions or two *b* functions at chain end. Thus, if we call j the total number of units, we have to deal with 3 different cases: j = 2i − 1; j = 2i; j = 2i + 1, i being the number of *aa* base units in the polycondensate. In addition, account has to be taken of the remaining *bb* type base molecules, the amount of which is not included in either category.

The total number of molecules N_t remaining when the fractions π_a and π_b of functions *a* and *b* have been already consumed amounts to:

$$N_t = \frac{N_0}{\pi_a + \pi_b}(\pi_a + \pi_b - 2\pi_a\pi_b) = N_0 \frac{1 + r - 2\pi_a r}{1 + r} \tag{2.20}$$

from which expression (2.9) for the P_n is easily deduced. As in the previous case of self-condensation of *ab* molecules, the principle of equal reactivity for each type of function is valid. Consequently, the probabilities p of reaction for a given function are equal, and identical to the overall extent of reaction π.

The distribution function (by number) of *aa* base units in the polycondensate molecules can thus be expressed as:

$$x_i = \frac{(1 - p_a)^2 p_a^{i-1} p_b^i}{p_a + p_b - 2p_a p_b} + \frac{2(1 - p_a)(1 - p_b) p_a^i p_b^i}{p_a + p_b - 2p_a p_b} + \frac{(1 - p_b)^2 p_a^{(i+1)} p_b^i}{p_a + p_b - 2p_a p_b} + \frac{p_a (1 - p_b)^2}{p_a + p_b - 2p_a p_b} \tag{2.21}$$

in which the first term refers to odd molecules with *aa* base units at both ends, constituted of i *aa* base units and (i − 1) *bb* base units; the second term takes account of molecules containing an even number of units (i *aa* and i *bb* units) and having one *a* and one *b* function at chain ends; the third term refers to odd macromolecules made

*) One must therefore consider other possible reactions in competition.

**) The calculation of the molecular weight distribution for a non-stoichiometric mixture of *aa* and *bb* ($r < 1$), when all *a* functions have reacted ($p_a = \pi_a = 1$) is far more complicated.

of i <u>aa</u> base units and (i + 1) <u>bb</u> base units, whereby the two chain ends are fitted with <u>bb</u> units. The last term of the above equation takes account of unreacted <u>bb</u> units, that are not considered in the preceding terms.

To calculate the weight distribution function of the macromolecules formed upon reaction between <u>aa</u> and <u>bb</u> base molecules it is necessary to introduce the weights of the base units W_a and W_b, which are not necessarily identical with the molecular weights of the base molecules that were reacted. For instance, the base units originating from the diol and from the diacid in poly(ethyleneterephthalate) are respectively $|O{-}CH_2{-}CH_2{-}O|$ and |CO— 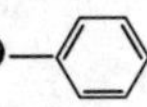 —CO|, as compared with the actual base molecules, ethylene glycol and terephthalic acid. The total weight of the base units formed upon polycondensation, at fractional conversion π_a and π_b of the <u>a</u> and of the <u>b</u> functions, is given by

$$W_t = \frac{\pi_b W_a + \pi_a W_b}{\pi_a + \pi_b} \tag{2.22}$$

Replacing again the degrees of conversion π by the probabilities of reaction p, the normalized weight distribution function of <u>aa</u> units can then be written as

$$w_i = (1 - p_a)^2 p_a^{i-1} p_b^i \cdot \frac{(iW_a + (i-1) W_b)}{p_b W_a + p_a W_b} + 2(1 - p_a)(1 - p_b)(p_a p_b)^i \cdot \frac{iW_a + iW_b}{p_b W_a + p_a W_b}$$

$$+ (1 - p_b)^2 p_a^{i+1} p_b^i \cdot \frac{iW_a + (i+1) W_b}{p_b W_a + p_a W_b} + \frac{p_a (1 - p_b)^2 W_b}{p_b W_a + p_a W_b} \tag{2.23}$$

In this expression again the first term refers to molecules containing *i* <u>aa</u> and (*i* − 1) <u>bb</u> units, the second to macromolecules with an even number of base units (*i* of each type), the third to species made of *i* units <u>aa</u> and (*i* + 1) <u>bb</u> units, the last one accounting for <u>bb</u> units left over.

The number average and the weight average molecular weights can thus be calculated:

$$\bar{M}_n = \frac{p_b W_a + p_a W_b}{p_a + p_b - 2 p_a p_b} \tag{2.24}$$

$$\bar{M}_w = \frac{1 + p_a p_b}{1 - p_a p_b} \left(\frac{p_b W_a^2 + p_a W_b^2}{p_b W_a + p_a W_b} \right) + \frac{4 p_a W_a W_b}{(1 - p_a p_b)(p_b W_a + p_a W_b)} \tag{2.25}$$

For a stoichiometric mixture of <u>aa</u> and <u>bb</u>, $\pi_a = \pi_b = \pi$ and thus, by our assumption, $p_a = p_b = p$, the above formulas reduce to the Schulz-Flory most probable distribution:

$$\overline{M}_n = W\frac{1}{1-p} = W\frac{1}{1-\pi}$$
$$\overline{M}_w = W\frac{1+p}{1-p} = W\frac{1+\pi}{1-\pi} \tag{2.26}$$

wherein W is the average weight of the units, $(W_{aa} + W_{bb})/2$ (or is the weighted average of all units if there are more than 2). For example, suppose a mixture is made of the following components:

1 mol hexamethylene diamine	$W_{aa,1} = 114$ g	$—NH(CH_2)_6NH—$
1 mol decamethylene diamine	$W_{aa,2} = 170$ g	$—NH(CH_2)_{10}NH—$
1 mol adipic acid	$W_{bb,1} = 112$ g	$—OC(CH_2)_4CO—$
1 mol sebacic acid	$W_{bb,2} = 168$ g	$—OC(CH_2)_8CO—$

Stoichiometric equivalence is thus provided (equal amounts of amine and carboxylic acid groups) and $W = (114 + 170 + 112 + 168)/4 = 141$ g.

2.3 Kinetics of Step-Growth Polymerization in Homogeneous Systems (not Diffusion Controlled)

In most cases, the kinetics of polycondensation derives directly from the kinetics of the organic reaction it is based on. Such reactions are generally bimolecular and the kinetic equations are written as:

$$-\frac{d[b]}{dt} = k[a][b] \qquad -\frac{d[a]}{dt} = k[a][b] \tag{2.27}$$

where $[a]$ and $[b]$ are the concentrations of *functions* $\underline{a}$ and $\underline{b}$ in the reaction medium. If the mixture of base molecules is in stoichiometric equivalence (or if the base molecule is of the $\underline{a}\underline{b}$ type) one can set $[a] = [b]$ and equation (2.27) can be integrated*) to:

$$\frac{1}{[a]} - \frac{1}{[a]_0} = kt \quad \text{or} \quad \frac{[a]_0 - [a]}{[a]} = [a]_0\, kt \tag{2.28}$$

$[a]_0$ being the initial number (molar concentration) of $\underline{a}$ functions. Since the number average degree of polymerization is defined as the ratio of the number of base molecules initially present ($\underline{a}\underline{a} + \underline{b}\underline{b}$, or $\underline{a}\underline{b}$) to the number of remaining molecules of all sizes at time t, one can also write, (if the mixture is stoichiometric):

$$\overline{P}_n = \frac{[a]_0}{[a]} = \frac{1}{1-\pi} \tag{2.10}$$

*) It is assumed that the total volume of the reaction mixture is constant, independent of conversion, whereas there could be a small shrinkage owing to higher density of polymer, as compared to monomers, and to loss of by-products (e.g. H_2O etc.).

and consequently the number average degree of polymerization is a linear function of time:

$$\bar{P}_n = 1 + [a]_0\, kt \tag{2.29}$$

Another way of obtaining the same result is to express the concentration $[a]$ in terms of extent of reaction π and constant *initial* concentration $[a]_0$.

Thus

$$[a]_0 \frac{d\pi}{dt} = k\,\{[a]_0\,(1 - \pi)\}^2 \tag{2.30}$$

or

$$\frac{d\pi}{(1 - \pi)^2} = k\,[a]_0\, dt \tag{2.31}$$

whence

$$\frac{1}{1 - \pi_2} - \frac{1}{1 - \pi_1} = k\,[a]_0\,(t_2 - t_1) \tag{2.32}$$

or

$$\Delta(\bar{P}_n) = k\,[a]_0\,\Delta t \tag{2.33}$$

This relation clearly illustrates the fact that high conversions (or long reaction times) are required to reach elevated values of $\bar{P}_n$, even under the best possible conditions: Stoichiometry, equal reactivity of all functions.

Underlying the above calculations is the assumption that any group $\underline{a}$ or $\underline{b}$ has the same reactivity, independent of the length of the chain to which it is attached. The same assumption allowed us to equate the probability p that a particular group $\underline{a}$ has reacted to the extent of conversion π. This condition is not always satisfied. For example, the primary alcohol of 1,2-dihydroxypropane has a higher reactivity than the secondary alcohol. Generally, however, the reactivity of a specific group does not depend on the length of the chain to which it is attached.

Yet another assumption is implicit in equations (2.27) through (2.33), namely that the equilibrium constant is so large that back-reaction can be neglected. To allow for the more general case the rate equations would be written:

$$\frac{-d[b]}{dt} = k_f[a][b] - k_r[c][d] \tag{2.34}$$

where k_f is the forward rate constant and k_r is the rate constant for the reverse reaction. The condition of equilibrium, obtained from setting $d[b]/dt$ in eq. (2.34) equal to zero, leads to:

$$\frac{[c]_e[d]_e}{[a]_e[b]_e} = \frac{k_f}{k_r} = K \tag{2.35}$$

where K is the equilibrium constant (cf. equation 2.1).

Earlier in this chapter we have made a distinction between *reversible* and *irreversible* polycondensations. The preceding analyis is based on the assumption of equal reactivity of the functions at chain end, and on the same forward rate constant k_f for all groups. However, if k_f is very large and at the same time the equilibrium constant K is very high, the rate constant for the reverse reaction $\underline{c} + \underline{d} \rightarrow \underline{a} + \underline{b}$ must be very small and the polycondensation might be diffusion controlled. In other words, if $\underline{a}$ and $\underline{b}$ can get together, they will form $\underline{c}$ and $\underline{d}$ at once. Thus introduction of the reagents and mixing them will be the central issue, not chemical kinetics. Under such conditions, neither the expression for the rate of conversion (eq. 2.27) nor that giving the degree of polymerization (eq. 2.29) will be appropriate. This is illustrated in the very fast, irreversible reaction of diamines with diacid chlorides in the *interfacial polycondensation.*

Conversely, a very large equilibrium constant does not a priori foretell failure of equations (2.27) and (2.29), for if the forward reaction rate constant k_f is low (as it is in the case of the reaction between a diisocyanate and a dialcohol), diffusion control may not intervene, and the true chemical reaction kinetics, as given in eq. (2.27) may prevail.

Cyclization reactions

Cyclization resulting from so-called *end-biting reactions,* i. e. reactions involving the $\underline{a}$ and $\underline{b}$ functions located at the chain ends of the same macromolecule have been excluded from consideration so far, but may occur. Since the degree of polymerization is computed by counting pairs of ends, a cycle having no ends will escape being counted. No error is introduced in the rate equations (2.27) and (2.28) but the degree of polymerization given by (2.29) would be in error.

Cyclizations can also arise from back-biting reactions, if a function $\underline{a}$ located at chain end is capable of reacting with a $\underline{c}$ link of the same chain. This is the intramolecular version of the reshuffling process already mentioned. Examples of such possible cyclizations include intramolecular transesterification upon polyesterification, or base catalyzed attack of a hydroxysilane end group onto a siloxane link in a polydimethylsiloxane.

Cyclizations usually can be neglected in first approximation, unless the polycondensation is stopped at an early stage. This arises from the fact that the probability of cyclization decreases as the chain length increases, as the chances for intramolecular reaction then vanish. Two cases may however yield measurable amounts of cyclic constituents: When the overall concentration of the reagents in the reaction medium is very low (high dilution polycondensation), and when at early stages of the reaction stable cycles are likely to be formed.

2.4 Network Formation by Step-Growth Polymerization

Assume a trifunctional reagent $\mathrm{a{-}\!\!<^{a}_{a}}$ is present in a step-growth polymerization together with $\underline{aa}$ and $\underline{bb}$ bifunctional base molecules. Functions $\underline{a}$ can only react with functions $\underline{b}$. Furthermore, let us call π_a and π_b the fractional conversions of $\underline{a}$ and $\underline{b}$ functions (which are equal to the probability for any $\underline{a}$ or $\underline{b}$ function to have undergone reaction, p_a and p_b) and assume that the reactivity of a function is independent of whether it is located on a bifunctional or on a trifunctional base molecule. The parameter ρ is defined as the ratio of the number of *a* functions belonging to trifunctional molecules to the total number of *a* functions, in the initial system.

The probability for a chain to start at a trifunctional base unit and to reach another trifunctional base unit after having added several *aa* and *bb* base units can be evaluated in a way very similar to that used in the calculation of molecular weight distributions in a linear polycondensate.

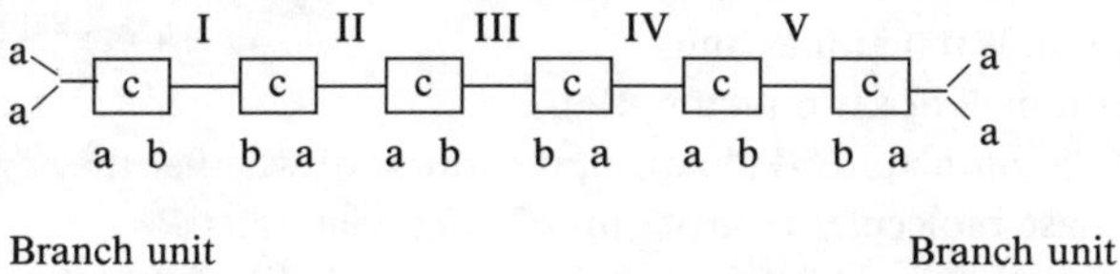

The probability for any *a* function to react with a *b* function is p_a. The probability for a *b* function to react with an *a* function belonging to a trifunctional base molecule is $p_b\,\rho$ and its probability to react with a bifunctional base molecule is consequently $p_b\,(1-\rho)$. In such a system the number of *base* units connected between two branch units (which are not counted) is always odd, say: $i = 2j + 1$ (I . . . V in above example)

$$P_i = p_a (p_b (1-\rho) p_a)^i \, p_b \, \rho \tag{2.36}$$

P_i is the molar distribution function (by number) of the chains connected at both ends with trifunctional branch-points. At high conversions, past the gelation point, this is also the number distribution function of the elastically effective network chains.

The gelation phenomenon that occurs upon formation of a network, involves the probability $\bar{\alpha}$ for a branching unit chosen *at random**) to be connected to another such unit by means of a chain, regardless of the number of base units of that connecting chain. This probability factor $\bar{\alpha}$ can be expressed as:

$$\bar{\alpha} = \sum_{j=0}^{\infty} p_a p_b \, \rho \, (p_a p_b (1-\rho))^j \tag{2.37}$$

$$\bar{\alpha} = \frac{p_a p_b \rho}{1-(p_a p_b (1-\rho))} = \frac{p^2 \rho r}{1-p^2 r (1-\rho)} \tag{2.38}$$

The criterion for gelation (network formation) is that a trifunctional branch unit selected at random is connected to *more* than 2 chains that lead to other such branch points; in turn each of those are connected with chains that will lead to further branch points, and so on. In other words, for trifunctional branch points, the value of $\bar{\alpha}$ required to obtain a network (i.e., to get gelation) can be expressed as

$$\alpha_c > \frac{1}{2} \tag{2.39}$$

*) Thus as $\bar{P}_n$ is the numerical average of all degrees of polymerization in a linear polycondensation, $\bar{\alpha}$ is the average of α_i for all values of *i*.

This expression can be generalized for f-functional branch points, and the *critical value* α_c at which gelation should occur is given by:

$$\alpha_c = \frac{1}{f-1} \tag{2.40}$$

By inserting this value α_c into the expression (2.38) one can calculate the conversion degree π_c (i.e. the proportion $(p_a)_c$ of $\underline{a}$ functions reacted) at the gel-point, for a system characterized by its stoichiometric ratio r and by the proportion ρ of $\underline{a}$ functions belonging to multifunctional units.

The above calculation includes two assumptions:

- The reactivity of all $\underline{a}$ functions is identical, regardless of whether they originate from bifunctional $\underline{aa}$ base molecules or from multifunctional units Ra_f.
- Cyclizations are neglected, which means that the possibility for a chain to be attached by its two ends to the same branch point (loop) is not considered.

Two limiting cases are of interest:

— If all $\underline{a}$ functions belong to multifunctional units, $\rho = 1$, and the expression reduces to

$$\alpha_c = \frac{1}{f-1} = \pi_c^2\, r \tag{2.41}$$

whereupon one obtains:

$$\pi_c = \left(\frac{1}{r(f-1)}\right)^{1/2} \tag{2.42}$$

— If the mixture is stoichiometric $r = 1$:

$$\alpha_c = \frac{1}{f-1} = \frac{\pi_c^2\, \rho}{1 - \pi_c^2(1-\rho)} \tag{2.43}$$

and the critical conversion degree, at the gel point, is given by:

$$\pi_c = \left(\frac{1}{\rho(f-2)+1}\right)^{1/2} \tag{2.44}$$

The values of π_c determined experimentally are generally very close to those calculated from the above expressions. Only when it is carried out in very dilute solution does the process involve some cyclization, whereby the gel point is delayed: crosslinking occurs for conversions π higher than the values calculated.

2.5 Examples of Step-Growth Polymerization Reactions

2.5.1 Polyesters

Polyesters can result from various types of reactions such as diacid + dialcohol, or di(acid chloride) + dialcohol, or from *transesterification processes.* Esterification generally requires catalysts such as toluene sulfonic acid or metal salts of various kinds*). As ester formation is a reversible process, elimination of water from the reaction medium is necessary to reach high conversion (and molecular weights high enough). Transesterification likewise is a reversible process which is used in industry to manufacture poly(ethylene terephthalate), and similar polyesters; in this case, an alcohol is eliminated.

The starting compound is dimethylterephthalate which is crystalline and easy to purify; it is reacted first with excess glycol, whereby methanol is distilled off, and bis(hydroxyethyl)terephthalate is formed. In a second step this compound undergoes transesterification on itself, whereby ethylene glycol is released, and removed under vacuum.

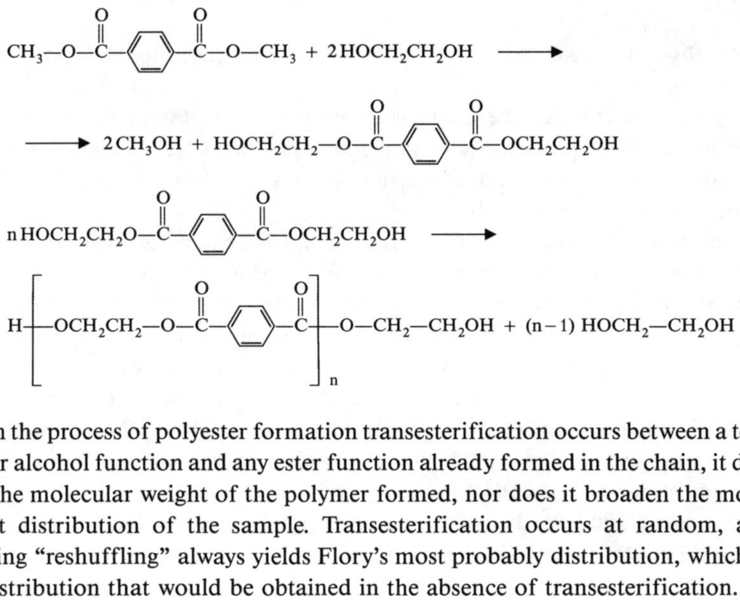

If in the process of polyester formation transesterification occurs between a terminal acid or alcohol function and any ester function already formed in the chain, it does not alter the molecular weight of the polymer formed, nor does it broaden the molecular weight distribution of the sample. Transesterification occurs at random, and the resulting "reshuffling" always yields Flory's most probably distribution, which is also the distribution that would be obtained in the absence of transesterification.

Adequate purification of the reagents is necessary to obtain the high conversion which is required to reach molecular weights high enough for the intended use of the polymers.

Side reactions are to be prevented in order to avoid any loss of functions during the course of the reaction: Decarboxylation, elimination, are such reactions, occurring upon polyesterification.

---

*) If no catalyst is present the reaction is second order in acid, as the acid plays the role of the catalyst. Assuming stoichiometric balance the reaction rate can be written as

$$-\frac{d[COOH]}{dt} = k[OH][COOH]^2 = k[COOH]^3$$

which upon integration yields

$$\frac{1}{[COOH]^2} - \frac{1}{[COOH]_0^2} = 2\,kt$$

It follows that the DP increases as the square-root of time $P_n^2 = 1 + 2[COOH]_0^2\,kt$

2.5.2 Polyamides

Polyamides*) result from the reaction between a diacid (such as adipic acid) and a diamine (such as hexamethylene diamine). This reaction is carried out in two steps. First the so-called AH-salt is made between these constituents:

$$\mathrm{HOOC(CH_2)_4COOH} + \mathrm{NH_2(CH_2)_6NH_2} \longrightarrow {}^{\ominus}\mathrm{O{-}\overset{O}{\overset{\|}{C}}{+}CH_2{+}_4\overset{O}{\overset{\|}{C}}{-}NH{+}CH_2{+}_6\overset{\oplus}{N}H_3}$$

The second step consists in heating this salt to temperatures high enough to yield amidification, whereupon water is relased.

$$\mathrm{H_3\overset{\oplus}{N}{+}CH_2{+}_6NH{-}\overset{O}{\overset{\|}{C}}{+}CH_2{+}_4\overset{O}{\overset{\|}{C}}{-}O^{\ominus}} \xrightarrow[260-270\,^\circ\mathrm{C}]{\Delta} \mathrm{{+}NH(CH_2)_6NHCO(CH_2)_4CO{+}_n} + 2n\,\mathrm{H_2O}$$

The temperature at which the second step occurs is of great importance, because of the high tendency of polyamides to undergo crystallization. To keep the reaction mixture homogeneous up to high conversion, the reaction has to be conducted above the melting point ($T_M^0 = 235\,^\circ C$).

Another procedure (non-equilibrium, irreversible) that can be used to make polyamides is the *interfacial polycondensation*, starting from the dicarboxylic acid chloride, dissolved in an organic solvent, and the diamine dissolved in water. The two non miscible layers are superposed and a polyamide film is formed at once at the interface serving as a separating membrane. As this film is pulled out, new polymer is instantly formed at the interface. This process requires that the water layer contains a base to remove HCl evolving from the reaction. It should be emphasized that this type of process is *diffusion controlled,* the extent of polycondensation (i.e. the molecular weight attained) largely depends upon the experimental conditions: Concentration of the reagents; means to keep the concentrations homogeneous in both phases throughout the process; rate at which the film is pulled out of the reaction medium, and so on.

2.5.3 Polyurethanes

Polyurethanes result from the irreversible, non-equilibrium reaction of diols with diisocyanates. Contrary to most other step growth reactions urethane formation does not involve release of a small molecules such as H_2O, HCl, CH_3OH, etc. The reaction can be written as

$$n\mathrm{OCN(CH_2)_6NCO} + n\,\mathrm{HO(CH_2)_6OH} \longrightarrow \mathrm{{+}(CH_2)_6{-}NH{-}CO{-}O(CH_2)_6{-}OCONH{+}_n}$$

Though the reaction is possible in the absence of catalysts, tertiary amines (DABCO)**), and metal salts (dibutyltin dilaurate) even more, accelerate the reaction

*) Some polyamides such as nylon-6 are made by ring-opening polymerization of lactams, considered in chapter 5.

**) DABCO: 1,4-Diazabicyclo[2,2,2]octane: 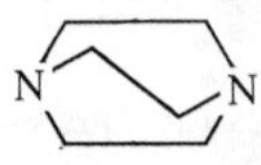

rate up to 1000 times. Among the diisocyanates that are commonly used in polyurethane synthesis the following should be quoted:

- Toluene diisocyanate (TDI), usually a mixture of toluene-2,4- and toluene-2,6-diisocyanate (80 : 20)
- Methylene diphenyldiisocyanate (MDI)
- Cyclohexane diisocyanate (CDI)
- Hexamethylene-1,6-diisocyanate (HMDI)
- The series of *"Desmodur"* are plurifunctional aromatic or aliphatic isocyanates that are used as crosslinkers in polyurethane foams especially.

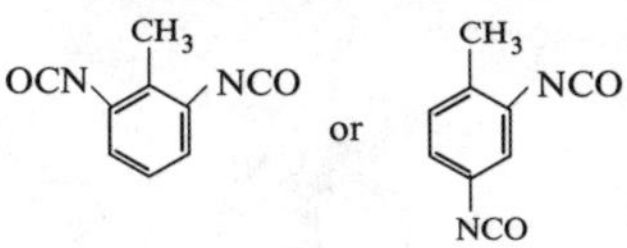

TDI

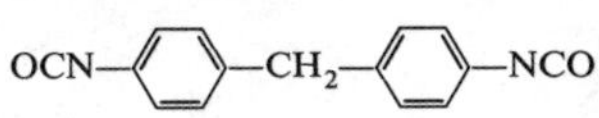

MDI

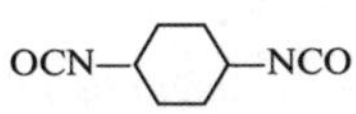

CDI

$$OCN{-}(CH_2)_6{-}NCO$$

HMDI

With toluene diisocyanate, the two isocyanate functions exhibit markedly different reactivities, and the rate expressions have to be reconsidered. The distribution of chain lengths is not very much affected by this difference in reactivity, if the reagents are in stoichiometric proportions.

Two side reactions can occur upon polyurethane synthesis: Formation of allophanate and of biuret linkages from an isocyanate function and an already existing urethane link. They not only affect stoichiometry, but they also introduce *branching*, yielding eventually *crosslinked* species.

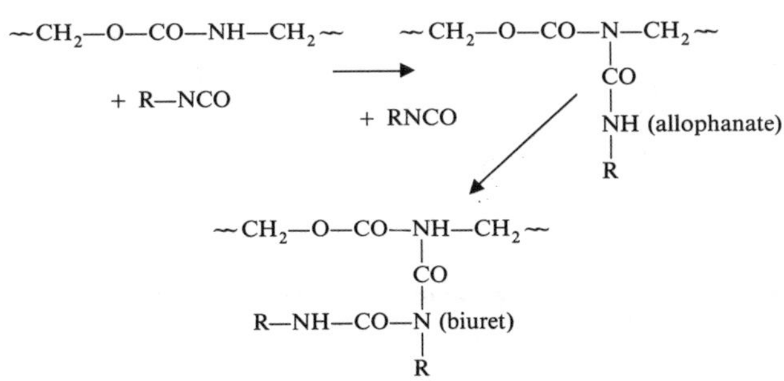

2.5.4 Polycarbonates

The polycarbonates have gained increasing importance in recent years, in the automotive industry especially. They result from the reaction of diphenols, such as bisphenol A, with phosgene. They can also be obtained by transesterification, starting from bisphenol A and diphenylcarbonate, whereby phenol is released.

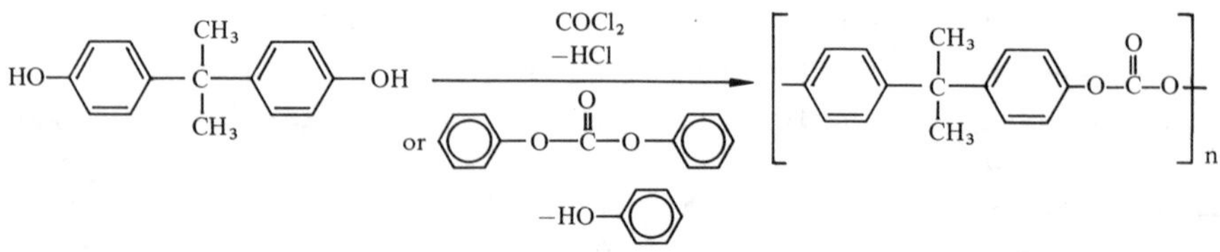

2.5.5 Polybenzimidazoles and Aromatic Polyimides

Imidazoles result from the reaction of aromatic ortho diamines with a carboxylic acid function, which generally proceeds in two steps as shown below:

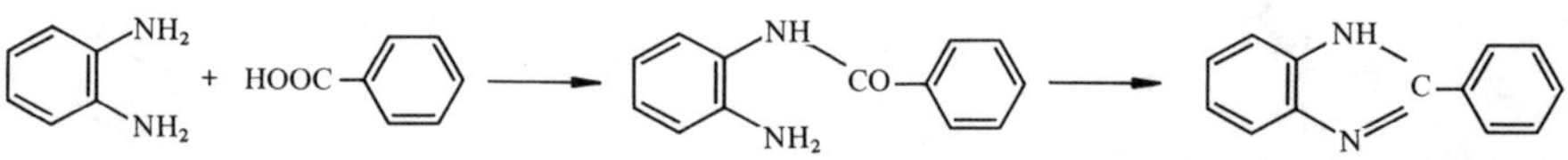

This reaction can be applied to polymer formation by reacting a diacid (isophthalic acid), or better its diester, with tetraamino diphenyl, whereby a tough thermostable polycondensate is formed.

Similarly, aromatic dianhydrides such as pyromellitic anhydride can be reacted with phenylene diamine to yield an aromatic polyamide; this polymer, in turn, can be subjected to intramolecular water elimination, to finally yield a polycondensate of high rigidity and exhibiting good thermal stability.

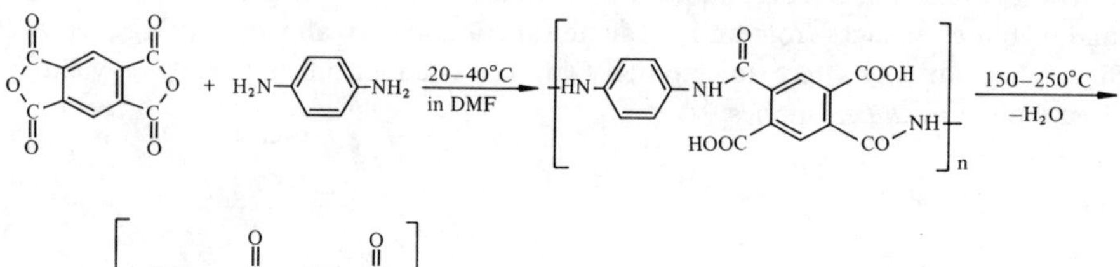

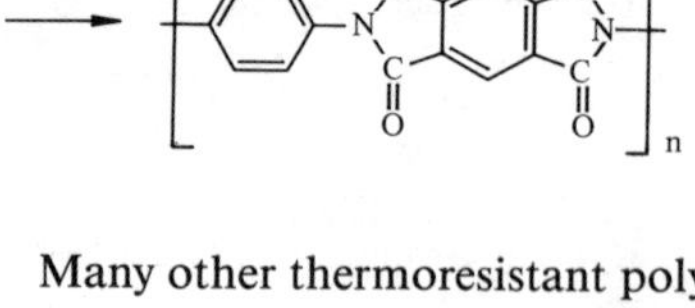

Many other thermoresistant polycondensates have been synthesized in recent years.

2.5.6 Crosslinked Polycondensates

Two classes of so-called *thermoset polycondensates* have been known for a long time. They result from the reaction of either phenol or urea with formaldehyde.

- Phenol-Formaldehyde Resins

Phenol has three reacting sites: The para and the two ortho positions with respect to the phenolic hydroxyl group. Depending on the molar ratio of phenol to form-

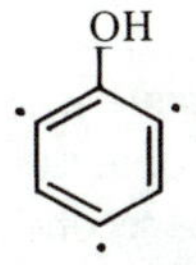

aldehyde and the type of catalyst used (acid, alkaline or Lewis acid catalysts are possible) a great variety of resins with different properties can be produced, ranging from low molecular weight *Novolaks* and *Resols* to highly crosslinked, non swellable, and non meltable polycondensates *(Resits)*. In technical manufacturing processes the resins are cured under pressure in a mold, and the shape of the sample and its size cannot be modified thereafter except by machining methods.

Owing to the fact that the reactive sites in phenol exhibit different reactivities it is also possible to carry out the reaction in two steps. The low molecular weight polycondensate resulting from the first step is highly branched, but still liquid, and fitted with many hydroxymethyl functions; the latter react, in a second step, at higher temperature either with each other to yield dimethylene ether bridges, or with still vacant sites on the benzene nuclei, to yield methylene bridges, whereby the thermoset polycondensate is formed.

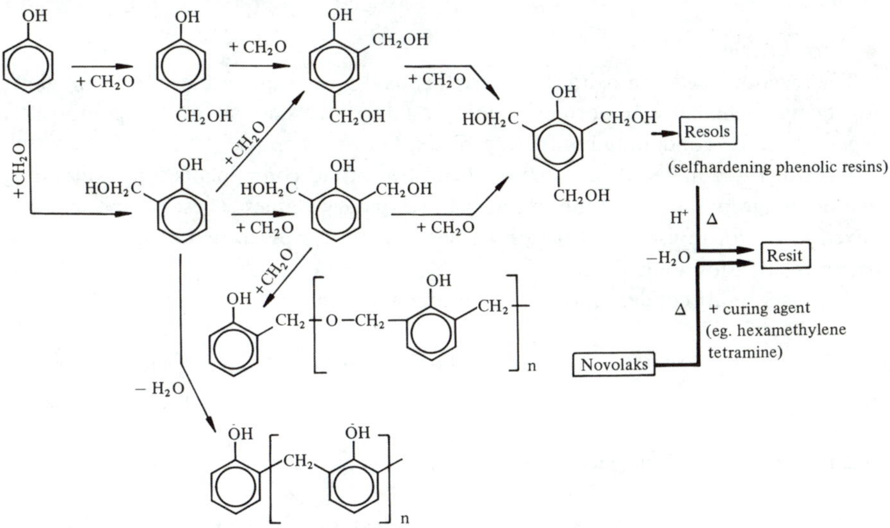

As shown above, the reaction of phenol with formaldehyde to phenolic resins is very complicated. The reaction scheme is strongly simplified because only two of the numerous reaction possibilities leading to phenolic resins are drawn in.

- Urea-Formaldehyde Resins

The same applies to the urea-formaldehyde polycondensates, where the four*) reactive sites of urea can contribute to the reaction with formaldehyde.

---

*) Four, because each H of urea can react with formaldehyde to give a methylol group:

$$O{=}C\begin{matrix} \diagup NH_2 \\ \diagdown NH_2 \end{matrix} + 4\,CH_2O \longrightarrow O{=}C\begin{matrix} \diagup N(CH_2OH)_2 \\ \diagdown N(CH_2OH)_2 \end{matrix}$$

- Polyurethanes

Crosslinked Polyurethanes have also gained numerous applications as bulk materials (rubber like) or as foams. They generally result from a two step polycondensation, the first involving only bifunctional reagents, one of them being in excess, and the second making use of the terminal functions of the percursor polymer to achieve crosslinking by addition of a tri- or multi-functional reagent, in appropriate proportion, to now achieve stoichiometry. This can be obtained by reacting a triol (such as trimethylol propane) with the isocyanate end groups of the precursor polyurethane, or by reacting a plurifunctional isocyanate (Desmodur™) onto the terminal hydroxyl functions of the precursor. To make *foams*, an excess of isocyanate functions has to be provided and a calculated amount of *water* is added during the crosslinking process, whereby CO_2 is released and trapped in the material as fine bubbles, and the isocyanate — thereby transformed to primary amine — rapidly reacts with the remaining isocyanate groups, to give urea linkages

$$\mathrm{R{-}NCO + H_2O \longrightarrow R{-}NH_2 + CO_2}$$

$$\mathrm{R{-}NH_2 + OCN{-}R \longrightarrow R{-}NH{-}CO{-}NH{-}R}$$

- Epoxy Resins

The reactions between oxirane (glycidyl, epoxy) groups and amines form the basis for many technically important coatings, adhesives, and structural composites, especially those containing reinforcing fibers, known as epoxies.

Most of these materials are formed from two liquid components, one containing oxirane endgroups and the other primary or secondary amines. These components are mixed just prior to use, and reaction ensues at room temperature, proceeding until it becomes diffusion limited.

The reaction of condensation is, for secondary amines:

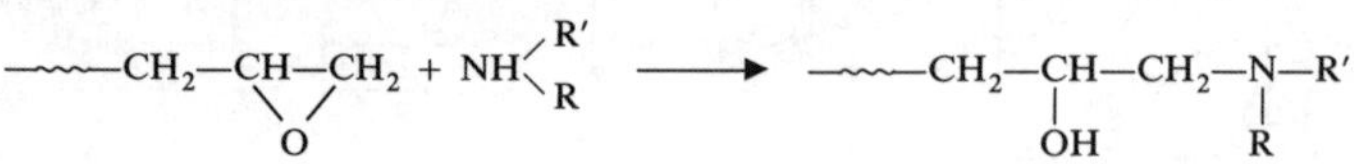

and for primary amines (each primary amine acts as a difunctional molecule):

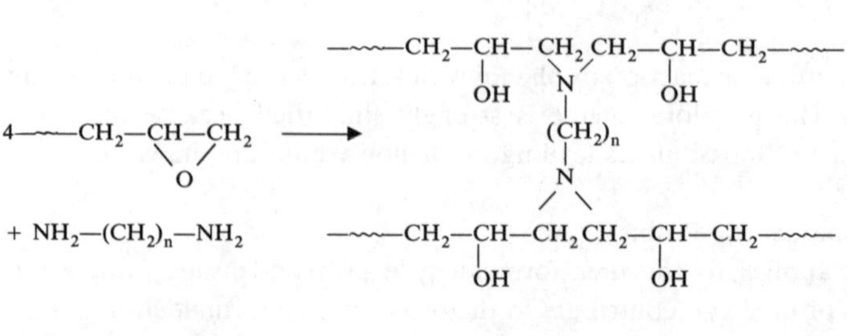

Thus dioxirane compounds (diepoxides) can be crosslinked by diamines (for example with 1,6-diaminohexane) which are tetrafunctional. The use of polyamines and polyimines will lead to gelation at very low conversions.

For composite structures, aromatic diamines are preferred, and the mixed resin after gelation is progressively heated to temperatures above the ever increasing glass transition to induce the greatest possible extent of reaction of the amine groups with the oxirane groups, so as to avoid subsequent sensitivity to moisture.

Another "epoxy" reaction is based on a single component mixture which is activated by raising the temperature to $\sim 150\,°C$. Schematically, the diepoxy compound generated from epichlorohydrin and "bisphenol A" (4,4′-dihydroxydiphenyl propane) has one or more secondary hydroxyl groups and can be represented as:

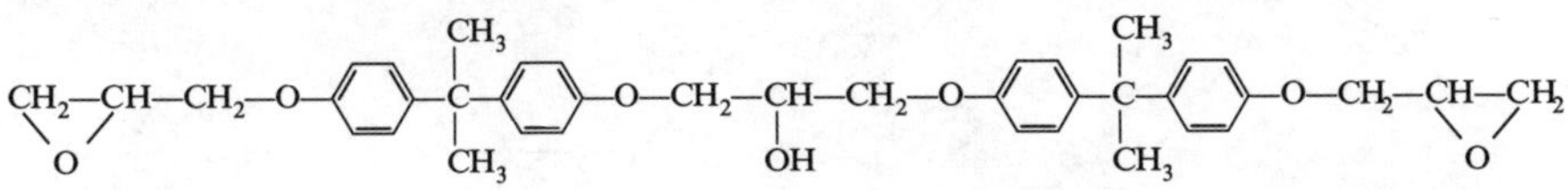

which reacts with an acid anhydride (like phthalic anhydride) at 150 °C to produce:

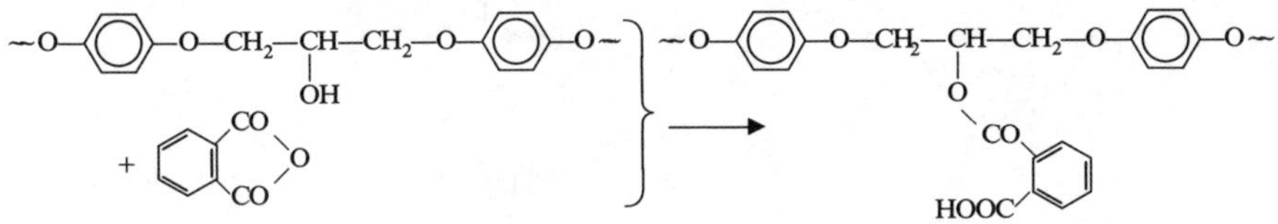

The carboxylic acid group thus formed at once reacts with the oxirane of another molecule.

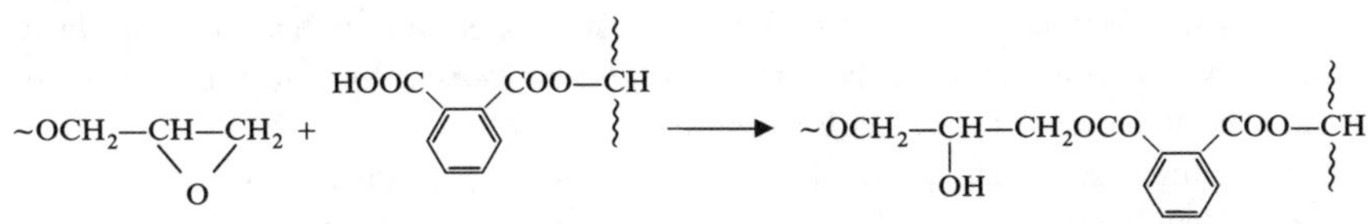

and a network is generated.

2.6 Special Cases of Polycondensation Reactions

In this section a few case are considered that still belong to polycondensation processes, though some of them are limiting cases.

2.6.1 Branched Uncrosslinked Polycondensates

Assume a difunctional base molecule $\underline{a}\underline{b}$ is reacted in the presence of some $a\langle{}^{b}_{b}$ trifunctional compound. The self condensation of $\underline{a}\underline{b}$ will yield linear chains, and every time a trifunctional species is incorporated a potential branch-point results. However, owing to the fact that $\underline{a}$ can *only* react with $\underline{b}$, crosslinking is impossible. The macromolecules formed are *tree-like*, and the number average molecular weight increases with conversion as in the case of $\underline{a}\underline{b}$ self-condensation, in spite of the fact that b̲ functions are in excess. Each molecule formed carries one single a̲ function and a number of b̲ functions equal to (l + 1), l being the number of trifunctional units incorporated in the macromolecule. In the example given below, l = 12 and the number of b̲ functions able to further react is equal to 13.

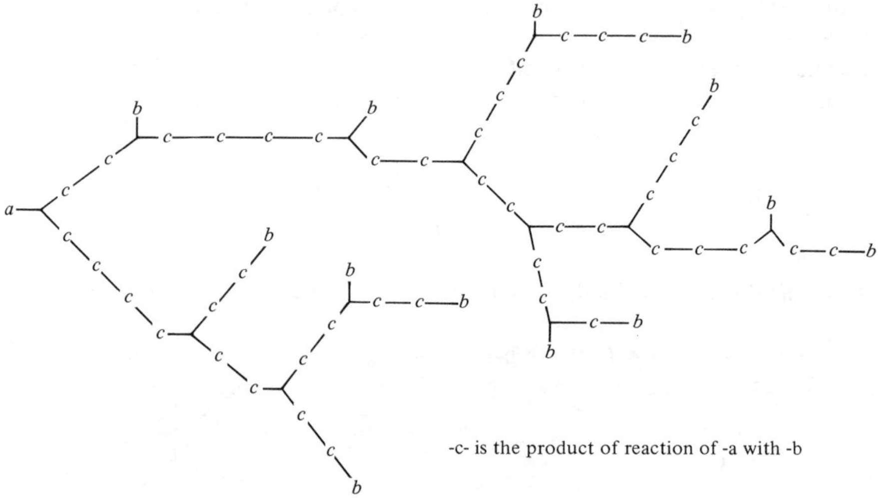

-c- is the product of reaction of -a with -b

2.6.2 Chain Extension Reactions

Step growth polycondensations have been widely applied to polymeric base molecules, too. These reactions are often referred to as *chain extension reactions* because they result in a large increase in molecular weight.

As an example one can quote the chain extension performed on α,ω-dihyroxy polymer molecules by means of a low molecular weight diisocyanate, such as MDI (methylene diphenyldiisocyanate) or HMDI (hexamethylene diisocyanate).

$$\text{HO—}\boxed{\text{polymer}}\text{—OH} + \text{OCN(CH}_2)_6\text{NCO} \longrightarrow$$

$$\text{+CO—NH—(CH}_2)_6\text{—NH—CO—O—}\boxed{\text{polymer}}\text{—O+}_n$$

The polymer can be poly(tetrahydrofurane), or poly(ethylene oxide), or a polyester fitted at both chain ends with alcohol functions. In all these cases the main purpose of chain extension is to increase the average molecular weight of the sample; this is done sometimes *upon processing* the polymer (Reaction Injection Molding).

There are, however, cases in which the molecular weight increase is not the chief aim of chain extension. It has been found of great interest to have polymer chains fitted with a succession of *"hard"* and *"soft" segments* whereby the material is strengthened, toughened, and sometimes made able to sustain higher temperatures. "Hard" segments are rather rigid, poorly soluble, sometimes crystalline. "Soft" segments are flexible, amorphous "liquid" chains, that have no mechanical strength of their own. The idea is to start from a polymer of the latter type, fitted with appropriately reactive functions at chain end, and to introduce it as one of the constituents in a polycondensation process. As an example one could quote poly(butylene oxide) (with 2 end-standing OH groups) reacting with an excess of a diisocyanate, whereby all chain ends get *capped* with isocyanate functions. Thereafter a diamine is added, in stoichiometric amount with respect to the total number of remaining available isocyanate functions. The poly(urea) linkages formed constitute the "hard" segments.

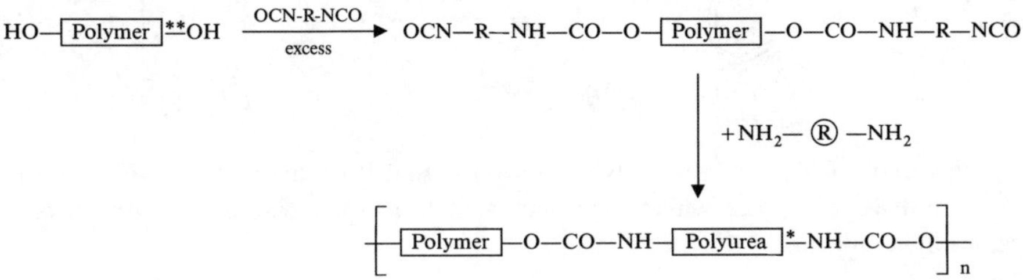

The hard segments*) associate into clusters which act as "physical" crosslinks leading to reversible elastic behavior, but by heating or use of solvents, the clusters can be dispersed and the material rendered plastic again (thermoplastic elastomers, see chapter 10).

2.6.3 Polyrecombination Reactions

When p-xylene is submitted to *pyrolysis* at temperatures of the order of 250 °*C* and the resulting gaseous compounds hit a surface which is cooled to temperatures below 30 °*C*, poly(p-xylylene) is deposited. This reaction can be considered as a polyrecombination between diradical species.

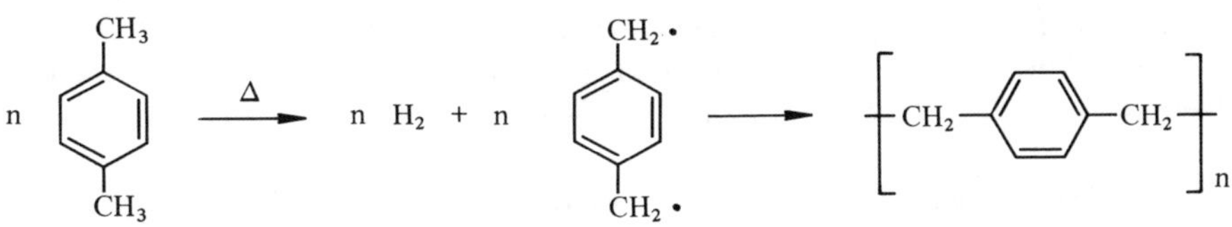

However it can also be considered as a free radical polymerization of a quinonic structure resulting from dehydrogenation of xylene.

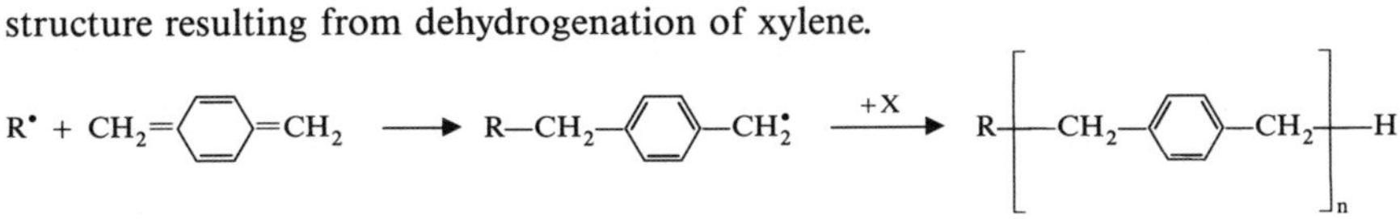

---

*) From the above this would be —R—NH—C(=O)—NH—Ⓡ—NH—C(=O)—NH—R—. The 4 hydrogen bonding nitrogen atoms of the two urea groups which are involved in crosslinking reactions lead to either high glass transition or melting temperature, thus resulting in a "hard" segment.

**) Polymer is a polyether or a polyester which is well above its glass transition or melting point (if any) at temperature of use.

More recently it was established that the initial step in the process is a dimerization reaction, and that poly(p-xylylene) results in fact from a ring opening polymerization of this cyclic dimer.

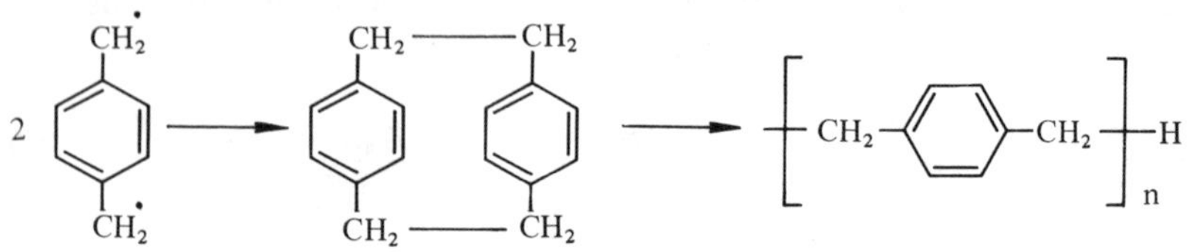

The specific feature of this polymerization is that it occurs upon condensing the vapor onto a cool surface, which might have something to do with propagation-depropagation equlibrium.

2.6.4 Ionenes

When a step growth polycondensation involves an alkylene diamine and an alkylene dihalide the polymer formed is a polyammonium salt, which becomes a polyamine upon treatment with a base:

$$Cl(CH_2)_6Cl + NH_2(CH_2)_4NH_2 \longrightarrow \left[\underset{Cl^{\ominus}}{\overset{\oplus}{N}H_2}-(CH_2)_4-\underset{Cl^{\ominus}}{\overset{\oplus}{N}H_2}-(CH_2)_6 \right]_n$$

$$\downarrow + OH^{\ominus}$$

$$\left[NH(CH_2)_4-NH-(CH_2)_6 \right]_n + n\,Cl^{\ominus} + n\,H_2O$$

In the example given cyclization does not constitute an alternative to polymer formation, as cycles with 8–12 atoms are unlikely to form. Instead of alkylene halides benzylic derivatives — such as dibromo (p-xylylene) — can be used to obtain polycondensates.

An interesting feature of this type of polycondensation is that if ditertiary amines are used together with an adequate alkylene dihalide, a real *polyelectrolyte* is obtained, the charge density depending upon the length of the base units chosen (as reflected by p and q):

$$Cl-(CH_2)_p-Cl + (CH_3)_2N-(CH_2)_q-N(CH_3)_2 \longrightarrow \left[(CH_2)_p-\underset{Cl^{\ominus}}{\overset{\oplus}{N}}(CH_3)_2-(CH_2)_q-\underset{Cl^{\ominus}}{\overset{\oplus}{N}}(CH_3)_2 \right]_n$$

2.6.5 Polyphenylene Oxides (PPO)

Polymers with phenylene oxide repeat units were first obtained by self condensation of the sodium salt of p-chlorophenol.

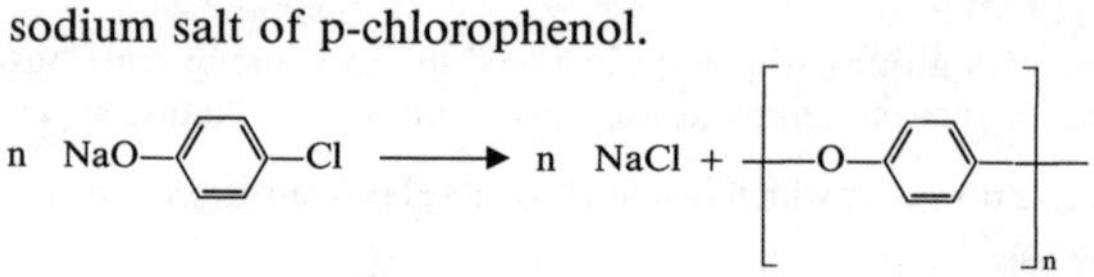

However this polymer is almost insoluble, and hard to purify.

A quite different procedure has been worked out for the synthesis of poly(2,6-dimethyl phenylene oxide), generally referred to as PPO: Controlled, mild oxidation of 2,6-dimethyl phenol in the presence of cupric salts, such as Cu^{++} $(NH_3)_2Cl_2^-$, can be obtained by bubbling oxygen slowly through a solution of the phenol.

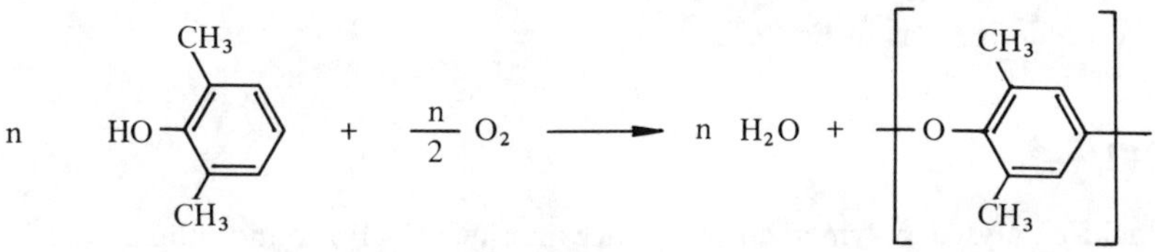

The precise mechanism of this oxidation is still under discussion, though it can be classified among step growth polymerzation processes, as any molecule present whatever its length, can still participate in further reactions.

Various other PPO's have been made. However, if the substituents in 2- and 6-position are too bulky, dimerization results — according to a different mechanism — and no polymer is obtained.

PPO's are soluble in some solvents, and they have found numerous applications as tough plastic materials e.g. "Noryl"™. They can be reacted relatively easily at the methyl group, and can be grafted subsequently.

Problem 2.1

One mole of bis(β-hydroxyethyl) terephthalate is heated under vacuum, and the glycol released is trapped quantitatively. What is the number average molecular weight of the poly(ethylene terephthalate) (PET) formed when 0.98 *moles* of glycol have been released? What is the expected polydispersity index?

Problem 2.2

Consider the case of synthesis of nylon-6,6 from 1,6-diaminohexane (hexamethylenediamine) and adipic acid, HOOC $(CH_2)_4$COOH. Two limiting cases are to be considered:

I. Exact stoichiometric equivalence, and extent of reaction $\pi = 0.995$

II. The mole ratio of the diacid to the diamine is 0.99 and all carboxylic acid groups are completely reacted.

a) Determine $\overline{P}_n$ and $\overline{P}_w$ for each case.

b) Determine the value of i at which w_i is a maximum (derive first a general expression to obtain this).

c) Determine the sum of the mole fractions and of the weight fractions of species of degrees of polymerization 1 through 5 inclusive.

d) To limit the degree of polymerization in polycondensation one can add a species $\underline{b}\,\underline{x}$ to $\underline{b}\,\underline{b}$ and $\underline{a}\,\underline{a}$ where $\underline{b}\,\underline{x}$ is monofunctional (only $\underline{b}$ is reactive, $\underline{x}$ is a "dead" end). If $N_{b,0} \geqq N_{a,0}$, show that the above case can be treated by the equation:

$$\bar{P}_n = \frac{1 + r}{1 - r + 2r(1 - p)}$$

if you redefine r as in Eq. 2.12.

Problem 2.3

In an acid catalysed polymerization of decamethylene glycol and adipic acid in equimolar proportions at 109.6 °C, the degree of polymerization $\bar{P}_n$ was followed by end group titration, and was found to increase from 8 at time zero to 50.0 at 300 *min*, all conditions being held constant.

a) Determine the rate constant k' as defined by:
$-\,d[a]/dt = k' \cdot [a] \cdot [b]$, where the concentrations are in $[\text{mol} \cdot l^{-1}]$ t in $[s]$. Note that $[a]/[a]_0 = N_a/N_{a,0}$ where N_0 = initial number of groups.

b) The equilibrium constant $K = \dfrac{[-C(O)O-][H_2O]}{[-COOH][-OH]}$ is of order unity at 109 °C.
What molar ratio of dissolved water of condensation to carboxyl end group concentration would lead to an *equilibrium* value of $\bar{P}_n = 50$ at 109 °C ($\bar{P}_n$ = degree of polymerization).

c) Suppose in the removal of water from the start of reaction 2% of the initially charged glycol were lost by steam distillation, but no adipic acid. What maximum degree of polymerization could be achieved?

Problem 2.4

The diisocyanate CHDI (1,4-trans cyclohexane diisocyanate) is to be reacted with a polyether tetrol, schematically

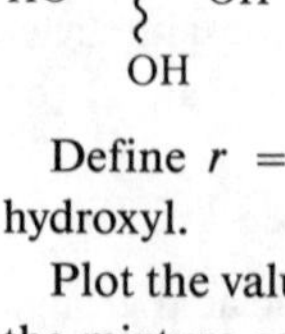

Define $r = [NCO]_0/[OH]_0$, the ratio of initial concentrations of isocyanate to hydroxyl.

Plot the value of π_{NCO} the fraction of NCO groups reacted, *below which* gelation of the mixture cannot take place, as a function of r for r: 1/4, 1/2, 3/4, 1, 2, 3, 4.

Here:

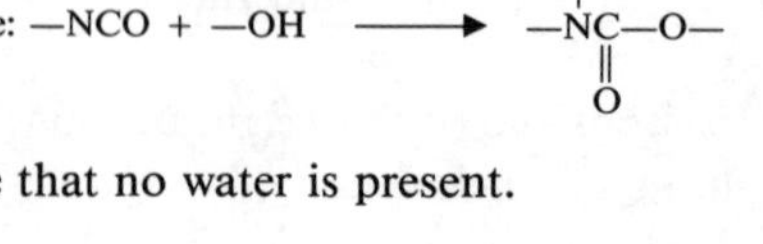

Assume that no water is present.

Problem 2.5

a) Reacting sebacic acid $HOOC(CH_2)_8COOH$ and hexamethylene diamine in stoichiometric amounts in an open batch reactor at 230 °C, what is the % weight loss at 99% conversion? What is the number average degree of polymerization?

b) If the equilibrium constant K of the polycondensation process is 100 at 230 °C, what concentration of water (mol/liter) would bring the reaction to equilibrium at 99% conversion? Assume that the density of the melt is 0.9 g/ml.

c) If the same reaction is now carried out at 230 °C with hot N_2 purge to keep the dissolved water concentration equal to zero, a conversion degree of 99% is attained in 6.2 hours. Deduce the forward rate constant, showing any assumptions necessary.

Problem 2.6

Nylon-11 and Nylon 6-10 (one mol of each) are melted together without loss or gain of water. The number average degree of polymerization of each of the nylons was 100. Calculate the number average and the weight average molecular weight of the final product.

Problem 2.7

Given a mixture of x moles of tetrafunctional molecules having A ends, y moles of bifunctional molecules having A ends, and z moles of bifunctional molecules carrying B ends. The only possible reaction is between A and B functions to give a link -ab- connecting two units. Call r the initial stoichiometric ratio between A and B functions, and ρ the proportion of A functions belonging to tetrafunctional molecules.

a) Find algebraically the range of values of r and ρ within which, after all A functions have reacted ($\pi_a = 1$), a network will have formed, and outside of which no gelation will occur.

b) Given the values of r and of ρ at what extent of conversion π_a will gelation occur? Calculate, for this value of π_a, the probabilities of sequence lengths to be formed between two successive branch points (see page 49).

Appendix 2A

Summations Often Needed in Calculation of Molecular Weight Distributions of Polymers:

1) $\sum_{1}^{n} p^{x-1}$ p being a constant: $0 < p < 1$

$x = 1, 2, 3 \ldots \infty$

$$\sum p^{x-1} = 1 + p + p^2 + \ldots + p^{n-1}$$

$$p\sum p^{x-1} = p + p^2 + p^3 + \ldots + p^n$$

By subtracting

$(1 - p)\ \Sigma p^{x-1} = 1 - p^n$

As $n \to \infty$, $p^a \to 0$ it follows:

$$\sum_{1}^{a} p^{x-1} = \frac{1}{1 - p} \tag{A.1}$$

2) $\sum_{1}^{a} p^{x-1}$ x and p defined as above

$\Sigma\ xp^{x-1} = 1 + 2p + 3p^2 + \ldots np^{a-1}$

$p\Sigma\ xp^{x-1} = p + 2p^2 + 3p^3 + \ldots + np^a$

Subtracting

$(1 - p)\ \Sigma\ xp^{x-1} = 1 + p + p^2 + p^3 + \ldots -p^a$

Taking into account the previous result, with $p^a \to 0$,

$$\sum_{1}^{n} xp^{x-1} = \frac{1}{(1 - p)^2} \tag{A.2}$$

3) $\sum_{1}^{n} x^2p^{x-1}$ x and p defined as above

$\Sigma\ x^2p^{x-1} = 1 + 4p + 9p^2 + 16p^3 + \ldots + n^2\,p^{n-1}$

$p\ \Sigma\ x^2\,p^{x-1} = p + 4p^2 + 9p^3 + 16p^4 + \ldots + n^2\,p^n$

subtract

$(1 - p)\ \Sigma x^2\,p^{x-1} = 1 + 3p + 5p^2 + 7p^3 + \ldots - n^2\,p^n$

$p(1 - p)\ \Sigma\ x^2\,p^{x-1} = p + 3p^2 + 5p^3 + 7p^4 + \ldots -n^2\,p^{n+1}$

subtract

$(1 - p)^2\ \Sigma\ x^2\,p^{x-1} = 1 + 2p + 2p^2 + 2p^3 + \ldots + n^2\,p^{n+1}$

$p(1 - p)^2\ \Sigma\ x^2\,p^{x-1} = p + 2p^2 + 2p^3 + 2p^4 + \ldots + n^2\,p^{n+2}$

subtract

$(1 - p)^3\ \Sigma\ x^2\,p^{x-1} = 1 + p - (n^2\,p^{n+2})$

$$\text{Thus } \Sigma\ x^2\,p^{x-1} = \frac{1 + p}{(1 - p)^3} \tag{A.3}$$

Similarly

$$\sum_{1}^{n} x^3p^{x-1} = \frac{1 + 4p + p^2}{(1 - p)^4} \tag{A.4}$$

Chapter 3
Free Radical Chain Polymerization

3.1 Introduction

Free radical chain polymerizations are by far the method most commonly used in industry to produce polymers. Polyethylene, polystyrene, poly(vinyl chloride) — the three polymer of major economic importance — but also poly(vinyl acetate), poly-(methyl methacrylate), poly(acrylonitrile) and many copolymers are manufactured by free radical polymerization.

There is, however, a great variety of such processes, depending on the characteristics of each system, whether it is homogeneous or hetereogeneous, whether the polymerization is carried out in bulk, in solution, in suspension or in emulsion, whether the polymer is soluble in its own monomer or not, whether the process involves one or several steps, etc. In this chapter, the discussion of the mechanisms of free radical polymerization, of the kinetics of the process, and of the consequences to be expected on molecular weight and molecular weight distribution will be limited to the case of homogeneous polymerization in solution. These considerations remain valid, to some extent, in bulk polymerizations and in suspension polymerizations, provided the polymer is soluble in its monomer. Emulsion polymerizations are quite different and will be considered in chapter 12.

The *propagating site* is a free radical, an unpaired electron at the last carbon atom of the growing chain. A free radical polymerization involves many successive steps of growth. In each of them a monomer molecule is added to the chain, whereby the radical site is reformed on the newly attached last unit of the chain. For example, an individual propagation (growth step) of the polymerization of a vinyl monomer $CH_2{=}CHR$ can be schematized as:

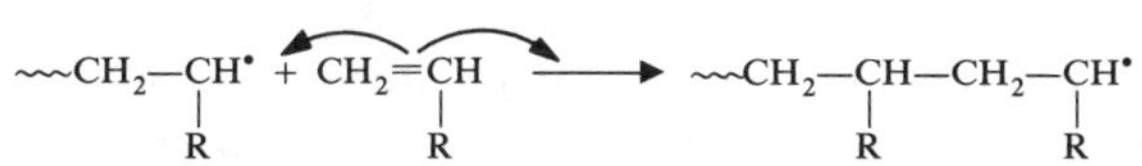

3.2 Initiators and Monomers Fitted for Free Radical Polymerization

3.2.1 Initiators and Initiation Processes

Substances that generate radicals by *homolytic scission* when heated or irradiated, or by any other means, may initiate free radical polymerizations, under certain conditions. The free radical initiators can be classified in several groups, of which the most important are:

a) Peroxides, organic or inorganic, like dibenzoyl peroxide (which is in fact a peranhydride, C_6H_5—CO—O—O—CO—C_6H_5), or potassium peroxodisulfate ($K_2S_2O_8$),
b) Hydroperoxides such as t-butylhydroperoxide: $(CH_3)_3C$—O—OH
c) Peresters like t-butyl perbenzoate: C_6H_5—CO—O—OtBu
d) Aliphatic azo compounds, such as azobisisobutyronitrile (AIBN):

$$(CH_3)_2\underset{\displaystyle C\equiv N}{\underset{|}{C}}-N=N-\underset{\displaystyle C\equiv N}{\underset{|}{C}}(CH_3)_2$$

Peroxides and azo compounds usually yield two *identical* radicals upon homolytic decomposition:

$$(CH_3)_2\underset{\displaystyle CN}{\underset{|}{C}}-N=N-\underset{\displaystyle CN}{\underset{|}{C}}(CH_3)_2 \longrightarrow N_2\uparrow + 2(CH_3)_2\underset{\displaystyle CN}{\underset{|}{C^\bullet}}$$

whereas peresters and hydroperoxides yield two *different* radicals which may not exhibit the same reactivity with respect to the monomer:

$$(CH_3)_3C-O-OH \longrightarrow (CH_3)_3C^\bullet + OH^\bullet$$

Not all free radicals can initiate the polymerization. Counter-examples are resonance stabilized or quite bulky radicals, like diphenylpicrylhydrazyl or triphenylmethyl, which would rather act as radical scavengers, and thus inhibit polymerization.

The activation energy of homolytic scission is rather high, which means that the rate of initiator decomposition is very much temperature dependent, as it will be seen later. A consequence is that for each initiator there is a domain of temperature that fits well for initiation: At lower temperatures the rate of formation of primary radicals is too slow; at higher temperatures the initiator decomposes too fast. The optimum temperature range is that leading to half life times*) of the order of 5 to 20 hours, compatible with the desired duration of a free radical polymerization process. Some data are displayed in Table 3.1.

UV irradiation is used sometimes to help cleave the initiator molecule. Thus the polymerization temperature can be lowered, but nothing is changed in the mechanism with respect to the thermal cleavage of the initiator.

Radical sites can also be created by a large variety of *oxidation-reduction ("redox") processes,* as in the examples quoted below:

$$Fe^{2+} + H_2O_2 \longrightarrow Fe^{3+} + \dot{O}H + \bar{O}H$$

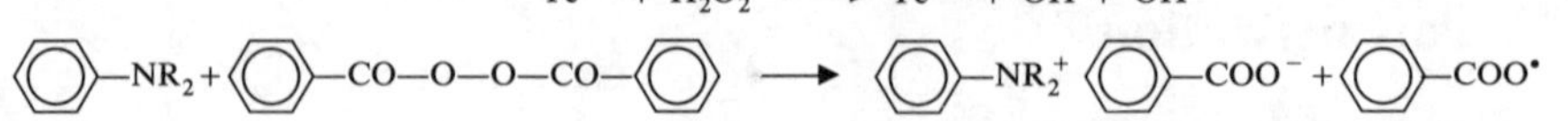

$$R-CH_2-OH + Ce^{4+} \longrightarrow R-\dot{C}H-OH + Ce^{3+} + H^+$$

$$S_2O_8^{2-} + S_2O_3^{2-} + H_2O \longrightarrow SO_4^{2-} + SO_4H^- + S_2O_3^{\bullet -} + \dot{O}H$$

*) The half-life time is the time necessary for half of the initiator molecules to have undergone homolytic decomposition (see § 3.4).

Table 3.1. Homolytic decomposition of various free radical initiators

Initiator	k_d (s^{-1})	at T (°C)	E_d (*Kcal/mol*)
AIBN	$0.85 \cdot 10^{-5}$	60	30.8
BPO	4.9×10^{-5}	85	29.7
Cumylperoxide	1.6×10^{-5}	115	40.7
t-BuOOH	0.45×10^{-5}	155	40.8

Redox initiators are commonly used in aqueous systems, and in emulsion polymerization processes, which require radical generation in the continuous phase. There are also redox systems operating in organic media.

Thermal Initiation and Photoinitiation

Some chain polymerizations can be started without added initiator. Thermal initiation has been applied for a long time to certain monomers (styrene, methylmethacrylate). The mechanisms involved are still controversial, in spite of detailed experimental search.

Irradiation of a monomer by UV light, in the presence of a suitable photosensitive agent, such as benzophenone, is also an efficient way to initiate free radical polymerizations. The initiation is usually attributed to free radicals originating from the decomposition of the light activated photosensitizer.

An alternate way is the photolysis of monomers by means of high energy irradiation (γ-rays, X-rays, or electron beams). These reactions generally produce radical species, e. g. by hydrogen transfer between two monomer molecules, as it was established by electron spin resonance (ESR):

$$2\,CH_2{=}CHR \xrightarrow[\text{High Energy Radiations}]{h\nu} CH_3{-}\dot{C}HR + CH_2{=}\dot{C}R$$

3.2.2 Monomers Suited for Free Radical Polymerization

Most of the vinylic, acrylic, and dienic monomers can undergo free radical polymerization. Free radical initiators are, however, unable to initiate ring opening polymerizations.

Here are a few examples of monomers that can be polymerized by free radical processes:

- *Ethylene* polymerizes only at elevated temperatures and very high pressures to yield highly branched macromolecules exhibiting a low degree of crystallinity. These polymers are referred to as low density polyethylenes (LDPE), in contrast to those generated by Ziegler-Natta processes, which are the linear, highly crystalline, high density polyethylene (HDPE).
- Propylene and its homologues yield only low molecular weight oligomers when polymerized with free radical initiators. The polymers are of no practical use.

• Styrene and substituted styrenes, dienes, vinyl pyridines, alkyl methacrylates and acrylates, acrylontrile, vinyl chloride and vinyl acetate are all well suited for free radical polymerizations. Acrylic and methacrylic acid, as well as acrylamide can be polymerized in water solution using water soluble initiators.

• Fluorinated monomers (vinylidene fluoride, trichlorofluoroethylene, tetrafluoroethylene) undergo free radical polymerization, although the polymers are crystalline, insoluble, and melt at temperatures far above those used in the process. The polymers are formed as a white powder.

A number of unsaturated monomers are insensitive to free radical polymerizations. Such is the case of:

• Monomers with symmetrically disubstituted unsaturations, like maleic anhydride, 2,3 dimethylbutene, fumaric esters.

• Monomers carrying two bulky substituents, like 1,1-diphenyl ethylene.

• Allylic monomers, because of the resonance stabilization of the radical sites: Transfer is dominant.

• α-methylstyrene and some other monomers because of thermodynamic reasons which will be discussed later: Propagation is reversible.

It should be noted that many of the monomers quoted as unable to undergo free radical polymerization are capable of copolymerizing with other monomers. Such monomers (α-methyl styrene, maleic anhydride, vinyl ethers, diphenyl ethylene) are not insensitive to the attack by radicals, they are merely unable to react with radicals derived from themselves.

Free radical polymerizations can be carried out in *solution* in an inert solvent, in the *bulk*, in *suspension,* or in *emulsion.* Polymerizations in solution are preferred in laboratories because of low viscosity and ease of handling. In industrial processes solvents are generally avoided because they are expensive, they require additional safety equipment, and recycling units. Bulk, suspension and emulsion polymerizations are therefore preferred in industry, and these methods will be considered in chapter 12.

3.3 Mechanism of Free Radical Polymerization

We shall now focus our attention to the polymerization of a vinylic monomer, initiated by a symmetrical free radical initiator and carried out in homogeneous solution in an inert solvent. The polymer to be formed is supposedly soluble in the reaction medium, but the viscosity increase due to polymer formation is not considered. As already mentioned, the kinetic equations that will be obtained are not necessarily valid if the above conditions are not satisfied, and they do not apply to emulsion processes.

3.3.1 Initiation

The initiation reaction is the attack of a monomer molecule by a primary radical originating from the initiator. This process involves two reactions:

- Decomposition of the initiator to form primary radicals:

$$C_6H_5CO{-}O{-}O{-}CO{-}C_6H_5 \longrightarrow 2\,C_6H_5COO^{\bullet}$$

or

$$I \longrightarrow 2\,R^{\bullet}$$

- The actual initiation reaction:

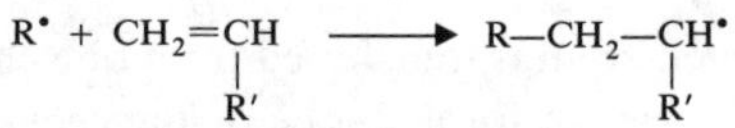

or

$$R^{\bullet} + M \longrightarrow RM^{\bullet}$$

Not all primary radicals formed will react with the monomer. In the above example of dibenzoyl peroxide, several other reactions are known to occur:

$$C_6H_5{-}COO^{\bullet} \longrightarrow C_6H_5^{\bullet} + CO_2\uparrow$$

$$C_6H_5^{\bullet} + C_6H_5COO^{\bullet} \longrightarrow C_6H_5COOC_6H_5$$

$$2\,C_6H_5^{\bullet} \longrightarrow C_6H_5{-}C_6H_5$$

Each molecule of biphenyl or of phenylbenzoate formed means loss of two primary radicals. In the case of AIBN, similarly, dimethyl succinonitrile is formed by recombination of two primary radicals, which otherwise could have initiated chains. The "initiator efficiency factor" f is defined as the probability for a primary radical to react with a monomer and to initiate a chain; thus $(1 - f)$ is the probability that it will react with another primary radical, to form a "dead" product*).

3.3.2 Propagation

This reaction is repeated hundreds to many thousands of times for each chain formed, as already stated; it can be written as:

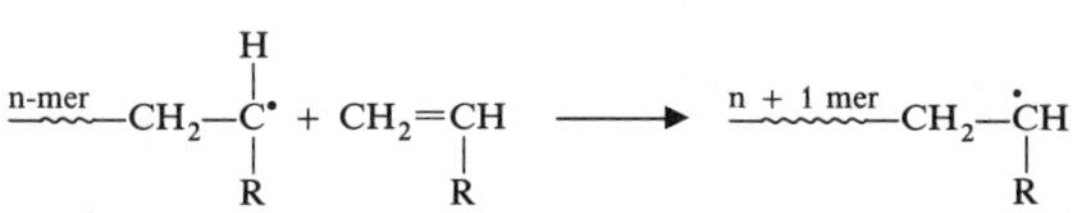

or

$$RM_n^{\bullet} + M \longrightarrow RM_{n+1}^{\bullet}$$

It is generally assumed that the rate constant of this reaction remains the same regardless of the length of the chain to which the radical site is attached.

---

*) This is not relevant to redox initiation, nor for any process involving more than one phase.

3.3.3 Termination

Growing radicals can react with each other in two different ways:

- By *recombination*, a homopolar bond is formed by pairing the single electrons of the free radical sites of two chains:

$$R{-}M_n^{\bullet} + R{-}M_p^{\bullet} \longrightarrow R{-}M_{n+p}{-}R$$

- By *disproportionation*, whereby a hydrogen atom is transferred from one chain to the other; two molecules of "dead" polymer are formed, one of them bearing a double bond at chain end.

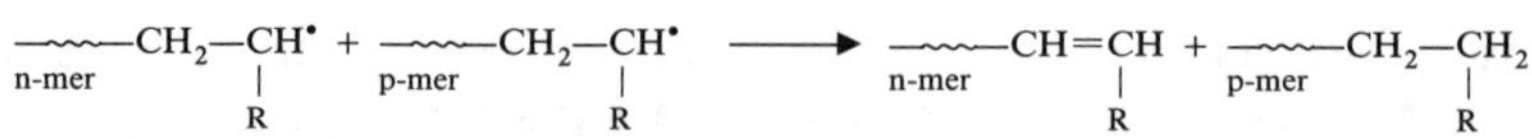

In both cases, the reaction involves mutual destruction of radicals, and kinetically these two processes cannot be distinguished. But their consequences as to what concerns the molecular weight and polydispersity of the polymers formed are different, as it will be seen.

3.3.4 Transfer Reactions

It can happen that a free radical site reacts with a molecule of solvent, or initiator, or monomer, or even with the polymer itself. In the last case branches are formed. The result of such reactions is that the growing site at chain end is removed. But another radical arises from the molecule which has reacted, and if it is able to add monomer it gives rise to another macromolecule, as shown on the following schemes:

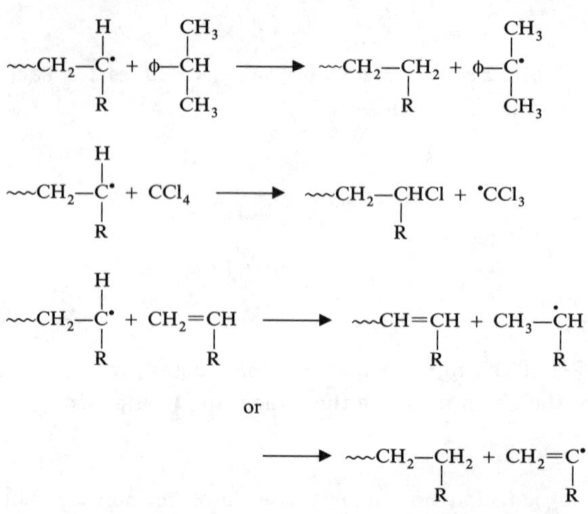

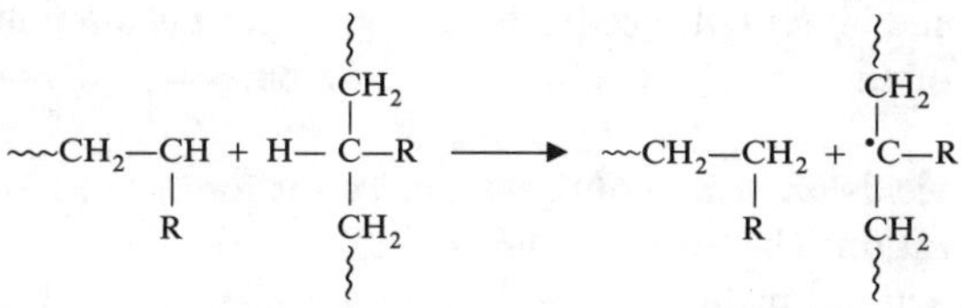

The total number of active radical sites is *not* modified by the occurrence of these "transfer reactions"; if the radicals formed upon transfer are active (with respect to the monomer), the kinetic consequences of transfer processes are of minor importance. But the occurrence of transfer processes strongly influences the molecular weight of the polymer formed. Instead of building *one* polymer molecule (or 1/2 if termination occurs by recombination) a given radical site can produce *several* polymer molecules during its lifetime, if transfer is involved. These effects will be discussed in greater detail in the next section.

Primary radicals can also get involved in transfer reactions. Conversely transfer to the initiator is usually very small. There are however exceptions, which will be discussed in section 3.10.

It can happen that radicals formed upon transfer are much less reactive than the radicals arising from the monomer. The polymerization process is then retarded or even inhibited.

The *kinetic chain length* is defined as the total number of growth steps (i. e. of monomer units added) that have occured during the life-time of a radical — from its formation to its termination by mutual annihilation with another radial. If the process does not involve transfer reactions the kinetic chain length is equal to the average degree of polymerization attained by the radicals when they terminate. If transfer has occured, each radical has produced several chains and the kinetic chain length is then usually far larger than the polymerization degree of the polymer formed.

3.3.5 Remarks

A few remarks have to be made on the mechanism of radical chain polymerizations:

1. The lifetime of each *individual* radical is very short. Initiation, the many steps of chain growth, possible transfer reactions, and mutual termination occur usually*) within a second or so. As a consequence, the concentration of radicals stays very low in the reaction medium, and in most cases its value is experimentally inaccessible.
2. The probability for any given transfer reaction to occur is usually low. But since the growth process can involve 100 to 1,000,000 steps, the probability of transfer being the same at each stage, it is impossible to discard the possibility of transfer, even when no specific transfer agents have been added.

*) If the medium becomes glassy or crystalline, radicals can become "frozen" in, and survive for much longer periods of time.

3. Transfer reactions are sometimes used specifically to obtain short chain polymers. A well defined amount of transfer agent is then added on purpose, to *"regulate"* the molecular weight.
4. Transfer to the polymer yields *branching* of the chain. This is for instance the case in vinyl acetate polymerization: The corresponding radicals have a marked tendency to give rise to transfer reactions, including onto its own polymer. However in most cases, especially when operating in solution and to low conversions, transfer to the polymer can be neglected to a first approximation, and the polymer molecules are expected to be linear.

3.4 Kinetics of Free Radical Polymerization

Initiation

Let us assume that the initiator I decomposes into two equally reactive primary radicals, with a rate constant k_d, according to

$$\mathrm{I} \xrightarrow{k_d} 2\mathrm{R}^\bullet \qquad -\frac{d[\mathrm{I}]}{dt} = k_d\,[\mathrm{I}] = \frac{1}{2}\,\frac{d[\mathrm{R}^\bullet]}{dt} \tag{3.1}$$

From equation (3.1) it follows that the concentration of the initiator in the reaction medium decreases exponentially:

$$[\mathrm{I}] = [\mathrm{I}]_0 \cdot e^{-k_d t} \tag{3.2}$$

If the half-life time of this initiator under the experimental conditions chosen is long as compared with the duration of the polymerization process, the initiator concentration can be assumed to be constant*).

The rate of formation of the actual active species (an unpaired electron on a monomer unit) can be written as:

$$\mathrm{R}^\bullet + \mathrm{M} \longrightarrow \mathrm{RM}^\bullet \text{ (or } \mathrm{M}_1^\bullet) \qquad \frac{d[\mathrm{M}_1^\bullet]}{dt} = 2\,k_d f[\mathrm{I}] \tag{3.3}$$

Equation (3.3) implicitely assumes that the decomposition of the initiator is the rate determining step. Here f is the probability factor for a primary radical to undergo reaction with a monomer M rather than to combine with another radical and form a "dead" product.

It is also possible to obtain equation (3.3) by assuming steady state conditions on the primary radicals. The rate of formation of the "effective" primary radicals $\mathrm{R}_e^\bullet$ can be written as:

$$\frac{d[\mathrm{R}_e^\bullet]}{dt} = 2k_d\,f[\mathrm{I}] \tag{3.3 a}$$

*) In fact the initiator concentration can also be maintained constant by continuously adding it to the reaction mixture at a rate equal to its decomposition rate.

Initiation involves consumption of these primary radicals by reaction with the monomer:

$$-\frac{d[R_e^{\bullet}]}{dt} = k_i [R_e^{\bullet}][M] \qquad (3.3\,b)$$

If both rates are the same the steady state concentration of effective primary radicals can be expressed as:

$$[R_e^{\bullet}] = \frac{2 k_d f[I]}{k_i [M]} \qquad (3.3\,c)$$

By introducing this value into equation (3.3 b) one obtains equation (3.3).

Propagation

The propagation steps can be written as follows, k_p being the rate constant of propagation:

$$R{-}M_n^{\bullet} + M \xrightarrow{k_p} R{-}M_{n+1}^{\bullet}$$

From this expression it follows that the rate of monomer consumption (neglecting the monomer consumed in the initiation process) is given by*):

$$R_p = -\frac{d[M]}{dt} = k_p[M^{\bullet}][M] \qquad (3.4)$$

where $[M^{\bullet}]$ stands for the steady state concentration of active sites, in the process of growth, *regardless of their size*, assuming equal reactivity for each of them:

$$[M^{\bullet}] = \sum_i [M_i^{\bullet}]$$

Termination

A termination reaction always involves two radicals, reacting either by *recombination* or by *disproportionation.* In the former case, a bond is formed between two growing radicals, whereby the polymer molecule is — on an average — twice the size of the radicals upon termination. In the latter case, a hydrogen atom is transferred from one macroradical to the other, whereby one of the polymer chains is saturated (M_i),

*) In bulk polymerizations of undiluted monomer, the meaning of [M] is unambiguous (unless more than one phase is present). When two or more monomers, or monomer and solvent, are part of the mixture, the global concentration of monomer [M] may not correspond to the local concentration around the radical chain end. The true concentration locally could be higher or lower. This question arises from phase equilibria phenomena: for example a single polymer type dissolved in a good solvent *S* and a non solvent *NS* can phase separate into polymer rich and polymer poor phases, with the *S/NS* ratio being conceivably several fold higher in the former than in the latter.

the other involves a terminal double bond ($M_{j\|}$); no change in size is involved, but none of the polymer species formed remains able to further react with monomer.

The termination reactions can be written as:

$$M_i^{\bullet} + M_j^{\bullet} \longrightarrow M_{i+j} \text{ (Recombination)}$$

$$M_i^{\bullet} + M_j^{\bullet} \longrightarrow M_i + M_{j\|} \text{ (Disproportionation)}$$

Kinetically the two types of termination are identical.

The rate of consumption of growing radicals is given by*):

$$-\frac{d[M^{\bullet}]}{dt} = k_t\,[M^{\bullet}]^2 \qquad (3.5)$$

If the number of growing sites is assumed to stay constant (over small time intervals) the rates of initiation and of termination are to be set equal. The "steady state equation" results:

$$k_t\,[M^{\bullet}]^2 = 2k_d\,f[I]$$

This equation can be solved for $M^{\bullet}$:

$$[M^{\bullet}] = \left(\frac{2k_d\,f[I]}{k_t}\right)^{1/2} \qquad (3.6)$$

Introducing this expression into equation (3.4), one obtains the general equation for the rate of polymerization:

$$R_p = -\frac{d[M]}{dt} = k_p\left(\frac{2k_d f[I]}{k_t}\right)^{1/2}\cdot[M] = \left(\frac{k_p^2}{k_t}\cdot R_i\right)^{1/2}\cdot[M] \qquad (3.7)$$

From this equation it follows that the rate of propagation (which is the overall rate of monomer consumption) is *first order* with respect to *monomer concentration* (proportional to [M]) and *one-half order* with respect to the *initiator concentration* (proportional to $[I]^{1/2}$).

The assumptions underlying this calculation are the following:

- k_p is independent of the length of the chain to which the growing sites are attached.
- The concentration of the active sites is constant, though the lifetime of each of them is extremely short. This implies that the initiator concentration stays constant, or does not change significantly during the process. The validity of this hypothesis is limited to low conversions, more precisely for process durations that are short as compared with the half-life time of the initiator.

*) In many books and articles, a factor of 2 appears in equation (3.5). This factor is kinetically unjustified, since obviously each radical can react with the $(n-1)$ other radicals present. The number of such events is $n(n-1)/2 \cong n^2/2$. This factor 1/2 cancels out the factor 2 implying that 2 radicals disappear simultaneously at each of these events.

- If this condition is not fulfilled account has to be taken of the initiator consumption during the process, by introducing equation (3.2) into the rate equation (3.7):

$$R_p = -\frac{d[M]}{dt} = k_p \left(\frac{2k_d f}{k_t}\right)^{1/2} [M]_t [I]_0^{1/2} e^{-\left(\frac{k_d t}{2}\right)} \tag{3.8}$$

To express the conversion as a function of time, equations (3.7) and (3.8) have to be integrated. Calling $[M_0]$ and $[I_0]$ the initial monomer and initiator concentrations, one gets:

- If [I] is held constant or does not vary over the period of time t (Fig. 3.1, 3.2):

$$\ln \frac{[M]_0}{[M]} = k_p \left(\frac{2k_d f}{k_t}\right)^{1/2} [I]^{1/2} t \tag{3.9}$$

- If the initiator consumption during the process is taken into account:

$$\ln \frac{[M]_0}{[M]} = 2k_p \left(\frac{2f}{k_d k_t}\right)^{1/2} [I]_0^{1/2} \left[1 - e^{-\left(\frac{k_d t}{2}\right)}\right] \tag{3.10}$$

Since the degree of conversion π is given by $\frac{[M]_0 - [M]}{[M]_0}$, equations (3.9) and (3.10) are suited to express the degree of conversion versus time, as:

$$\ln \frac{[M]_0}{[M]} = -\ln(1 - \pi) \qquad (3.9^a), (3.10^a)$$

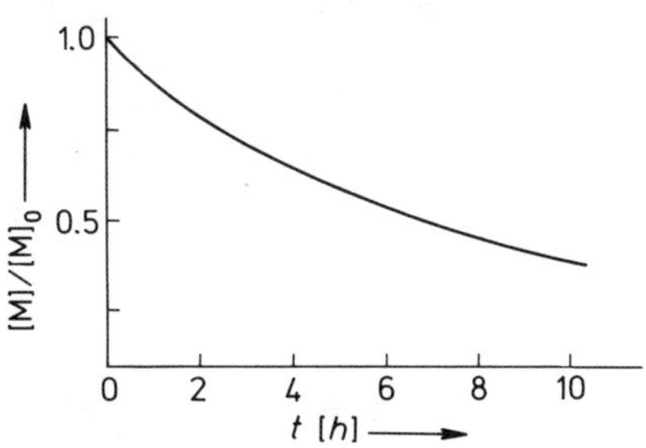

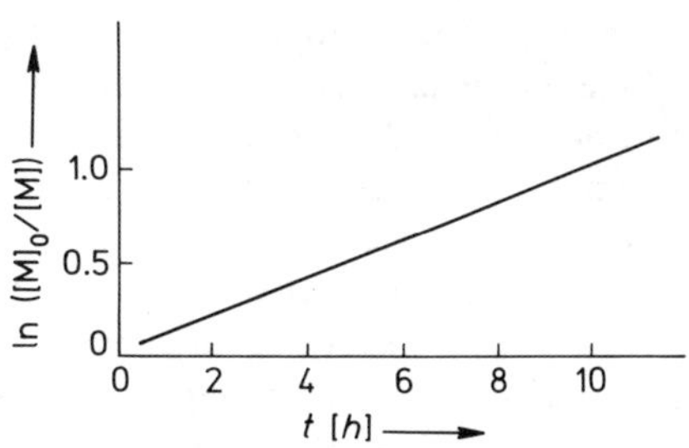

Fig. 3.1. Monomer consumption in a free radical polymerization of styrene

Fig. 3.2. First order plot of the free radical polymerization shown in Fig. 3.1 ([i] is assumed to be constant.)

Since the kinetics of polymerization are not much affected by transfer processes, the above equations usually hold, whether transfer reactions occur or not, provided the process is carried out in a single phase: Polymer soluble in its own monomer, or in the solvent used.

3.5 Kinetic Chain Length, Average Degree of Polymerization and Molecular Weight Distribution

The *kinetic chain length* ν is defined as the average number of steps of growth per effective radical*). It is given by the ratio of the propagation rate to the initiation rate. Taking into account the steady-state conditions, it is expressed as:

$$\nu = \frac{R_p}{R_i} = \frac{k_p[M]}{k_t[M^\bullet]} \tag{3.11}$$

Replacing $[M^\bullet]$ by its value taken from equation (3.6), the kinetic chain length is given by:

$$\nu = \frac{k_p[M]}{(2k_d f k_t[I])^{1/2}} = \frac{k_p^2[M]^2}{k_t R_p} \tag{3.12}$$

In the absence of transfer processes, ν is equal to the number average degree of polymerization $\overline{P}_n$ of the *radicals* formed. If termination occurs exclusively by disproportionation ν is equal to $\overline{P}_n$ of the polymer, whereas if termination occurs by recombination of two growing chains, the degree of polymerization obtained $\overline{P}_n$ is twice the value of ν. In general the two ways of termination coexist, and one can write:

$$P_n = 2\,a\,\nu \qquad \text{with: } 1/2 \leqslant a \leqslant 1$$

the parameter a characterizing the proportion of chains which have undergone termination by recombination: It is equal to one if all of them do, and to 1/2 if termination occurs by disproportionation only. It should also be emphasized that the expression for ν contains the actual monomer concentration [M]. The above equations therefore refer to the *instantaneous* kinetic chain length and to the *instantaneous* degree of polymerization. As the monomer concentration decreases the average number of growth steps per radical also decreases, implying that the degree of polymerization is lowered correspondingly.

Equation (3.12) can also be obtained in a more formal way. Let us consider a polymer radical, and assume it can either react with a monomer and grow, or disproportionate by reaction with another radical, in which case it vanishes. If p is the probability of growth and $(1 - p)$ the probability of chain termination:

$$p = \frac{k_p[M]}{k_p[M] + k_t[M^\bullet]} \qquad 1 - p = \frac{k_t[M^\bullet]}{k_p[M] + k_t[M^\bullet]} \tag{3.13}$$

The probability for a given radical to undergo $(i - 1)$ successive growth steps before terminating is given by:

---

*) This definition is unambiguous for bulk or solution polymerization if account is taken of f. It has no meaning in emulsion polymerization, and will be discussed in chapter 12.

$$x_i = p^{i-1}(1 - p) \tag{3.14}$$

The degree of polymerization of this species is i (because one monomer unit arises from initiation). x_i is also the fraction of *radicals* which have reached a degree of polymerization i. Therefore, the number average degree of polymerization is given by:

$$\bar{P}_n = \sum_{i=1}^{\infty} ix_i = (1 - p)\sum_{i=1}^{\infty} ip^{(i-1)} \tag{3.15}$$

or, after summation:

$$\bar{P}_n = \frac{1}{1 - p} \tag{3.16}$$

For values of p close to unity, this expression is identical with equation (3.12)

$$\bar{P}_n = \frac{k_p[M]}{k_t[M^\bullet]} = \frac{k_p^2[M]^2}{k_t R_p} \tag{3.12}$$

Let us now calculate the weight average degree of polymerization of the radicals upon termination. The weight fraction w_i of species that have undergone $(i - 1)$ steps of growth, and contain i monomer units is given by:

$$w_i = \frac{ix_i}{\sum_{i=1}^{\infty} ix_i}$$

$\bar{P}_w$ is defined as:

$$\bar{P}_w = \sum_{i=1}^{\infty} iw_i = \frac{\sum_{i=1}^{\infty} i^2 x_i}{\sum_{i=1}^{\infty} ix_i} = \frac{\sum_{i=1}^{\infty} i^2 x_i}{\bar{P}_n}$$

The summation leads to:

$$\bar{P}_w = \frac{1 + p}{1 - p} \tag{3.17}$$

For values of p close to unity, the instantaneous polydispersity index $\frac{\bar{P}_w}{\bar{P}_n}$, characterizing the size distribution of *macro-radicals,* is very close to 2. Obviously, if termination occurs by disproportionation, this is also the polydispersity index of the *polymer.* If, on the contrary, recombination is the only possible reaction between radicals, the polydispersity index is reduced*): Its value is equal to $(1 + p)$ if termination occurs by disproportionation only, and to $\left(1 + \frac{p}{2}\right)$ if it occurs by re-

*) This arises from the fact that recombination is a *random* process, involving each time two radicals of *any* size. A large radical can react with a short one, whereby the molecular weight distribution becomes narrower.

combination only. As p is close to unity, under the assumption of extensive chain growth, the instantaneous polydispersity index is close to 2, resp. to 1.5, provided the radicals do not undergo transfer reactions.

If the process is not limited to low conversions, the polydispersity vastly exceeds the values calculated above. This is because the concentrations of monomer and of initiator intervene in the expressions of the probability of growth p. Thus, for each small conversion interval — during which [M] and [I] can be considered constant — one such distribution of molecular weights applies to the macromolecules formed during that interval. Each such individual distribution is shifted with respect to the preceding one, thus accounting for the shift of p arising from the consumption of monomer and initiator. The true overall molecular weight distribution of the sample is the convolution of all individual instantaneous distributions, and it is usually characterized by polydispersity indices far higher than 2.

3.6 Influence of Transfer Reactions on the Degree of Polymerization

Transfer reactions do not markedly change the rate of reaction provided the radicals formed upon transfer efficiently reinitiate the polymerization, with a rate constant k_i' similar to k_i:

$$\text{Transfer:} \quad R-M_n^{\bullet} + AH \xrightarrow{k_{tr}} RM_n-H + A^{\bullet}$$

$$\text{Reinitiation:} \quad A^{\bullet} + M \xrightarrow{k_i'} AM^{\bullet}$$

Conversely, the occurence of transfer strongly affects the molecular weight of the polymers obtained. It can be stated again that the degree of polymerization is given by the average number of growth steps a radical has undergone from initiation to *either* transfer *or* termination. Effectively, P_n is given by the ratio of the rate at which monomer units are being formed in the polymer chain, to the rate at which pairs of permanent ends are being created (one such pair defining one macromolecule). This can be expressed as:

$$\bar{P}_n = \frac{k_p[M]}{a k_t[M^{\bullet}] + k_{tr,M}[M] + k_{tr,S}[S] + k_{tr,I}[I]} \tag{3.18}$$

in wich a is defined as before, and characterizes the type of termination.

Introducing the ratio C_T of the rate constant of transfer to the rate constant of propagation, usually referred to as "transfer constant" of a given species T

$$C_S = \frac{k_{tr,S}}{k_p} \quad C_I = \frac{k_{tr,I}}{k_p} \quad C_M = \frac{k_{tr,M}}{k_p} \tag{3.19}$$

it follows that:

$$\frac{1}{\overline{P_n}} = \frac{a k_t R_p}{k_p^2 [M]^2} + C_M + C_S \frac{[S]}{[M]} + C_I \frac{[I]}{[M]} \tag{3.20}$$

The first term of eq. (3.20) gives the $\overline{P_n}$ value that would result if no transfer had occurred (Fig. 3.3).

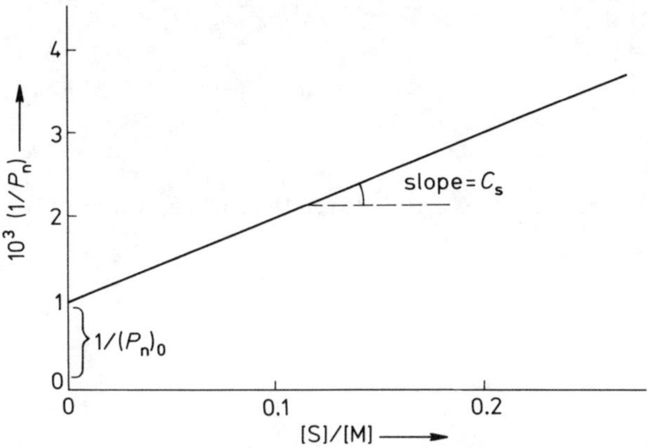

Fig. 3.3. Plot of the reciprocal degree of polymerization of polystyrene polymerized in the presence of a transfer agent (CCl_4), as a function of the mole ratio of CCl_4 to monomer. $(P_n)_0$ = DP of polystyrene obtained in the absence of CCl_4. C_S = transfer constant of CCl_4. Here C_M and C_I are negligibly small

As it has been done before, the distribution of molecular weights within a sample can be obtained by considering the probability p of chain growth at any stage, as opposed to all other events (transfer and termination). The number of growing radicals involving i monomer units can be calculated, at steady state, from the equation:

$$-\frac{d[M_i^\bullet]}{dt} = k_p [M]([M_i^\bullet] - [M_{i-1}^\bullet]) + k_t [M_i^\bullet] \sum_i [M_i^\bullet] + \sum_j k_{tr,j} [M_i^\bullet][Q_j] = 0 \tag{3.21}$$

Q_j being the various substances able to yield transfer reactions. Setting:

$$U = \sum_j k_{tr,j} [Q_j]$$

it follows that:

$$[M_i^\bullet] = \frac{k_p [M]}{k_p [M] + k_t [M^\bullet] + U} [M_{i-1}^\bullet] \qquad \text{for } i \geqslant 2 \tag{3.22}$$

or

$$[M_i^\bullet] = p[M_{i-1}^\bullet], \qquad \text{with } p = \frac{k_p [M]}{k_p [M] + k_t [M^\bullet] + U}$$

The concentration of $[M_1^\bullet]$ radicals (arising from initiation) is given by:

$$\frac{d[M_1^\bullet]}{dt} = R_i - k_p[M_1^\bullet][M] - k_t[M_1^\bullet][M^\bullet] - U[M_1^\bullet] + U[M^\bullet] = 0 \tag{3.23}$$

Recalling that $R_i = R_t$, this expression yields:

$$[M_1^\bullet] = \frac{k_t[M^\bullet] + U}{k_p[M] + k_t[M^\bullet] + U}[M^\bullet] \tag{3.24}$$

which can be written as:

$$[M_1^\bullet] = (1 - p)[M^\bullet]$$

Thus the following recurrence relationships exist between the concentrations of the growing chains:

$$\begin{aligned} [M_i^\bullet] &= p[M_{i-1}^\bullet] \\ [M_{i-1}^\bullet] &= p[M_{i-2}^\bullet] \\ &\text{-------------------} \\ [M_1^\bullet] &= (1 - p)[M^\bullet] \end{aligned}$$

with:

$$[M^\bullet] = \Sigma\,[M_i^\bullet]$$

The result is that the molar concentration of M_i radicals is given by:

$$[M_i^\bullet] = p^{i-1}(1 - p)[M^\bullet] \tag{3.25}$$

The fraction of radicals which have reached a degree of polymerization i is thus expressed by the same equation as above:

$$x_i = \frac{[M_i^\bullet]}{[M^\bullet]} = p^{i-1}(1 - p)$$

It has to be remembered that this is the *instantaneous* distribution of the radicals in the reaction medium. The probability of growth for any radical is namely a function of the concentrations [M], [Q], and [I] (the latter determining $[M^\bullet]$). All these concentrations are subject to variation with time.

Ideal and non-ideal Transfer Agents

As already mentioned, transfer agents are sometimes added on purpose in a polymerization in order to lower the average molecular weight of the polymer to be formed. In such instances all other transfer reactions can be disregarded as the compound added is chosen for the high value of its transfer constant. Equation 3.20 then reduces to:

$$\frac{1}{P_n} = \frac{a\,k_t R_p}{k_p^2[M]} + C_T\frac{[T]}{[M]}$$

Such processes have been applied to the synthesis of telomers or of ω-functional polymers. The so-called "telogen" is a functional transfer agent, and any transfer will result in the functionalization of a chain end. Only the polymer molecules arising from primary radicals will not carry endstanding functions. Thiols, various trichloromethyl or bromodichloromethyl compounds are efficient transfer agents (see section 3.5). However the efficiency of a transfer agent is not the same with different monomer systems, and the temperature also plays a role, as it can be anticipated.

Of special interest is the case of transfer agents exhibiting a transfer constant C_T close to unity. In this "ideal" case, the average length of the polymer chains to be formed is determined solely by the mole ratio of monomer to transfer agent. Furthermore, as monomer and transfer agent are consumed at the same rate, there can be no shift in molecular weight in the polymer as conversion progresses. The molecular weight distribution is not shifted towards lower polymerization degrees, as conversion progresses. In this case the polydispersity index to be expected for the polymer is equal to 2, as in any "most probable" (Flory-Schulz) distribution.

There are however only very few cases known in which a compound exhibits a transfer constant close to unity. Furthermore, the choice of polymerization degree in such cases is rather limited, ranging roughly from 1 to 100. If the [T]/[M] ratio is chosen smaller than 10^{-2}, the other terms in equation (3.20) can no more be disregarded.

As C_T is lower than 1, the monomer and the transfer agent are not consumed at the same rate: monomer is converted faster. Consequently the [T]/[M] ratio will increase and the molecular weight distribution curves tend to shift to lower polymerization degrees.

3.7 Determination of Rate Constants in Homogeneous Free Radical Polymerization

In free radical polymerizations, the rate constants k_p and k_t cannot be measured independently. The rate of initiator consumption, the overall rate of polymerization, and the number average molecular weight of the polymer obtained are experimentally accessible. To make the best use of these experimental data, two parameters are usually considered:

1. Parameter k_p^2/k_t

The term $k_p\,(2f\,k_d/k_t)^{1/2}$ in equation (3.7) can be expressed as:

$$\frac{2fk_d k_p^2}{k_t} = \frac{R_p^2}{[I][M]^2}$$

Introducing: $R_i = 2k_d f\,[I]$ *one gets:*

$$\frac{k_p^2}{k_t} = \frac{R_p^2}{R_i[M]^2}$$

Knowing R_p for a given monomer concentration [M] — and determining R_i from the rate of the initiator consumption and from correlative inhibitor consumption (to obtain f) in a separate experiment carried out under the same conditions — the parameter k_p^2/k_t can be calculated rather accurately.

The same parameter can be obtained from the number average degree of polymerization (in absence of any transfer), provided the termination reaction is clearly identified, and characterized by the value of a.

$$\frac{k_p^2}{k_t} = \frac{R_p}{[M]^2} \cdot a \cdot \bar{P}_n \qquad (3.27)$$

If transfer reactions occur, more elaborate equations are required.

2. Parameter k_p/k_t

The average life-time of the radicals, τ, is defined as the ratio of their steady-state concentration to the rate of their consumption:

$$\tau = \frac{[M^\bullet]}{k_t [M^\bullet]^2} \qquad (3.28)$$

This expression can be rewritten as:

$$\tau = \frac{k_p [M]}{k_t R_p} \quad \text{or} \quad \frac{k_p}{k_t} = \frac{\tau R_p}{[M]} \qquad (3.29)$$

Methods to determine the life time of the radicals originate chiefly from photopolymerization experiments, in which the reaction medium is submitted to alternate short periods of illumination and darkness. It is then possible to determine τ, and hence the ratio $\frac{k_p}{k_t}$, for given experimental conditions.

Knowing both k_p^2/k_t and k_p/k_t, the rate constants k_p and k_t can be evaluated.

Here are some orders of magnitude merely meant to help understand the kinetics of radical polymerization:

• The life time τ is of the order of	$10^{-1} - 10\,[s]$
• The rate constants of propagation	$10^2 - 10^4\,[l \cdot mole^{-1}\, s^{-1}]$
• The rate constants of termination	$10^6 - 10^8\,[l \cdot mole^{-1}\, s^{-1}]$
• The rate constants of initiator decomposition	$10^{-4} - 10^{-6}\,[s^{-1}]$
• The equilibrium radical concentration	$10^{-7} - 10^{-9}\,[mole \cdot l^{-1}]$
thus $\frac{k_p}{k_t}$ is of the order of	$10^{-4} - 10^{-6}$ (dimensionless)
and $\frac{k_p^2}{k_t}$	$1 - 10^{-4}\,[l \cdot mole^{-1} \cdot s^{-1}]$

From these figures it follows that 10^2 to 10^4 monomer molecules are added per second to a growing radical.

3.8 Temperature Dependence of Rate Constants and of Degrees of Polymerization

It is well known that the temperature dependence of each individual rate constant obeys an Arrhenius-type relationship:

$$k_p = A_p \, e^{-E_p/RT} \tag{3.30}$$

where A_p is the entropy factor (also called "pre-exponential factor") and E_p is the activation energy of the reaction.

Since the overall rate of polymerization involves three different rate constants, the temperature dependence of the rate of polymerization includes three activation energies, the values of which are markedly different.

Starting from the general equation (3.7) for the instantaneous rate of monomer consumption, derived earlier:

$$R_p = -\frac{dM}{dt} = k_p \left(\frac{2k_d f}{k_t} \right)^{1/2} [I]^{1/2} [M]$$

Calling E_p, E_t and E_d the activation energies for propagation, for termination, and for the homolytic decomposition of the initiator, respectively (and assuming that the efficency factor f is not temperature dependent) one can write:

$$\ln \left[2^{1/2} k_p \left(\frac{k_d}{k_t} \right)^{1/2} \right] = \ln \left[2^{1/2} A_p \left(\frac{A_d}{A_t} \right)^{1/2} \right] - \frac{E_p - \frac{E_t}{2} + \frac{E_d}{2}}{RT}$$

and therefore:

$$\ln R_p = \ln \left[2^{1/2} A_p \left(\frac{A_d}{A_t} \right)^{1/2} \right] + \ln [(f[I])^{1/2} [M]] - \frac{E_R}{RT} \tag{3.31}$$

with

$$E_R = E_p + \frac{E_d - E_t}{2},$$

where E_R is the *"overall" activation energy* for the rate of polymerization.

The E_p values are of the order of 5 to 10 [*Kcal/mole*], the E_t values range from 2 to 5 [*Kcal/mole*], whereas the activation energies for the decomposition of the initiator are far higher, of the order of 30 to 40 [*Kcal/mole*] (Tab. 3.2). Thus, E_R is of the order of 20 [*Kcal/mole*], which means that a 10 °*C* increase of the temperature will provide a doubling of the rate of polymerization approximately.

The same type of calculation applies to the degree of polymerization. If no transfer processes take place, the number average degree of polymerization is given by the expression (3.12) derived earlier:

$$\bar{P}_n = \frac{k_p [M]}{(2k_d k_t f[I])^{1/2}} \tag{3.12}$$

As one can write:

$$\ln \frac{k_p}{(k_d k_t)^{1/2}} = \ln \left[\frac{A_p}{(A_d \cdot A_t)^{1/2}} \right] - \frac{E_p - \frac{E_d}{2} - \frac{E_t}{2}}{RT},$$

the temperature dependence of the instantaneous degree of polymerization can be expressed as:

$$\ln \bar{P}_n = \ln \left[\frac{A_p}{(2A_d A_t)^{1/2}} \right] + \ln \left[\frac{[M]}{(f[I])^{1/2}} \right] - \frac{E_{DP}}{RT} \tag{3.32}$$

with

$$E_{DP} = E_p - \frac{E_d + E_t}{2}$$

Table 3.2. Data on polymerization of various monomers at 60 °C by free radical initiators

Monomer	k_p *l mol*$^{-1}$ *s*$^{-1}$	k_t *l mol*$^{-1}$ *s*$^{-1}$	E_p *Kcal/mol*	E_t *Kcal/mol*
Vinyl chloride	12 300	$5 \cdot 10^4$	3.7	4.2
Vinyl acetate	3 700	$6 \cdot 10^7$	7.3	3.2
Acrylonitrile	2 000	$1.6 \cdot 10^7$	3.9	3.7
Methyl acrylate	2 090	$1.8 \cdot 10^7$	7.1	5
Methyl methacrylate	705	$5 \cdot 10^7$	4.7	1.2
Styrene	145	$5.8 \cdot 10^7$	7.3	1.9

The "overall" activation energy for the polymerization degree, E_{DP}, is of the order of -15 [*Kcal/mole*], which means that upon increasing the temperature, the degree of polymerization of the obtained polymer will decrease rather rapidly*⁾.

The orders of magnitude quoted above apply only to *chemically initiated* free radical chain polymerizations. In the case of *thermal initiation*, the activation energy for the initiation reaction is only about 10 to 15 [*Kcal/mole*], and in the case of *photochemical initiation* $E_d = 0$. Therefore in this latter case, the $\bar{P}_n$ value should increase (and not decrease) with increasing temperature.

3.9 Trommsdorff-Effect

It was shown that the concentration of active species (radicals) obeys steady state conditions, provided the initiator concentration can be considered constant.

$$[M^\bullet] = \left(\frac{2fk_d[I]}{k_t} \right)^{1/2}$$

and the rate of propagation is given by

*⁾ In addition, transfers are more likely to occur at elevated temperature and this can further decrease the molecular weight, i. e., C_M, C_S, and C_I increase with temperature.

$$R_p = -\frac{d[M]}{dt} = k_p[M^\bullet][M]$$

However, if a polymerization is carried out in the *bulk*, or in *concentrated solution* to high conversion, the viscosity of the reaction medium increases as the reaction proceeds. This increase of viscosity does not affect the rate of decomposition of the initiator (monomolecular process), nor does it affect significantly the chain propagation reaction, since the diffusion rate of the small monomer molecules is hardly slowed down, unless the glass transition is approached. But it results in a drastic decrease of the *rate of termination**), which involves two large macromolecular species in a viscous surrounding. It follows that steady-state conditions can no longer be fulfilled, and that $[M^\bullet]$ starts to increase. This departure from steady-state conditions has two main consequences:

- An increase of the polymerization rate
- An increase of the average molecular weight of the polymer

As the polymerization degree increases the viscosity of the medium increases even more. The increase of the polymerization rate will produce an increase in the rate of heat evolved from the reaction. The high viscosity of the medium can prevent this heat from being transferred in a proper way. A temperature increase is a possible consequence, speeding up the whole process, which may get out of control and lead to an explosion.

Fig. 3.4 illustrates the increase of the polymerization rate and of the polymerization degree of a polymer formed in the bulk as a function of time. In stage I conversion is

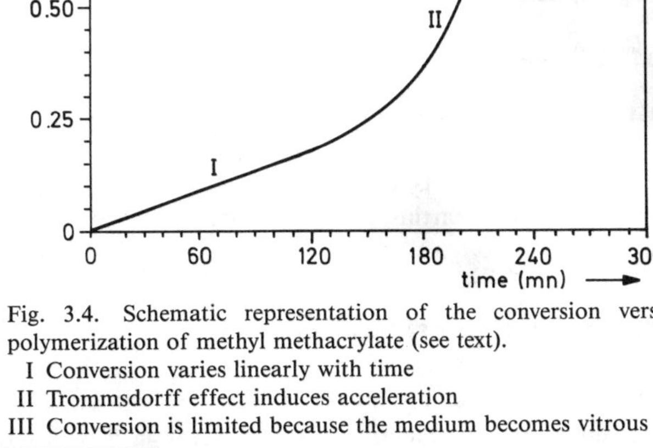

Fig. 3.4. Schematic representation of the conversion versus time in a classical bulk polymerization of methyl methacrylate (see text).
I Conversion varies linearly with time
II Trommsdorff effect induces acceleration
III Conversion is limited because the medium becomes vitrous

*) "k_t" decreases drastically — not because of a decrease in chemical reactivity of terminal radicals, but because of inaccessibility of a reaction partner (bimolecular reaction).

linear in time. In stage II the acceleration of the process is clearly due to Trommsdorff effect. In stage III the process slows down because the reaction mixture is becoming vitrous (glass transition temperature higher than polymerization temperature), thus finally depressing the diffusion of the monomer to the active chain end.

Bulk polymerizations of monomers which act as solvents for their own polymers have to be carried out with great care. If the polymer is insoluble in its own monomer, the danger for the reaction to get out of control is much less: The polymer precipitates out of solution to form a slurry, and it does not contribute to an increase of the viscosity of the reaction medium. Even though further growth is not prevented, since trapped radicals are still present and monomer can still diffuse into the slurry, the rate will become slower rather than faster as the reaction proceeds. This will be discussed in detail in Chapter 12.

3.10 Telomerization

Free radical polymerization processes involving functional transfer agents have found applications for the synthesis of so-called *telomers*, which are functional oligomers. The telogen is a compound which carries a chemical function and exhibits a high transfer constant. Upon reacting a suitable monomer in the presence of a telogen with an initiating system generating free radicals, the probability of transfer is enhanced, and almost all the molecules formed carry at chain end the function originating from the telogen. The more effective the transfer agent, and the higher its concentration, the lower the average molecular weight of the telomer.

There are even cases in which adduct formation predominates, i. e. a telomer with polymerization degree 1 is obtained. Redox initiation is preferred in telomerizations, although it is not required.

In some cases the adducts formed in a first run can be used in turn as telogens in a second step. An example is given below:

$$CCl_4 + CF_2 = CFCl \xrightarrow{\text{(Free rad, Initiator)}} CCl_3{-}CF_2{-}CFCl_2$$

$$CCl_3{-}CF_2{-}CFCl_2 + n\ CH_2{=}CHCl \longrightarrow CFCl_2{-}CF_2{-}CCl_2{-}(CH_2{-}CHCl)_n{-}Cl$$

Telomers have found a number of applications, especially as low molecular weight functional polymers. They are also used as starting points for further synthesis.

3.11 "Living" Free Radical Polymerizations

In classical free radical polymerizations the initiator serves exclusively to generate primary radicals. Transfers arising from the reaction of a growing radical with an initiator molecule can usually be disregarded. Termination involving a primary radical is also quite unlikely. The primary radicals react with a monomer to give an active site and subsequently propagation steps take place until the growing radical terminates (with another such macroradical) or gets involved in a transfer reaction.

In the so-called "iniferter" systems the reaction scheme is quite different. It involves six main reactions:

- Initiation by homolytic decomposition of the initiator (symbolized here as R—R)

$$\text{R—R} \longrightarrow 2\text{R}^\bullet$$

followed by attack of the primary radical onto a monomer molecule:

$$\text{R}^\bullet + \text{M} \longrightarrow \text{R M}_1^\bullet$$

- Propagation proceeds as usually in many steps by attack of a radical site onto monomer

$$\text{R—M}_i^\bullet + \text{M} \longrightarrow \text{R—M}_{i+1}^\bullet$$

- Chain transfer to initiator:

$$\text{R—M}_i^\bullet + \text{R—R} \longrightarrow \text{R—M}_i\text{—R} + \text{R}^\bullet$$

whereby the newly formed primary radical $R^\bullet$ can generate a new chain.

- Primary radical termination:

$$\text{R—M}_i^\bullet + \text{R}^\bullet \longrightarrow \text{R—M}_i\text{—R}$$

If the probability of primary radical termination is high, it means that bimolecular termination (between two macroradicals) can be neglected. In a similar fashion, if the probability of transfer to the initiator is high enough, all other transfer reactions can be disregarded. Such a process would imply that the molecular weights formed are low. However, another reaction has to be considered, namely:

- Reinitiation, by homolytic cleavage of the M-R bond at chain end of "dead" polymer molecules:

$$\text{R—M}_i\text{—R} \longrightarrow \text{R—M}_i^\bullet + \text{R}^\bullet$$

In other words, the primary radical termination reaction is reversible, which means that the whole process involving iniferter-type initiators and monomers can be considered as insertion reactions of monomer units between the radical sites arising from the initiator.

A consequence of the above statements is that the polymers appear to be living, as in the case of anionic polymerizations carried out in aprotic solvents. Molecular weight is proportional to conversion. Further addition of monomer results in an increase of the length of the existing macromolecules. Nevertheless, the molecular weight distribution is usually rather broad, as a result of slow initiation.

Specific initiators for iniferter polymerizations include some thiocarbamates and bulky azo compounds such as:

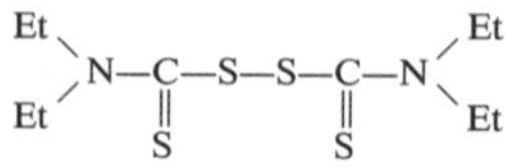

N,N′ Tetraethyldithiocarbamate

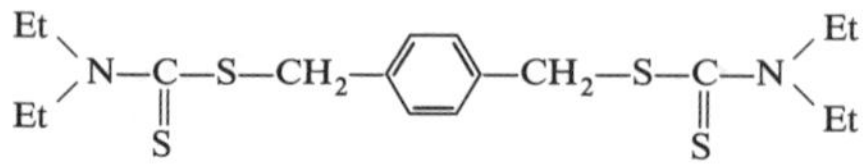

Xylylene bisdiethylthiocarbamate

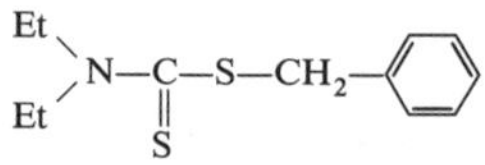

Benzyl N-diethylthiocarbamate

(The former two behave as difunctional initiators.)

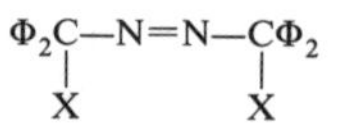

with X = Et, CN, OΦ

The pseudo-living character of these radical polymerizations has led to applications in macromolecular synthesis, which are not accessible in classical free radical processes. An example is given by block copolymer synthesis:

First monomer A is polymerized to yield a poly A chain fitted at its end with a thiocarbamate site. Next, monomer B is added and the mixture is irradiated with UV to cleave the C—S bond, whereby its polymerization is initiated. The block copolymer yield is rather high, although some homopoly B is present in the final product. Diblock or triblock copolymers can be obtained, depending upon whether the initiator used is mono- or bifunctional. This method of block copolymer synthesis is of special interest for systems that cannot be obtained anionically.

Problem 3.1

Assuming the overall probability for chain growth in a free radical polymerization in the presence of a transfer agent is $p = 0.98$, and that termination occurs only by disproportionation, calculate and draw the distribution curve (by number and by weight) of the polymerization degrees of the polymer formed initially. Calculate also the number and weight averages of the DP.

Problem 3.2

a) Estimate the initial rates of bulk polymerization of undiluted monomer at 60 °C, using azobisisobutyronitrile (AIBN) at a 10^{-3} *molar* concentration, for methyl acrylate, methyl methacrylate, styrene and vinyl acetate. Assume that the efficiency factor f is 0.5 for all monomers.

b) Estimate the steady state radical concentration for each monomer under the conditions of part a).
c) Use the listed values of the activation energies E_d, E_p, E_t to predict the initial rate of polymerization of methyl methacrylate at 40 °*C* and at 80 °*C*.
d) Estimate the initial kinetic chain lengths of the polymerizations referred to in part a), and the number average molecular weights, assuming that transfer to the monomer is negligible and that termination occurs by coupling for methyl acrylate and styrene, and by disproportionation for the two other monomers (these assumptions are not quite justified).
e) Assuming that the AIBN concentration is kept constant by make-up throughout the polymerization of methyl acrylate, calculate the molecular weights $\overline{M}_n$ of the polymer formed initially, of the polymer formed at 95% conversion, and the temporal average from 0 to 95% conversion.

Problem 3.3

a) Knowing the rate constant of decomposition of AIBN at 60 °C and the corresponding activation energy, calculate the half life time of this initiator at 50, 60 and 80 °C.
b) AIBN is used to initiate the polymerization of methyl methacrylate in benzene at 60 °*C*. The initial concentrations are $[I]_0 = 0.02$ $[mol \cdot l^{-1}]$ and $[M]_0 = 2$ $[mol \cdot l^{-1}]$. Assume the efficiency factor *f* is 0.5, and termination occurs solely by disproportionation. Calculate
   — the initial rate of polymerization
   — the time at which 1/10 of the monomer has been converted into polymer
   — the conversion degree π after 10 hours of polymerization. It should be checked whether or not the initiator consumption can be neglected,
   — the rate of polymerization after 10 hours of reaction.
   Also determine the molecular weight $\overline{M}_n$ of the polymer formed initially, assuming the absence of transfer of any kind. What would be the value of $\overline{M}_n$ if the reaction medium contained 0.3 $[mol \cdot l^{-1}]$ of chlorotrimethyl silane as transfer agent, the transfer constant of which is $C_T = 144 \times 10^{-4}$.

Data needed for problems 3.2 and 3.3

Initiator AIBN at 60 °C $k_d = 0.85 \cdot 10^{-5}\ s^{-1}$ $E_d = 30\ [Kcal \cdot mol^{-1}]$

		Propagation		Termination	
Monomers	Density [*g/ml*]	k_p (60 °C) $[l \cdot mol^{-1} \cdot s^{-1}]$	E_p $[Kcal \cdot mol^{-1}]$	$k_t \cdot 10^{-6}$ (60 °C) $[l \cdot mol^{-1} \cdot s^{-1}]$	E_t $[Kcal \cdot mol^{-1}]$
Methyl acrylate	0.95	2 090	7.1	18	5
Methyl methacrylate	0.94	705	4.7	50	1.2
Styrene	0.91	145	7.3	58	1.9
Vinyl acetate	0.93	3 700	7.3	58	3.2

— Volume expansion of bulk monomers is neglected
— Arrhenius' equation is: $k = A \exp(-E/RT)$

Problem 3.4

In a free radical polymerization, attack of a growing chain radical on a hydrogen atom of a *mer* (unit) in a previously completed macromolecule produces a radical site at that point on the macromolecule, while the abstracted hydrogen terminates the attacking chain. This radical site can then begin a long side chain by the successive addition of monomer. This "long chain branching" is kinetically different from the "back-biting" process in which the growing chain radical reaches back on its chain 5 to 6 carbon atoms to abstract a hydrogen atom.

In the case at hand, (long chain branching), the attacked mer becomes a branching mer or branching unit.

Derive an expression relating the branching density, ρ, to the polymer transfer constant C_p and the fraction conversion π, where these and other terms are defined as follows:

[B] = *mol* branch units/*litre*
$[M]_0$ = initial monomer concentration in [*mol/l*]
[M] = monomer concentration at some later time, $= (1 - \pi)\,[M]_0$
[P] = *mole* of what were monomer units, now part of chain, $= \pi\,[M]_0$
π = fraction of original monomer converted to polymer
ρ = branching density, $= [B]/([B] + [P]), \cong [B]/[P]$
C_p = $k_{tr,P}/k_p$ where $k_{tr,P}$ is defined by:

$$d[B]/dt = k_{tr,P}\,[M^\bullet][P]$$

The order of magnitude of C_p ranges from 10^{-7} to 10^{-3}.
Show that the value of $\bar{P}_n$ is not affected by polymer transfer.

Problem 3.5

Consider the polymerization of styrene in toluene solution, initiated by di-tert.butylperoxide. For a solution containing 0.04 mole of initiator and 2 mole of monomer per liter, the *initial* rates of initiation R_i and of propagation R_p are found to be $1.6 \cdot 10^{-10}$ and $6.4 \cdot 10^{-7}$, respectively.

a) Calculate the values of $(f \cdot k_d)$ and of $(k_p/k_t^{1/2})$.

b) Calculate the initial kinetic chain length, λ, under the assumption that no transfer occurs.

c) The transfer constant of styrene, under the conditions chosen, is $C_M = 0.85 \cdot 10^{-4}$. To what extent does this transfer reaction affect the molecular weight of the polymer? It is assumed that termination occurs by disproportionation only.

d) The molecular weight of this polymer is too high. A value of 40000 would be appropriate. To get this value, CCl_4 is added to the reaction medium. The transfer constant C_T of CCl_4 is $90 \cdot 10^{-4}$. What amount (expressed in g/l) of this compound is needed to get this value of $\bar{M}_n$.

e) Under the conditions described above the polymerization is much too slow. To enhance the rate, the polymerization is carried out at 100 °C. Calculate the corresponding value of the *initial* rate of polymerization, using the values of the activation energies given in the Table.

f) Calculate the conversion attained after 5 hours of reaction, at 100 °C, neglecting the changes of concentration due to volume expansion. (The initiator concentration can be considered constant over the whole time.)

Problem 3.6

Consider the bulk isothermal homogeneous free radical polymerization of a vinyl monomer by a homolytically cleaving initiator:

$[M]_0 = 10$ mol/l; $k_p = 100\ l/mol \cdot s$; $[M^\bullet] = 10^{-8}\ mol/l$ (steady state); $k_t = 2 \cdot 10^7$; $C_M = 0$; termination is assumed to occur exclusively by recombination. The number average degree of polymerization $\overline{P}_{no}$ of the polymer formed initially is 1000.

Derive an expression for the time averaged $\overline{P}_n$ in terms of π, the fraction of monomer converted to polymer. Determine the value of $\overline{P}_n$ for $\pi = 0.9$ and for $\pi = 0.99$. State the assumptions made.

Chapter 4
Free Radical Copolymerization

4.1 Introduction

Free radical chain copolymerizations are widely used in industry to produce copolymers, the properties of which are often intermediate between those of the corresponding homopolymers. These processes are, however, rather complicated. In many cases large fluctuations in composition within the samples are involved. A kinetic approach is necessary to understand these phenomena. We shall consider here the case of homogeneous free radical copolymerization, implying the presence of a solvent, or solubility of the polymeric species formed in the monomer mixture they arise from.

If a mixture of two monomers is polymerized by means of a free radical initiator, the following reactions have to be considered:

- *Initiation:* The primary radicals originating from the homolytic decomposition of the initiator may react with either of the two monomers:

$$\mathrm{I} \longrightarrow 2\,\mathrm{R}^{\bullet} \begin{cases} \mathrm{R}^{\bullet} + \mathrm{A} \longrightarrow \mathrm{RA}^{\bullet} \\ \mathrm{R}^{\bullet} + \mathrm{B} \longrightarrow \mathrm{RB}^{\bullet} \end{cases}$$

- *Propagation:* Four types of propagation steps contribute to chain growth, either types of radicals $A^{\bullet}$ or $B^{\bullet}$ being able to react with either of the two monomers A and B:

$$\begin{aligned}
\sim\!\!\sim\mathrm{A}^{\bullet} + \mathrm{A} &\xrightarrow{k_{aa}} \sim\!\!\sim\mathrm{A}^{\bullet} \\
\sim\!\!\sim\mathrm{A}^{\bullet} + \mathrm{B} &\xrightarrow{k_{ab}} \sim\!\!\sim\mathrm{B}^{\bullet} \\
\sim\!\!\sim\mathrm{B}^{\bullet} + \mathrm{B} &\xrightarrow{k_{bb}} \sim\!\!\sim\mathrm{B}^{\bullet} \\
\sim\!\!\sim\mathrm{B}^{\bullet} + \mathrm{A} &\xrightarrow{k_{ba}} \sim\!\!\sim\mathrm{A}^{\bullet}
\end{aligned}$$

- *Termination* can occur by disproportionation or by recombination, and it can involve both types of radicals. Three termination reactions, with assumedly different rate constants, are thus involved, as recombination and disproportionation are kinetically undistinguishable.

$$\begin{aligned}
\sim\!\!\sim\mathrm{A}^{\bullet} + {}^{\bullet}\mathrm{A}\sim\!\!\sim \qquad & k_{taa} \\
\sim\!\!\sim\mathrm{A}^{\bullet} + {}^{\bullet}\mathrm{B}\sim\!\!\sim \qquad & k_{tab} \\
\sim\!\!\sim\mathrm{B}^{\bullet} + {}^{\bullet}\mathrm{B}\sim\!\!\sim \qquad & k_{tbb}
\end{aligned}$$

The initiation step involves only a single monomer unit, and it will not be discussed any further. But the propagation reactions determine the actual composition of the copolymer formed, as well as the distribution of A and B units along the chain. The cross-termination reactions will be considered in connection with the overall kinetics of chain copolymerization.

4.2 Copolymerization and Compositional Heterogeneity

The rates of consumption of each of the two monomers involved can be expressed as:

$$-\frac{d[A]}{dt} = k_{aa}[A^\bullet][A] + k_{ba}[B^\bullet][A]$$
$$-\frac{d[B]}{dt} = k_{bb}[B^\bullet][B] + k_{ab}[A^\bullet][B] \tag{4.1}$$

In free radical polymerizations, *stationary conditions* are fulfilled over small time intervals, as shown earlier. In the case of radical chain copolymerization, this implies that the same condition applies also to each individual type of radical*). This can be equated as:

$$k_{ab}[A^\bullet][B] = k_{ba}[B^\bullet][A] \tag{4.2}$$

and is generally referred to as *Bodenstein's rule.*

Taking this equation into consideration, and dividing eqs. (4.1), one by the other, one obtains:

$$\frac{d[A]}{d[B]} = \frac{k_{aa}[A^\bullet][A] + k_{ba}[B^\bullet][A]}{k_{bb}[B^\bullet][B] + k_{ab}[A^\bullet][B]} = \frac{k_{aa}[A^\bullet][A] + k_{ab}[A^\bullet][B]}{k_{bb}[B^\bullet][B] + k_{ba}[B^\bullet][A]} \tag{4.3}$$

d[A] and d[B] being the amounts of monomer A and of monomer B that have been converted into polymer during time interval dt. The ratio d[A]/d[B] thus characterizes the instantaneous composition of the copolymer.

We can set:

$$r_a = \frac{k_{aa}}{k_{ab}} \qquad r_b = \frac{k_{bb}}{k_{ba}} \tag{4.4}$$

to define the *radical reactivity ratios* (often called erroneously monomer reactivity ratios). r_a characterizes the reactivity of the $A^\bullet$ radical with respect to the two monomers present, A and B. If $r_a > 1$, homopolymeric growth is preferred; if $r_a = 0$ the $A^\bullet$ radical can only react with monomer B, which implies that monomer A cannot homopolymerize. Introducing r_a and r_b into equation (4.3) leads to:

---

*) If this condition would not be fulfilled, the process would rapidly evolve into a homopolymerization of one of the monomers, leaving the other almost untouched.

$$\frac{d[A]}{d[B]} = \frac{[A]}{[B]} \cdot \frac{r_a[A] + [B]}{r_b[B] + [A]} = \frac{r_a\dfrac{[A]}{[B]} + 1}{r_b\dfrac{[B]}{[A]} + 1} \tag{4.5}$$

This so-called *"instantaneous copolymer composition equation"* can also be written in a different form, upon introducing the mole fractions of A and B in the monomer mixture f_a and f_b:

$$f_a = \frac{[A]}{[A] + [B]} \qquad f_b = 1 - f_a$$

and the mole fractions of units A and B in the copolymer formed *instantaneously* F_a and F_b:

$$F_a = \frac{d[A]}{d[A] + d[B]} \qquad F_b = 1 - F_a$$

Equation (4.5) can then be written as:

$$F_a = \frac{r_a f_a^2 + f_a f_b}{r_a f_a^2 + r_b f_b^2 + 2 f_a f_b} \tag{4.6}$$

Given the reactivity ratios r_a and r_b, this relation enables one to calculate the instantaneous mole fraction of A in the *copolymer* formed from a mixture of monomers A and B, in which the mole fraction of A is f_a.

In most cases, F_a is different from f_a, which means that one of the monomers is consumed more rapidly than the other. Consequently, the composition of the monomer mixture varies as the reaction proceeds, and the composition of the copolymer formed also varies with conversion. This is the origin of the *compositional heterogeneity* (or fluctuations in composition) observed in copolymers obtained to high conversions*).

Skeist's differential equation expresses these considerations in a more quantitative way. If [M] is the total monomer concentration at time t and d[M] the amount of monomer (A and B) converted to polymer during time interval dt, one can write:

$$[A] = f_a[M] \quad \text{and} \quad d[A] = F_a\, d[M]$$

The proportion of A incorporated into the polymer during dt results in a shift by df_a of the A content of the monomer mixture, as it can be expressed by:

$$f_a[M] - (f_a - df_a)([M] - d[M]) = F_a\, d[M] \tag{4.7}$$

Skeist's differential expression results (neglecting the second order infinitesimal):

---

*) Only if the process is limited to very small conversions, or if make-up monomer is continuously supplied in the correct ratio of [A] to [B] can one consider the copolymer formed to be homogeneous in composition. In the latter case equation (4.8) is incorrect, since it is a material balance based on the assumption that input is zero.

$$\frac{d[M]}{[M]} = \frac{df_a}{F_a - f_a} \tag{4.8}$$

where $\frac{d[M]}{[M]}$ is the conversion increment $d\pi$ occurring during time interval dt. In this expression F_a should be replaced by its value from equation (4.6); integration of this equation is not easy. Analytical expressions have been given by several authors.

4.2.1 Plots of the Instantaneous Copolymer Composition

The effect of the actual values of the radical reactivity ratios r_a and r_b on the copolymerization process has to be examined now. The composition of the copolymer formed instantaneously is plotted as a function of the composition of the monomer mixture it arises from. Time (or conversion) are not considered in these diagrams of instantaneous copolymer composition: F_a is plotted versus f_a. Their main use is to illustrate the tendency of a system to give yield to compositional heterogeneity. Two examples are shown on Figure 4.1.

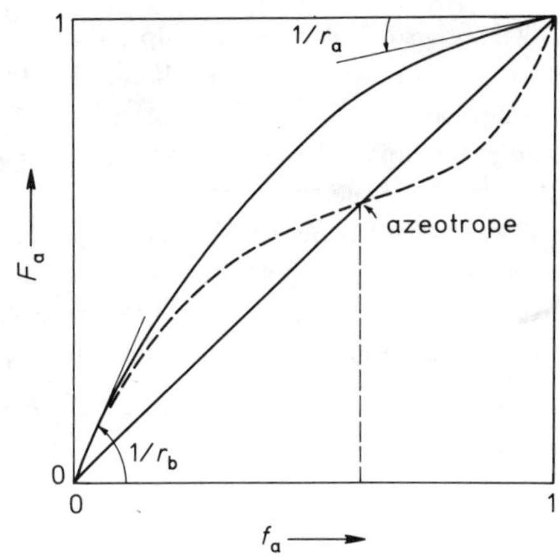

Fig. 4.1. Examples of instantaneous compositions of copolymers (F_a versus f_a diagram); ——— without azeotrope; – – – with azeotrope

It can be established easily that the slopes of the curves for $f_a = 0$ and for $f_a = 1$ are the reciprocals of the radical reactivity ratios:

$$\left(\frac{dF_a}{df_a}\right)_{f_a=0} = \frac{1}{r_b}; \qquad \left(\frac{dF_a}{df_a}\right)_{f_a=1} = \frac{1}{r_a}$$

The distance between the actual curve $F_a(f_a)$ and the first diagonal can be considered as a measure of the tendency of the system to involve compositional heterogeneity.

If the curve is *above* the diagonal, it means that the copolymer formed instantaneously is richer in A than the monomer mixture it originates from: A is consumed more rapidly than B, and the A content of the monomer mixture progressively decreases.

Consequently, the A content of the polymer formed instantaneously will also decrease, following the instantaneous copolymerization composition diagram, as indicated by the arrow on Fig. 4.1.

If, on the contrary, the A content of the copolymer formed instantaneously is lower than the A content of the monomer mixture it originates from, then the A content of the monomer mixture (and that of the copolymer formed instantaneously) will increase as conversion progresses.

It may happen that for a given composition of the monomer mixture the curve of the instantaneous composition of the copolymer *crosses* the first diagonal. This implies the following condition to be fulfilled, easily obtained from equation (4.5), by setting $d[A]/d[B] = [A]/[B]$,

$$\frac{r_a[A] + [B]}{r_b[B] + [A]} = 1 \quad \text{or} \quad \frac{[A]}{[B]} = \frac{1 - r_b}{1 - r_a} \tag{4.10}$$

which can also be expressed as:

$$F_a = \frac{1 - r_b}{2 - r_a - r_b} \tag{4.10a}$$

The cases of *"azeotropy"* imply that both reactivity ratios of the system are either smaller than unity, or larger than unity. In fact, there are numerous examples of the former case, but none of the latter case. If both $r_a < 1$ and $r_b < 1$, the azeotropy is said to be "unstable". This means that when the A content of the monomer mixture is slightly above the azeotropic composition, the composition of the monomer mixture will be drawn further off the azeotropic composition. The same is true if the initial A content is slightly below that of the azeotrope. The arrows on the diagram are meant to visualize that effect. Only at the precise location of the azeotropic composition does the composition stay the same until total conversion.

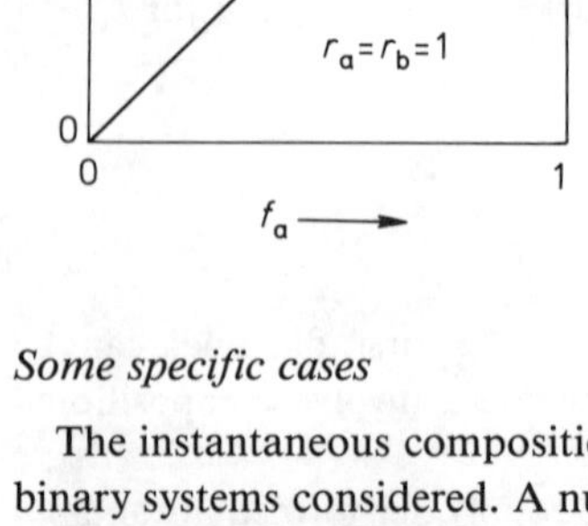

Fig. 4.2. Instantaneous composition diagram, $r_a = r_b = 1$

Some specific cases

The instantaneous composition diagram F_a vs. f_a reflects the characteristics of the binary systems considered. A number of specific cases involving vinylic monomers A and B will be studied now, in relation with the values of the radical reactivity ratios r_a and r_b defining the system.

Case 1 : $r_a = r_b$

In this trivial case, illustrated in Fig. 4.2, the instantaneous composition of the copolymer coincides with that of the monomer mixture over the entire composition range. Consequently, no fluctuations in composition are to be expected. There are a number of binary monomer systems that approach closely this case, in which the affinity of each of the radicals for both monomers is the same. This condition also implies a truly random placement of the A and B units in the copolymer.

Case 2 : $r_a \cdot r_b = 1$

This case is referred to as *ideal copolymerization,* as the radicals $A^\bullet$ and $B^\bullet$ exhibit the same relative reactivity for monomers A and B:

$$\frac{k_{aa}}{k_{ab}} = \frac{k_{ba}}{k_{bb}}$$

In this case, illustrated by Fig. 4.3, the instantaneous composition equation reduces to:

$$\frac{d[A]}{d[B]} = r_a \frac{[A]}{[B]} \quad \text{or} \quad F_a = \frac{r_a f_a}{r_a f_a + f_b}$$

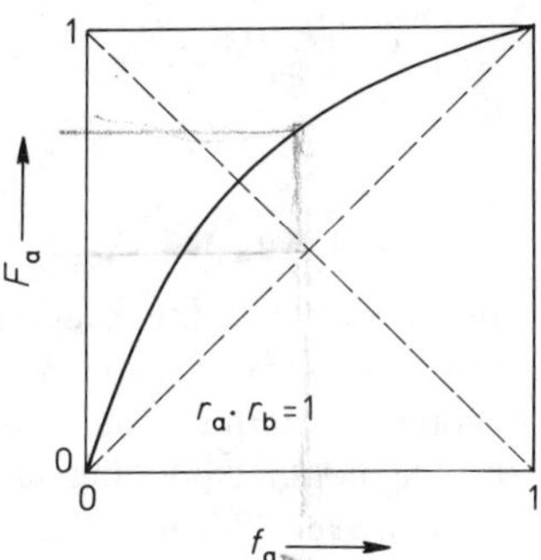

Fig. 4.3. Instantaneous composition diagram, ideal case

The plot of F_a vs f_a is symmetrical with respect to the second diagonal. The more r_a departs from unity, the larger the compositional heterogeneity at large conversions. The placement of the A and B units along the chain obeys Bernouillian statistics, because it is commanded by a single parameter. The very important pair styrene-butadiene approximates an ideal copolymerization.

Case 3 : $r_b = 0$

In this case monomer B does not homopolymerize, as $k_{bb} = 0$. If monomer B is in excess, an alternating copolymer is formed, regardless of the proportion of A in the mixture. For mixtures that are rich in A, the shape of the curve (Fig. 4.4) depends on the values of r_a. If $r_a < 1$ an azeotrope is found, and the tendency to alternation remains rather large. If $r_a > 1$, the A content of the copolymer tends to increase with f_a.

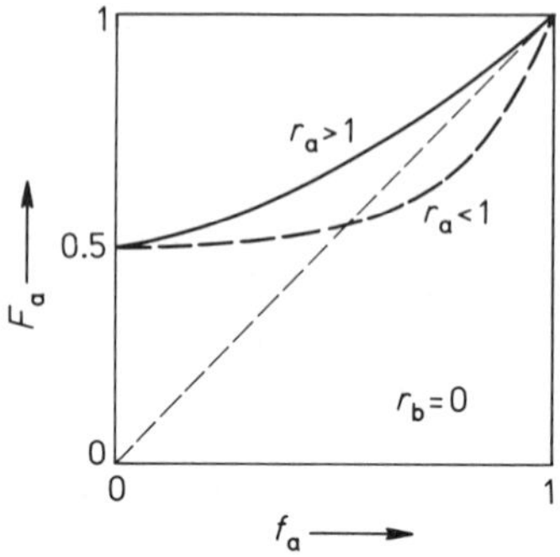

Fig. 4.4. Instantaneous composition diagram, $r_b = 0$

4.2.2 Variation of Copolymer Composition versus Conversion

To investigate how the instantaneous composition of the copolymer varies with conversion it is necessary to integrate Skeist's equation. An analytical expression has been given by Meyer and Lowry, but it is not very easy too deal with it:

$$\pi = 1 - \frac{[M]}{[M]_0} = 1 - \left(\frac{f_a}{f_{a0}}\right)^{\alpha} \left(\frac{f_b}{f_{b0}}\right)^{\beta} \left(\frac{f_{a0} - \delta}{f_a - \delta}\right)^{\gamma} \tag{4.11}$$

with:

$$\alpha = \frac{r_b}{1 - r_b}; \quad \beta = \frac{r_a}{1 - r_a}; \quad \gamma = \frac{1 - r_a r_b}{(1 - r_a)(1 - r_b)}; \quad \delta = \frac{1 - r_b}{2 - r_a - r_b}$$

In the ideal case the solution to equation (4.8) becomes more simple.

It is far more efficient to apply an incremental calculation process, set on a computor program. Given the initial molar A-content of the monomer mixture $(f_a)_0$ it is easy to calculate the corresponding initial molar A-content of the copolymer $(F_a)_0$ from equation (4.6). Skeist's differential equation can now be used to calculate the compositional shift df_a of the monomer mixture for a given (small) conversion increment $d\pi = d[M]/[M]$. Next, the incremented value of the A-content of the monomer mixture, $(f_a)_1 = (f_a)_0 - df_a$, is introduced into equation (4.6) to calculate the new instantaneous value of the A-content of the copolymer $(F_a)_1$; both values f_a and F_a are introduced into Skeist's equation for another conversion increment $d\pi$, and so on.

It is obviously necessary to choose the conversion increments small enough to be able to accurately follow the shift in monomer composition, which in turn commands the fluctuations in composition within the copolymer sample.

Having the values of f_a and of F_a as a function of conversion π, it is possible to evaluate the "average" composition of the copolymer formed — i.e. the A-content analytically found — upon integration of the instantaneous composition over the conversion interval chosen:

$$\langle F_a \rangle = \frac{1}{\pi} \int_0^{\pi} F_a(\pi)\, d\pi$$

There is, however, a much easier way to get the overall composition of the copolymer, as it is given, at any conversion, by the expression:

$$\langle F_a \rangle = \frac{(f_a)_0 - (1 - \pi)\,(f_a)_\pi}{\pi}$$

This equation arises from the fact that the proportion of monomer A remaining in the monomer mixture at any conversion is given by the ratio of the mole concentration of unconsumed monomer A to the total remaining monomer:

$$(f_a)_\pi = \frac{(f_a)_0 - \langle F_a \rangle_\pi\, \pi}{1 - \pi}$$

In the diagram shown on Figure 4.5. f_a, F_a and $\langle F_a \rangle$ are plotted as a function of conversion degree π, to illustrate the above development. For any given conversion degree, the "average" (or overall) composition of the copolymer is given by the value of $\langle F_a \rangle$ for that precise value of π. The amplitude of the compositional heterogeneity extends from the value $(F_a)_0$ — instantaneous composition of the copolymer at zero conversion — to the value $(F_a)_\pi$ observed at that conversion.

A second example is given in Fig. 4.6., illustrating the case in which an azeotropic composition exists. Here, the plots of F_a versus π are given for various initial compositions of the monomer mixture. At the value of $(F_a)_{az}$ corresponding to the azeotropic composition, no compositional heterogeneity can occur. On either side of the azeotrope the variation of the composition with conversion is slow and fluctuations in composition remain small, up to rather high conversions. As was already discussed, the composition is always moving away from the azeotrope.

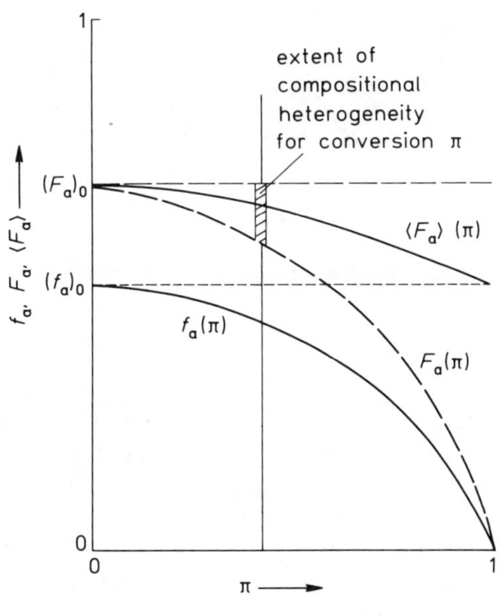

Fig. 4.5. Schematic representation of the variation with conversion π of the instantaneous A-content of the monomer mixture f_a; of the instantaneous composition of the copolymer F_a and of the integrated A-content of the copolymer formed to π: $\langle F_a \rangle$

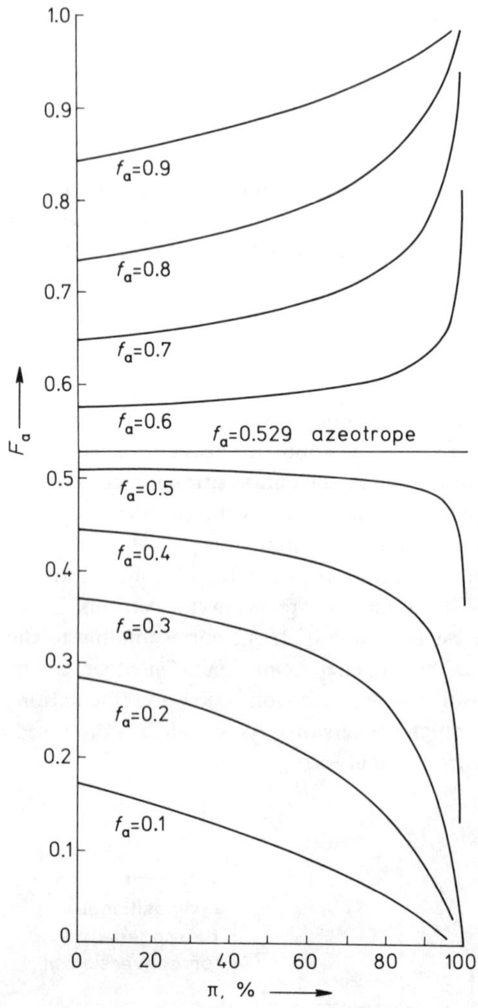

Fig. 4.6. Instantaneous copolymer composition F_a plotted versus conversion π for various values of the initial A-content of the monomer mixture f_a (A = styrene; B = MMA); $r_a = 0.52$, $r_b = 0.47$

4.3 Determination of the Radical Reactivity Ratios

4.3.1 Methods

To determine the radical reactivity ratios, copolymerization experiments are carried out to low conversion, in order to avoid bias originating from fluctuations in composition. The "integrated" composition $\langle F_a \rangle$ of the copolymer can then be set equal to its "initial" composition $(F_a)_0$ and considered with reference to the initial composition of the monomer mixture, $(f_a)_0$.

Several linearization procedures of the instantaneous composition equation have been developed for the determination of r_a and r_b:

$$\frac{dA}{dB} = \frac{[A]}{[B]} \frac{r_a[A] + [B]}{r_b[B] + [A]} \tag{4.5}$$

yields:

$$r_b = \frac{[A]}{[B]} \left\{ \frac{d[B]}{d[A]} \left(1 + r_a \frac{[A]}{[B]} \right) - 1 \right\} \tag{4.12}$$

which can also be written as:

$$r_b = \frac{f_a}{1 - f_a} \cdot \frac{1 - 2F_a}{F_a} + \frac{f_a^2}{(1 - f_a)^2} \cdot \frac{1 - F_a}{F_a} \cdot r_a = - G + Hr_a \tag{4.13}$$

From the experimental values of f_a and F_a one can calculate G and H. One can then plot either G versus H, or r_b versus r_a. In the former case (*Fineman-Ross plots)*, each experiment gives one point (G, H) and a straight line is expected to pass through these points. The slope of this line is r_a and the ordinate intercept is r_b (Fig. 4.7).

In the latter case (Mayo-Lewis plot: r_b versus r_a), each experiment results in a straight line. The best intersection between all these lines gives the values of r_a and r_b (Fig. 4.8).

The major criticism that can be expressed against such linearization methods is that r_a and r_b do not play symmetrical roles in equation (4.13). Thus it happens that exchange of A and B do not yield identical results. The method of *Yerzeliev* and *Roskin* introduces some improvements in that respect. Setting:

$$k = \frac{[B]}{[A]} \cdot \frac{d[A]}{d[B]} \quad \text{and} \quad x = \frac{[A]}{[B]}$$

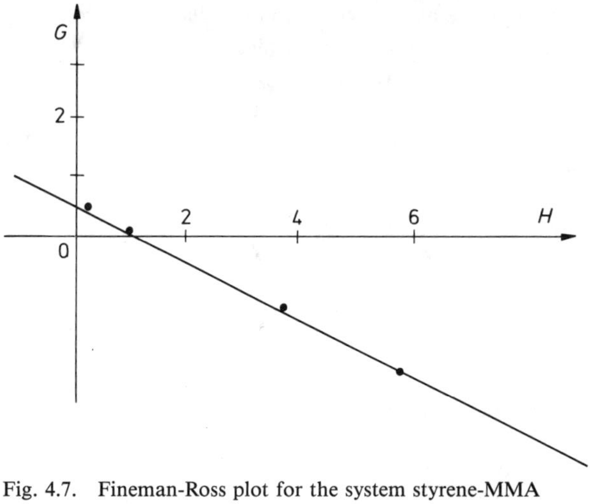

Fig. 4.7. Fineman-Ross plot for the system styrene-MMA

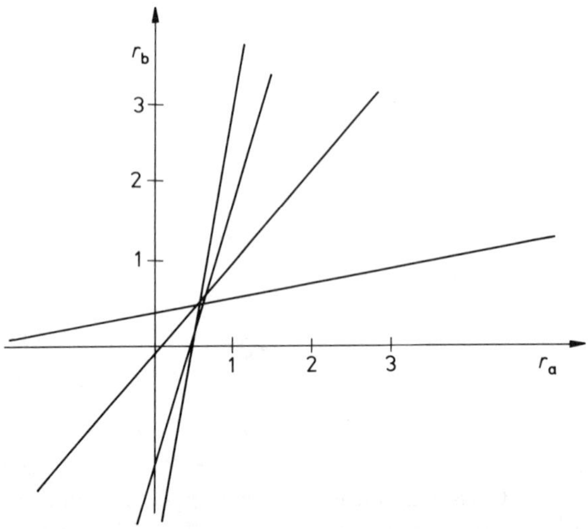

Fig. 4.8. Mayo-Lewis plot for the system styrene-MMA; (styrene: $r_a = 0.52$; methyl methacrylate: $r_b = 0.48$)

equation (4.12) can be written as:

$$kr_b + kx = r_a x + 1 \tag{4.14}$$

Now upon dividing both members of this equation by $\sqrt{kx}$ one obtains equation (4.15) in which r_a and r_b do play symmetrical roles:

$$\sqrt{kx} - \frac{1}{\sqrt{kx}} = r_a \sqrt{\frac{x}{k}} - r_b \sqrt{\frac{k}{x}} \tag{4.15}$$

Exchange of A and B should introduce no changes as to the values of the reactivity ratios, and the uncertainty on both values should be the same.

The method most commonly applied to determine radical reactivity ratios is that of Kelen and Tüdös. This procedure also aims at preventing the non-symmetrical character of the composition equation from affecting the values of r_a and r_b determined experimentally. Setting:

$$G = \frac{f_a}{f_b} \cdot \frac{2F_a - 1}{F_a} \qquad H = \frac{f_a^2}{f_b^2} \cdot \frac{1 - F_a}{F_a}$$

and

$$\eta = \frac{G}{\alpha + H} \qquad \xi = \frac{H}{\alpha + H}$$

the instantaneous composition equation can be written as:

$$\eta = \xi \left(r_a + \frac{r_b}{\alpha} \right) - \frac{r_b}{\alpha} \tag{4.16}$$

In fact the introduction of constant α has the advantage of limiting the domain of variation of η and of ξ to the interval (0, 1), instead of (0, ∞) for G and H. Furthermore an exchange between A and B does not affect the values. By plotting η versus ξ one obtains a straight line and r_a and $-\frac{r_b}{\alpha}$ are the values of η for $\xi = 1$ and for $\xi = 0$, respectively.

More elaborate treatments have been proposed, involving iterations (*computer calculcations*); however, if account is taken of the experimental accuracy on the compositional data, the above methods are adequate in most cases.

4.3.2 Experimental Procedures

Mixtures of A and B in known proportions are placed in small reaction vessels and set to polymerize in the presence of a free radical initiator. As soon as the conversion degree has reached 5–10% the reaction is stopped, the copolymer is precipitated, washed and dried carefully, and its overall composition is determined.

In recent years *gas chromatography* has been used increasingly, to measure the residual monomer concentrations [A] and [B] at various times, during one single copolymerization experiment. More information is provided than with a whole set of separate experiments carried out to low conversion, over the entire range of compositions. Gas chromatography is not only well suited to determine the radical reactivity ratios, it is also used to study copolymerization kinetics, and to evaluate the magnitude of fluctuations in composition within a sample.

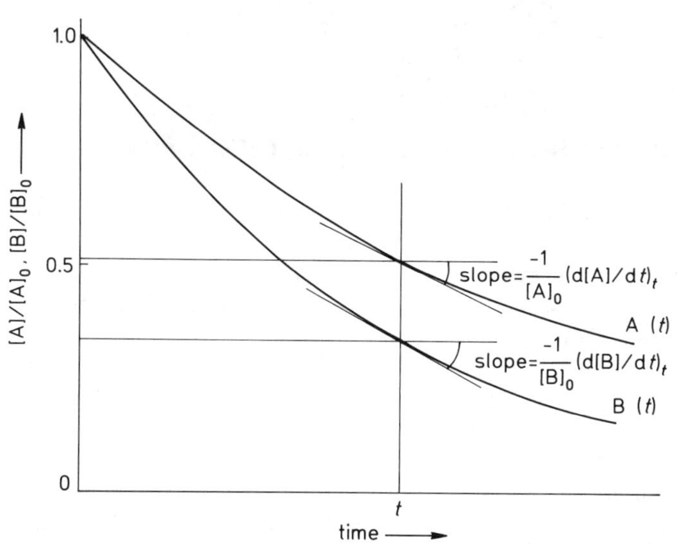

Fig. 4.9. Schematic representation of a plot of monomer composition in a free radical copolymerization, versus time. The remaining monomer concentrations $[A]_t$ and $[B]_t$ are obtained from gas chromatographic measurements. The slopes of the curves allow calculation of the instantaneous copolymer composition

To determine the values of [A] and of [B] as a function of time an internal standard is used. Its concentration should be kept constant, its retention time should be close to that of the monomers, but no overlap should occur. From the normalized peak areas it is possible to determine accurately the residual monomer concentrations as a function of reaction time. Fig. 4.9 shows one such plot. The slopes of these curves at any given time yield the instantaneous rates of monomer conversion $d[A]/dt$ and $d[B]/dt$. the instantaneous monomer ratio [A]/[B] at time t can be related with the instantaneous ratio of monomer units $d[A]/d[B]$ (ratio of the slopes). Thus F_a and f_a can be calculated at various reaction times, and inserted in the above equations which will be solved by any of the graphical methods described above. The chief interest of such a method is that it yields the instantaneous values of monomer and copolymer composition, and is therefore not limited to low conversions.

Note that the chromatographic data give also access to the overall rate of copolymerization.

Table 4.1. Examples of reactivity ratios in binary free radical copolymerization

Monomer — System A	B	r_a	r_b
Styrene	Butadiene	0.75	1.3
Styrene	Methylmethacrylate	0.52	0.47
Styrene	Vinyl acetate	50	0.01
Styrene	Diethyl fumarate	0.3	0.07
Styrene	Maleic anhydride	0.05	0.0
Vinyl acetate	Vinyl chloride	0.23	1.68
Methyl M. A.	Vinyl acetate	20	0.015

4.4 Factors Affecting the Copolymerization Reactivity

Three factors affect the ability of a monomer to take part in a copolymerization process:

- *Steric factors.* Bulky monomers, and some 1,2-disubstituted monomers do not homopolymerize, though they can add to radicals of another kind. Consequently their radical reactivity ratios are usually close to zero.
- *Resonance stabilization of the radical site* results in the decrease of its reactivity and of its ability to attack another monomer molecule. Such is the case of styrene or vinylpyridine or acrylic ester radicals. This parameter is taken into consideration in the Alfrey Price scheme by the factor Q.
- *The polarity of the double bond* also plays a role. An electron donating substituent polarizes the double bond to some extent, and an electron withdrawing substituent does it in the reverse manner. A monomer carrying an electron donating substituent will therefore have a marked tendency to add preferentially to a radical derived from a monomer carrying an electron withdrawing substituent, and vice versa. It may even happen that a *charge transfer complex* is formed between monomers A and B, prior

to any polymerization. This accounts for the high degree of alternation which is observed in such systems.

Two examples of monomer systems in which the substituent effects are in opposite directions, and in which the resonance stabilization of the radicals is quite different are given below:

Example 1

system 1 $\begin{cases} \text{B} = \text{styrene (resonance stabilized); } r_b = 50 \\ \text{A} = \text{vinyl acetate (unstabilized); } r_a = 0.01 \end{cases}$

$k_{ab} = 370{,}000$, $k_{ba} = 3.2$ — cross addition rate constants
$k_{bb} = 176$ — styrene homopolymerization rate constant
$k_{aa} = 3{,}700$ — vinyl acetate homopolymerization rate constant

Example 2

system 2 $\begin{cases} \text{B} = \text{methyl methacrylate (electron withdrawing substituent); } r_b = 20 \\ \text{A} = \text{vinyl acetate (electron donating substituent); } r_a = 0.015 \end{cases}$

$k_{ab} = 247{,}000$, $k_{ba} = 37$ — cross addition rate constants
$k_{aa} = 3{,}700$ — rate constant for vinyl acetate homopolymerization
$k_{bb} = 734$ — rate constant for methyl methacrylate homopolymerization

In both cases the equilibrium concentration of the $A^\bullet$ radicals is very low, and in spite of the fact that $k_{aa} \gg k_{bb}$, the copolymer contains more B units than the monomer mixture it originates from.

This feature has found some industrial applications in the homopolymerization of monomers such as vinyl chloride or vinyl acetate. If the rate of polymerization rises too much and tends to get out of control, a small injection of styrene suffices to slow it down, because stabilized styrene radicals are formed at the expense of highly reactive vinyl chloride or vinyl acetate radicals.

4.5 *Q*-*e*-Scheme

In order to rationalize the copolymerization abilities of unsaturated monomers, it was attempted to find some constants which would characterize the reactivity of each radical or monomer from their *resonance* and *polarizability parameters*. As for *Hammett's rules* meant to characterize the reactivities of substituted benzene rings, it was postulated that the reaction rate constants between radical $A^\bullet$ and monomers A or B should obey the following equations:

$$
\begin{aligned}
k_{aa} &= P_a Q_a \exp(-e_a^2) \\
k_{ab} &= P_a Q_b \exp(-e_a e_b)
\end{aligned}
\tag{4.17}
$$

Here P_a is a measure of the resonance stabilization of radical $A^{\bullet}$, and characterizes its reactivity; Q_a and Q_b characterize similarly the resonance in monomers A and B; e_a and e_b are related to the factors affecting the polarities of A and B. It is assumed that the polarity of the substituent affects monomer and radical in the same way and to the same amount.

Thus, after *Alfrey* and *Price*, the reactivity ratios can be expressed as:

$$
\begin{aligned}
r_a &= \frac{Q_a}{Q_b} \exp[-e_a(e_a - e_b)] \\
r_b &= \frac{Q_b}{Q_a} \exp[-e_b(e_b - e_a)]
\end{aligned}
\tag{4.18}
$$

Styrene is taken as a standard, its Q and e values being set to 1 and -0.8, respectively. Lists of values of Q and e are available. Using the sets of values for two monomers, it is possible to calculate approximate values of their radical reactivity ratios r_a and r_b. This is of interest for pairs of monomers which have not been investigated before. In some cases the calculated values are not far from the true values, but, in general, this method remains *rather qualitative*.

4.6 Kinetics of Radical Copolymerization

It is of interest to study the kinetics of free radical copolymerization with reference to the corresponding homopolymerization rates. Since the total number of active sites is stationary, the initiation and termination rates can be set equal:

$$R_i = k_{taa}[A^{\bullet}]^2 + k_{tbb}[B^{\bullet}]^2 + k_{tab}[A^{\bullet}][B^{\bullet}] \tag{4.19}$$

The steady-state condition also applies to each type of radical (*Bodenstein's rule*) for short reaction times, as it was shown before.

$$k_{ab}[A^{\bullet}][B] = k_{ba}[B^{\bullet}][A] \tag{4.2}$$

Equation (4.19) implies that termination is chemically controlled, exclusively. The overall rate of monomer consumption can be expressed from equation (4.1), taking Bodenstein's rule into account:

$$R_p = -\frac{d[A] + d[B]}{dt} = k_{aa}[A^{\bullet}][A] + k_{bb}[B^{\bullet}][B] + 2k_{ab}[A^{\bullet}][B] \tag{4.20}$$

From equation (4.2), it follows that $[B^{\bullet}]$ can be expressed as a function of $[A^{\bullet}]$:

$$[B^\bullet] = [A^\bullet]\,\frac{k_{ab}}{k_{ba}}\,\frac{[B]}{[A]}$$

One obtains:

$$R_p = \frac{[A^\bullet]}{[A]}\left[k_{aa}[A]^2 + 2k_{ab}[A][B] + k_{bb}[B]^2 \cdot \frac{k_{ab}}{k_{ba}}\right] \qquad (4.21)$$

and $[A^\bullet]$ can be expressed by means of eq. (4.19), to yield:

$$[A^\bullet] = \frac{R_i^{1/2}}{\left[k_{taa} + k_{tbb}\left(\frac{k_{ab}}{k_{ba}}\,\frac{[B]}{[A]}\right)^2 + k_{tab}\left(\frac{k_{ab}}{k_{ba}}\,\frac{[B]}{[A]}\right)\right]^{1/2}} \qquad (4.22)$$

Replacing in equation (4.21) $[A^\bullet]$ by its value shown in equation (4.22) and setting

$$\delta_a = \left(\frac{k_{taa}}{k_{aa}^2}\right)^{1/2} \;;\quad \delta_b = \left(\frac{k_{tbb}}{k_{bb}^2}\right)^{1/2} \;;\quad \phi = \frac{k_{tab}}{(k_{taa}\cdot k_{tbb})^{1/2}}$$

the following instantaneous copolymerization rate equation is obtained:

$$R_p = \frac{r_a[A]^2 + 2[A][B] + r_b[B]^2}{[r_a^2\,\delta_a^2[A]^2 + 2\phi\,\delta_a\delta_b r_a r_b[A][B] + r_b^2\delta_b^2[B]^2]^{1/2}} \cdot R_i^{1/2} \qquad (4.23)$$

The δ parameters are the familiar $\left(\frac{k_p^2}{k_t}\right)^{-1/2}$ parameters characterizing the homopolymerization of each of the monomers involved.

The parameter φ characterizes the cross termination rate constant with respect to the geometric mean of the termination rate constants for the corresponding homopolymerization of A and of B. Since the cross additions ($A^\bullet$ + B, or $B^\bullet$ + A) are often faster than the homoaddition reaction (the radical reactivity ratios being then lower than unity), the copolymerization might be exprected to be faster than the corresponding homopolymerizations. However, this is not the case, because the cross termination reaction is much faster than the homotermination process ($\phi > 1$). More precisely, the greater the tendency towards alternation, the higher the value of φ. Consequently, the steady-state concentration of radicals in a copolymerization of A and B is much lower than in the homopolymerizations of A and of B, carried out under the same experimental conditions.

The above treatment is based on the assumption that termination is *not* controlled by diffusion, i. e., is a true chemical kinetic rate constant. While this is correct at low conversions (low viscosity media), it may not be correct at conversions exceeding 15% – 30%. Since chain length, and thus viscosity, is widely variable, it is impossible to cite an exact concenctration or percent conversion, beyond which diffusion controls.

Furthermore, it was found that, for some systems investigated, φ is a function of copolymer composition. This problem was handled using a single termination rate constant k_t and assuming that the value of k_t depends upon the instantaneous composition of the copolymer formed. The steady state equation can be written as:

$$R_i = k_t([A^\bullet] + [B^\bullet])^2 \tag{4.24}$$

and the rate equation results:

$$R_p = \frac{R_i^{1/2}\,(r_a[A]^2 + 2[A][B] + r_b[B]^2)}{k_t^{1/2}\left[\dfrac{r_a[A]}{k_{aa}} + \dfrac{r_b[B]}{k_{bb}}\right]} \tag{4.25}$$

with

$$k_t = F_a k_{taa} + F_b k_{tbb} \tag{4.26}$$

The latter relation enables one to calculate k_t from the corresponding termination rate constants in homopolymerization and the molar composition parameters of the copolymer F_a (and $F_b = 1 - F_a$). As the latter varies with conversion, comparisons should be limited to low conversions. However, this method can explain the observed variation of the copolymerization rate versus composition of the feed as satisfactorily as the preceding method.

4.7 Sequence Distributions in Radical Copolymerization

The sequence distribution of monomer units A and B along the chain is also governed by the kinetic parameters of the system.

Assume a growing chain has its active site located on a (terminal) A unit. The probability that the next unit to be added is again an A unit is given by:

$$p_{aa} = \frac{k_{aa}[A]}{k_{aa}[A] + k_{ab}[B]} = \frac{r_a f_a}{r_a f_a + f_b} \tag{4.27}$$

Obviously, the probability of formation of an AB diad is:

$$p_{ab} = 1 - p_{aa} = \frac{f_b}{r_a f_a + f_b}$$

Furthermore, the probability of finding an AB diad *at any particular place* in the copolymer chain is given by the probability of finding an A unit (namely F_a) multiplied by the probability of occurrence of an AB diad at this point. Thus

$$P_{ab} = F_a \cdot p_{ab} = F_a \frac{f_b}{r_a f_a + f_b} \tag{4.28}$$

It can be easily demonstrated that diads or triads can be counted either way along the chain, by setting:

$$P_{ab} = F_a \frac{f_b}{r_a f_a + f_b} \quad \text{and} \quad P_{ba} = F_b \frac{f_a}{r_b f_b + f_a}$$

If $P_{ab} = P_{ba}$, it follows that:

$$\frac{F_a f_b}{r_a f_a + f_b} = \frac{F_b f_a}{r_b f_b + f_a}$$

or:

$$\frac{F_a}{F_b} = \frac{f_a(r_a f_a + f_b)}{f_b(r_b f_b + f_a)} \equiv \frac{d[A]}{d[B]} = \frac{[A]}{[B]} \cdot \frac{r_a[A] + [B]}{r_b[B] + [A]} \tag{4.5}$$

which is precisely the instantaneous composition equation.

Let us now consider the *probabilities of triad occurrence* along the chain. As the probabilities of occurrence of each individual diad are independent, the probability of formation of any given triad is the product of the probabilities for the constituting diads to occur. As a result, it follows:

$$\left.\begin{aligned} P_{aaa} &= F_a \cdot p_{aa}^2 & P_{bbb} &= F_b \cdot p_{bb}^2 \\ P_{abb} &= F_a \cdot p_{aa} \cdot p_{ab} & P_{bba} &= F_b \cdot p_{bb} \cdot p_{ba} \\ P_{abb} &= F_a \cdot p_{ab} \cdot p_{bb} & P_{baa} &= F_b \cdot p_{ba} \cdot p_{aa} \\ P_{aba} &= F_a \cdot p_{ab} \cdot p_{ba} & P_{bab} &= F_b \cdot p_{ab} \cdot p_{ba} \end{aligned}\right\} \tag{4.29}$$

with

$$p_{aa} + p_{ab} = 1 \qquad p_{ba} + p_{bb} = 1$$

The sum of the eight triad probabilities P is of course unity.

The probability of alternation in a copolymer chain is the sum of the probabilities of occurrence of ABA and BAB triads. It is therefore given by:

$$P_{alt} = F_a p_{ab} p_{ba} + F_b p_{ab} p_{ba} = p_{ab} \cdot p_{ba}$$

and it can be expressed as:

$$P_{alt} = \frac{f_b}{r_a f_a + f_b} \cdot \frac{f_a}{r_b f_b + f_a} \tag{4.30}$$

When both r_a and r_b are *small* (or equal to zero), the probability of alternation is high. The proportion of alternation in a copolymer can be used as a check for the r_a and r_b values of the system.

For a given system, alternation reaches its maximum value for a composition that can be calculated upon differentiating equation (4.30) and setting $\frac{dP_{alt}}{df_a}$ equal to zero, whereby one obtains:

$$\frac{[A]}{[B]} = \left(\frac{r_b}{r_a}\right)^{1/2} \quad \text{or} \quad f_a = \frac{r_b^{1/2}}{r_a^{1/2} + r_b^{1/2}}$$

This monomer feed composition is precisely that yielding a 1:1 copolymer (with 50% of monomer units of each type).

The probability P_{na+} *of finding homopolymeric sequences* in a copolymer having at least n A units connected together is given by:

$$P_{na+} = F_a p_{aa}^n$$

which can be written as:

$$P_{na+} = F_a \left\{\frac{r_a f_a}{r_a f_a + f_b}\right\}^n \tag{4.31}$$

It follows that, unless r_a is by far greater than unity, the formation of long homopolymeric sequences AA . . . A in a binary copolymer chain is unlikely to occur. The formation of block copolymers (involving both sequences of A units *and* sequences of B units) is impossibly since in no reported case both radical reactivity ratios r_a and r_b are larger than unity. Thus homogeneous radical chain copolymerization cannot yield block copolymers*).

Bernouillian distributions of the A and B units**) along the chain can be attained only in two special cases:

- When both radical reactivity ratios r_a and r_b are equal to one, which is a trivial limiting case, but is illustrated by a few systems.
- In the case of ideal copolymerization ($r_a \cdot r_b = 1$) since both radicals $A^\bullet$ and $B^\bullet$ exhibit then the same relative reactivities towards the two monomers present.

In these cases the distribution of A and B units along the chain is a one-parameter distribution, since

$$F_a = p_{aa} = p_{ba} \quad \text{and} \quad F_b = p_{bb} = p_{ab} = 1 - F_a$$

In all other copolymerization systems (those involving an azeotrope included), two independent parameters are required — at least — to account for the distribution of A and B units along the chain.

4.8 Penultimate Effects

It sometimes happens that the reactivity ratios are not found to be constant over the entire range of monomer feed composition. This can be interpreted in various manners.

---

*) If solid homopolymer B in a glassy or crystalline state is exposed to monomer A and is irradiated by electrons, *grafting* of polymer-A chains to poly-B will occur.

**) A Bernouillian distribution is a distribution in which an event is not conditioned by previous events. One single parameter suffices to characterize it.

One way to account for the experimental facts is to assume that the penultimate chain unit affects the reactivity of the active site located on the last unit of the growing chain. In other words, the $A^\bullet$ radical would exhibit different reactivities depending upon whether the preceding unit is A or B. Such an assumption would lead to 8 rate constants of growth (instead of 4) and to 4 radical reactivity ratios (instead of 2). The composition equation then gets rather complicated. Very precise and reproducible measurements are required to check the values obtained, even though penultimate effects usually do not show up simultaneously on both types of radicals $A^\bullet$ and $B^\bullet$.

Alternate explanations have been proposed to account for the apparent variation of r_a and/or r_b with the composition of the monomer feed: They may arise from *preferential solvation* of the polymer chain by one of the two monomers. The actual composition of the monomer feed in the vicinity of the growing sites may be different from the analytical overall composition. They can also arise from *strong interactions* between A and B monomer units, prior to their addition to the growing site.

4.9 Crosslinking Free Radical Copolymerization

The free radical copolymerization of a monomer A with a bis-unsaturated monomer D is of special interest as it yields *crosslinked* species which have found a wide variety of applications. Such systems are rather complicated. When the process starts the system is binary and it can be treated as a regular free radical copolymerization. However, as soon as some D units have been added to radical sites, pendent double bonds are formed (scheme), which exhibit their own reactivity. From there on the system involves three different polymerizable constituents: A monomer, D monomer and the pendant unsaturations: The system has become ternary.

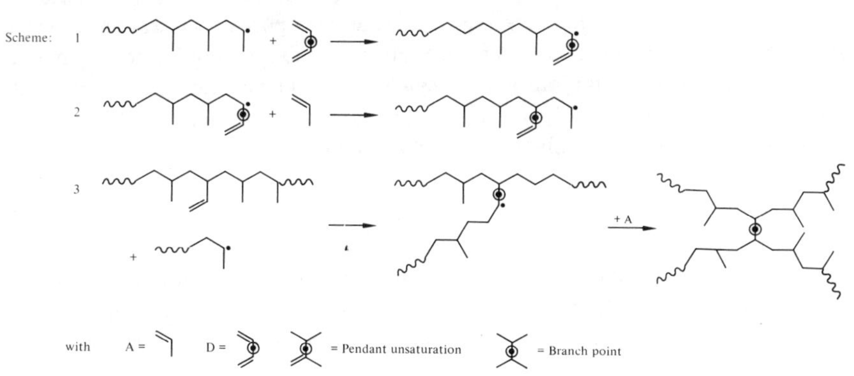

Furthermore, the attack of a pendant unsaturation by a growing radical results in formation of a tetrafunctional branch point, connecting two chains. As discussed earlier (chapter 2), once each chain is connected with an average of two other chains by means of tetrafunctional branch points, gelation ensues. A polymer network is formed. At the gel point, the weight average molecular weight of the polymer reaches infinity.

The probability of branch points to be formed depends primarily upon the proportion of the diunsaturated monomer D in the initial mixture. The reactivity of the pendant unsaturations (as compared to those of monomers A and D) with respect to the radicals also influences the course of the reaction. This accounts for the fact that, for the same ratio of [D] to [A], the gel point can be attained at widely varying conversions.

4.10 Some Specific Cases

4.10.1 Polyalkylene Sulfones

Free radical copolymerization of ethylene and sulfur dioxide does not require high pressures and temperatures, and it yields essentially alternating copolymers, generally referred to as polyethylene sulfones.

$$n\,SO_2 + n\,CH_2{=}CH_2 \longrightarrow \left[CH_2{-}CH_2{-}SO_2 \right]_n$$

This case is interesting as it shows how two compounds can easily copolymerize while neither of them would homopolymerize under the conditions chosen ($r_a = 0$, $r_b = 0$). The kinetics of the process are complicated owing to the fact that propagation-depropagation equilibria are involved.

Beside ethylene, a number of other monomers give alternating copolymers with SO_2: Propylene, butene, and vinyl ethers. They have electron donating substituents and are therefore likely to give some kind of complex with SO_2 that is strongly electropositive. This favors the tendency to alternation.

Ethylene can also be copolymerized with carbon monoxide under higher temperatures and pressures, but the tendency to give alternating copolymers is less pronounced in this case. The CO units are still isolated, though, as CO does not homopolymerize. By choosing adequate molar ratios of ethylene and CO one can get ketone units distributed at will along the polymer backbone, and this has been applied in a search for "disposable", degradable polymers (see chapter 9).

4.10.2 Cyclopolymerization

It was shown earlier that polymers arising from diunsaturated monomers, such as p-divinylbenzene, or ethylene dimethacrylate are tightly crosslinked, and that they induce crosslinking upon copolymerizing with monomers such as styrene or methyl methacrylate.

There are however cases in which diunsaturated monomers would give rise to *cyclic* units rather than to crosslinks. Such monomers include unconjugated diunsaturated monomers in which intramolecular chain propagation can lead to 5 or 6 membered rings along the chain. Typical examples include:

- Diallylammonium salts, that yield fully soluble, unsaturation free polyelectrolyte cyclopolymers according to

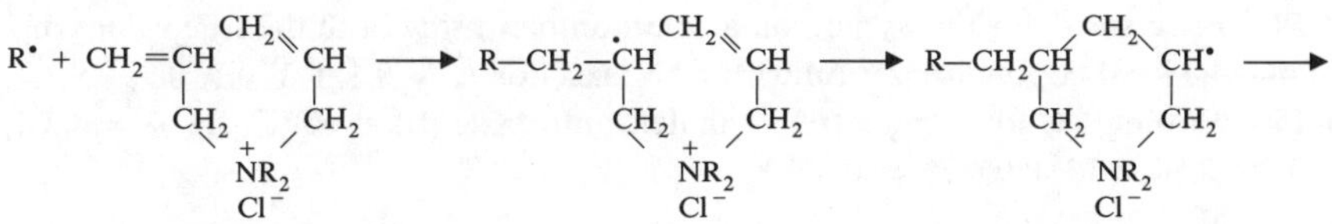

- Acrylic and methacrylic anhydrides, that also yield uncrosslinked cyclopolymers

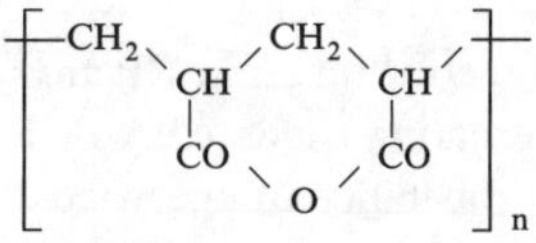

- 1,5-hexadiene and 1,6-heptadiene and derivatives.

When cycles larger than 6-membered are to be formed, the probability of cyclization is lower and pendant unaturations are formed, which eventually react with other growing radicals, yielding crosslinking. However, even in such cases a fraction of the diunsaturated monomer gives yield to intramolecular chain propagation and does not contribute to crosslinking. In the case of ortho diallylphthalate, 40% of the monomer units are involved in intramolecular chain propagation (cyclization): the others contribute to intermolecular chain propagation, yielding pendant double bonds, that eventually lead to crosslinking.

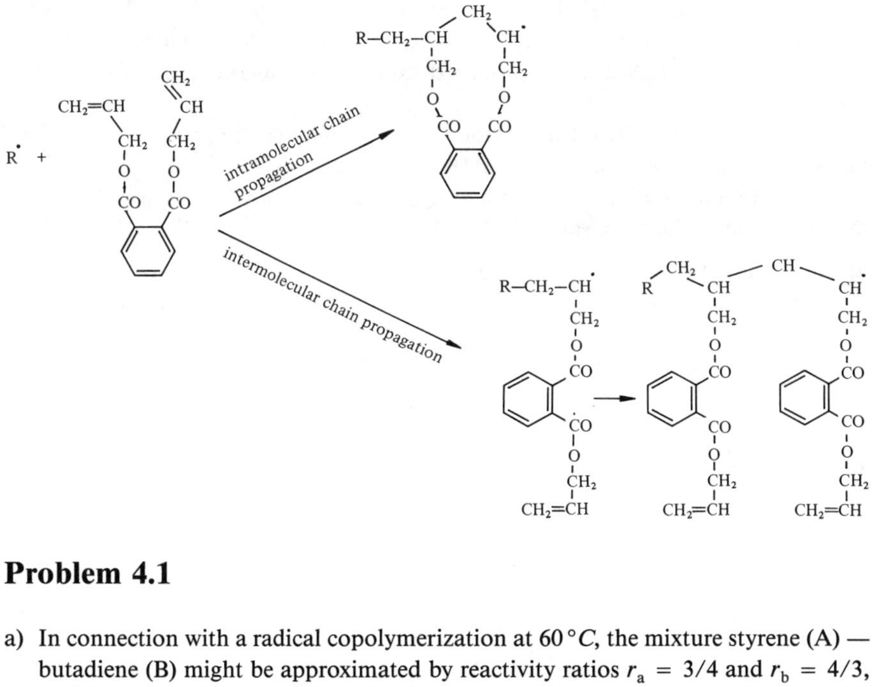

Problem 4.1

a) In connection with a radical copolymerization at 60 °C, the mixture styrene (A) — butadiene (B) might be approximated by reactivity ratios $r_a = 3/4$ and $r_b = 4/3$, and thus as an "ideal" copolymerization. The true values are: $r_a = 0.78$; $r_b = 1.39$.

Plot F_a versus f_a for this system on a common plot using both the true values of r and the ideal copolymerization with $r_a = 3/4$, for $f_a = 0.1, 0.3, 0.5, 0.7, 0.9$.

b) Plot F_a versus f_a for styrene (A) — maleic anhydride (B) at 60 °*C*; for $f_a = 0.05$, 0.25, 0.50, 0.75, 0.95 ($r_a = 0.02$; $r_b = 0$).

Problem 4.2

The reactivity ratios for the system styrene (A) and diethyl fumarate (B) are $r_a = 0.3$ and $r_b = 0.07$. The probability of formation of alternating triads ABA or BAB are $p_{aba} = p_{ab}p_{ba}$, and $p_{bab} = p_{ba}p_{ab}$. Likewise, the probability of the triad AAA is $p_{aa}p_{aa}$. These probabilities obviously depend on the molar ratio x of the concentrations of monomers A and B.

a) Show that the value of x which leads to the highest proportion of alternating triads is 0.483.
b) Using the above value of x calculate p_{aba}, p_{aaa}, p_{aab}, p_{abb} the sum of which is unity, and p_{bab}, p_{bbb}, p_{bba}, p_{baa} the sum of which is unity.
c) Repeat the calculation of (b), but for $x = 2$.

Problem 4.3

A copolymer containing 20 mol-% vinyl acetate (A) ($r_a = 0.23$) and 80 mol-% vinyl chloride (B) ($r_b = 1.68$) is to be produced from an appropriate monomer composition, at 60 °*C*.

Assume that a partially filled reaction vessel contains 100 *moles* of the appropriate mixed monomer. This is initiated by AIBN.

The more rapidly consumed monomer will be fed continuously so as to maintain this monomer composition constant.

a) What should the monomer composition be?
b) After 100 *moles* of polymer have been produced, how many total moles are in the vessel?
c) Determine the triad probabilities of this polymer.

Problem 4.4

At 60 °*C*, the kinetics of homopolymerization is determined by the rate constants below:

for styrene:	$k_p = 176\ [l \cdot mol^{-} \cdot s^{-1}]$
(a)	$k_t = 3.6 \times 10^7\ [l \cdot mol^{-1} \cdot s^{-1}]$
for butadiene:	$k_p = 100\ [l \cdot mol^{-1} \cdot s^{-1}]$
(b)	$k_t = 36 \times 10^7\ [l \cdot mol^{-1} \cdot s^{-1}]$

At given $r_a \cong 3/4$, $r_b \cong 4/3$, and $\phi \cong 1$, determine AIBN concentration necessary to produce, at 60 °C, kinetic chain lengths of 5000 in the copolymerization of styrene with butadiene, to produce a polymer with $F_a = 0.20$ and $F_b = 0.80$. Assume no solvent will be used and that the conversion will be kept to low values, so that monomer composition does not change much.

Problem 4.5

In the homogeneous polymerization of monomer A with monomer B, the radical reactivity ratios are: $r_a = 2$ and $r_b = 1/2$, and the initial concentrations are: $[A]_0 = [B]_0 = 5$ mol/l.

Determine:

a) The extent of conversion required to reduce f_a to 1/10 of its initial value.
b) The instantaneous value of F_a at this value of conversion.
c) The integrated value $\langle F_a \rangle$ of the A-content of the copolymer, at that conversion.

[Note that in this case of ideal copolymerization the equations are simplified.]

Chapter 5
Anionic Polymerizations

5.1 Introduction

Anionic polymerizations have been known for a long time, and they have been studied extensively over the past three decades. Their industrial interest is limited, because they usually require solvents, but their development is related to the possibilities they can offer to synthesize well defined macromolecules. Few other methods are as versatile as anionic polymerizations to control molecular weights and molecular weight distributions, and to give access to functional polymers, to block copolymers, to branched species, and others more (see chapter 10).

Anionic polymerizations proceed via metalorganic sites: *Carbanions (or oxanions),* with their metallic counterions for electrical neutrality. Carbanions are nucleophiles, consequently the monomers best fitted for anionic polymerization are those bearing an electroattractive substituent, to polarize the double bond in the proper way. Anionic growth reactions (propagation) proceed by nucleophilic attack of a carbanionic site onto a monomer, whereby the carbanionic site is restored at chain end:

The same is true for the ring opening polymerization of cyclic monomers containing some function sensitive to nucleophiles, such as oxiranes, thiiranes, lactones, etc.

Early examples of anionic polymerizations include:

- The polymerization of styrene in liquid ammonia, initiated by sodium amide ($NaNH_2$).
- The polymerization of dienes initiated either by metallic sodium (Buna) or with butyl lithium (BuLi).
- The ring opening polymerization of oxiranes (ethylene oxide) initiated by potassium alkoholates, and that of caprolactam with sodium.
- The polymerization of monomers such as cyanoacrylates by means of weak bases, in acrylic glues.

The influence of solvent on the nucleophilic reactions involved in anionic growth reactions can be expected to be high. If the solvent is able to release a proton, it may react with the active site, as proton donating compounds generally do with organometallic derivatives. It is therefore essential to distinguish between anionic polymerizations carried out in proton donating solvents, and those carried out in the absence of any proton donating compounds, which are the so-called *"living" anionic polymerizations*. Termination reactions by mutual annihilation, characteristic of free radicals, have no counterpart in ionic polymerization.

5.2 Anionic Polymerizations Carried out in Protic Media

Anionic polymerizations carried out in protic media are characterized by a competition between chain growth (reaction of the active site with the monomer) and chain transfer involving a solvent molecule. If the new anion formed upon transfer is unable to reinitiate (by reaction with the monomer) no polymerization can take place. There are however cases in which the anions originating from the solvent upon transfer can react with the monomer. The polymerization of styrene in liquid ammonia initiated by sodium amide is such a case, and it will be examined below.

5.2.1 Reaction Scheme

The general reaction scheme can be established as follows:

- Dissociation of sodium amide into its ions:

$$NaNH_2 \rightleftarrows Na^{\oplus} + NH_2^{\ominus} \qquad \text{(equilibrium constant } K_D\text{)}$$

- Initiation, assuming that the $NH_2^{\ominus}$ ions are the only initiating species:

$$NH_2^{\ominus} + M \longrightarrow NH_2M_1^{\ominus} \qquad \text{(rate constant of initiation } k_i\text{)}$$

- Growth (chain propagation):

$$NH_2\text{—}M_i^{\ominus} + M \longrightarrow NH_2M_{i+1}^{\ominus} \qquad \text{(rate constant of growth } k_p\text{)}$$

- Transfer to the solvent:

$$NH_2M_i^{\ominus} + NH_3 \longrightarrow NH_2M_iH + NH_2^{\ominus} \qquad \text{(rate constant of transfer } k_{tr}\text{)}$$

Here k_i, k_p, k_{tr} are the rate constants of initiation, of propagation and of transfer to solvent, and K_D stands for the equilibrium constant of the mass action law for the electrolytic dissociation of sodium amide:

$$K_D = \frac{[NH_2^{\ominus}][Na^{\oplus}]}{[NaNH_2]} \qquad (5.1)$$

Provided no other ions are present one can set $[Na^{\oplus}] = [NH_2^{\ominus}]$, and $[NH_2^{\ominus}]$ can be expressed as:

$$[NH_2^{\ominus}] = \sqrt{K_D \cdot [NaNH_2]} \qquad (5.2)$$

5.2.2 Kinetics

The rate of initiation can be written as:

$$R_i = \frac{d[M^\ominus]}{dt} = k_i [NH_2^\ominus][M] \tag{5.3}$$

Once created, the active sites can add monomer molecules, and the rate of chain growth is given by:

$$R_p = -\frac{d[M]}{dt} = k_p [M^\ominus][M] \tag{5.4}$$

where $[M^\ominus]$ stands for the molar concentration of the active sites *of any size* in the reaction medium:

$$[M^\ominus] = \sum_{i=1}^{\infty} [M_i^\ominus]$$

The rate of transfer of the active sites onto the solvent is expressed as:

$$R_{tr} = -\frac{d[M^\ominus]}{dt} = k_{tr} [M^\ominus][NH_3] \tag{5.5}$$

Assuming steady state conditions*) for the active sites, one can set $R_i = R_{tr}$ and solve for $[M^\ominus]$:

$$[M^\ominus] = \frac{k_i [NH_2^\ominus][M]}{k_{tr} [NH_3]} \tag{5.6}$$

Replacing $[NH_2^\ominus]$ by its value taken from equation (5.2) and inserting equation (5.6) into the rate equation (5.4) one obtains:

$$R_p = -\frac{d[M]}{dt} = k_p \frac{k_i}{k_{tr}} K_D^{1/2} \frac{[NaNH_2]^{1/2}}{[NH_3]} [M]^2 \tag{5.7}$$

This is one of the few examples of polymerization rates that are second order in monomer concentration. The variation with time of the monomer concentration in the reaction medium can be obtained by integration of equation (5.7) to:

$$\frac{1}{[M]} - \frac{1}{[M]_0} = k_p \frac{k_i}{k_{tr}} \frac{(K_D [NaNH_2])^{1/2}}{[NH_3]} \cdot t \tag{5.8}$$

which can also be expressed in terms of conversion degree π, defined as:

$$\pi = \frac{[M]_0 - [M]}{[M]_0}$$

$$\frac{1}{1-\pi} = \frac{[M]_0}{[M]} = 1 + k_p \frac{k_i}{k_{tr}} \frac{(K_D [NaNH_2])^{1/2}}{[NH_3]} [M]_0 t \tag{5.9}$$

---

*) This assumption is justified, as the $NH_2^\ominus$-ions formed upon transfer participate again in the dissociation equilibrium of the initiator.

The rate of conversion depends upon the initial monomer concentration, contrary to the case of first-order polymerization rate equations, in which the fraction of monomer converted to polymer at time t is independent of the initial monomer concentration $[M]_0$.

The number average degree of polymerization of the polymer arising from such a process is equal to the average number of growth steps a carbanion undergoes before it transfers with NH_3. In other words, it is given by the ratio of the rate of growth to the rate of transfer:

$$\bar{P}_n = \frac{k_p[M]}{k_{tr}[NH_3]} \tag{5.10}$$

The molecular weight distribution exhibited by a polymer obtained anionically in a protic solvent will now be considered in more detail.

5.2.3 Molecular Weight Distribution of Polymers Obtained by Anionic Polymerization in Protic Media

As steady-state can be assumed (within small time intervals), it is possible to evaluate the polydispersity of the polymers obtained – as done previously for radical polymerizations — by evaluating the rates of formation and of consumption of the active species of polymerization degree i, and expressing then $[M_i^\ominus]$ as a function of $[M_{i-1}^\ominus]$.

For $i \geqslant 2$ this expression can be written as:

$$\frac{d[M_i^\ominus]}{dt} = k_p[M_{i-1}^\ominus][M] - k_p[M_i^\ominus][M] - k_{tr}[M_i^\ominus][NH_3] = 0 \tag{5.11}$$

Therefore:

$$[M_i^\ominus] = \frac{k_p[M]}{k_p[M] + k_{tr}[NH_3]} \cdot [M_{i-1}^\ominus] \tag{5.12}$$

And for $i = 1$:

$$\frac{d[M_1^\ominus]}{dt} = -k_p[M_1^\ominus][M] - k_{tr}[M_1^\ominus][NH_3] + k_i[NH_2^\ominus][M] = 0 \tag{5.13}$$

Since the steady state condition implies that:

$$k_i[NH_2^\ominus][M] = k_{tr}[M^\ominus][NH_3]$$

it follows:

$$[M_1^\ominus] = \frac{k_{tr}[NH_3]}{k_p[M] + k_{tr}[NH_3]}[M^\ominus] \tag{5.14}$$

Calling p the probability for chain growth to occur, $(1 - p)$ being then the probability of transfer:

$$p = \frac{k_p[M]}{k_p[M] + k_{tr}[NH_3]}; \qquad 1 - p = \frac{k_{tr}[NH_3]}{k_p[M] + k_{tr}[NH_3]} \tag{5.15}$$

one obtains:

$$[M_i^{\ominus}] = p[M_{i-1}^{\ominus}]$$
$$[M_{i-1}^{\ominus}] = p[M_{i-2}^{\ominus}]$$
$$- - - - - - - - - - -$$
$$[M_1^{\ominus}] = (1 - p)[M^{\ominus}]$$

which can be written as:

$$[M_i^{\ominus}] = p^{i-1}(1 - p)[M^{\ominus}] \tag{5.16}$$

The fraction of macromolecules that have reached a polymerization degree *i* results:

$$x_i = \frac{[M_i^{\ominus}]}{[M^{\ominus}]} = p^{(i-1)}(1 - p) \tag{5.17}$$

This expression is the same as that derived for free radical polymerization. Consequently, the molecular weight distribution is Flory's most probable with:

$$\bar{P}_n = \frac{1}{1 - p} \qquad \bar{P}_w = \frac{1 + p}{1 - p} \qquad \frac{M_w}{M_n} = 1 + p \tag{5.18}$$

and the polydispersity index comes close to 2 as p tends to 1.

This result is valid only for small conversions. The probability of chain growth decreases as the concentration of monomer in the reaction medium decreases. Thus, the overall molecular weight distribution is the convolution of all individual instantaneous distributions, as in the case of free radical polymerizations. It should be recalled that each site may give yield to many polymer molecules, owing to the transfer reactions, and that the rate of polymerization in such cases is second order with respect to monomer concentration.

5.2.4 Ring Opening Polymerization in Protic Media

Another case of interest is the ring opening polymerization of ethylene oxide (oxirane) initiated with an alcoholate, in the presence of excess alcohol, acting as transfer agent. The reaction scheme can be written as follows:

Initiation

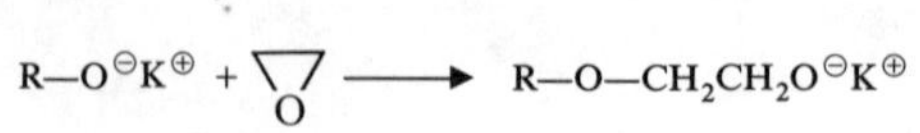

Propagation

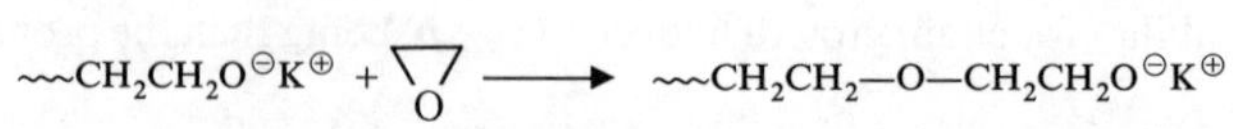

Transfer

$$\sim\!\!\sim CH_2\text{—}CH_2\text{—}O^{\ominus}K^{\oplus} + ROH \longrightarrow \sim\!\!\sim CH_2CH_2OH + RO^{\ominus}K^{\oplus}$$

The main difference with respect to the preceding example is that *any* alcohol function can be involved in transfer: The excess alcohol present, as well as any polymer chain that has already undergone transfer and therefore carries a terminal alcohol function. Assuming that the reactivity of all OH functions in the transfer process is the same, regardless of the length of the chains, it follows that the number average degree of polymerization of the poly(ethylene oxide) formed is determined by the molar ratio of monomer and alcohol involved, as if all alcohol molecules had been metallated initially. However, the polymerization rate takes only into account metallated sites that participate in the propagation and not the excess alcohol functions present.

5.3 Anionic Polymerization in Aprotic Media: "Living" Polymers

5.3.1 Reaction Scheme

Let us consider now the polymerization of a monomer dissolved in an *aprotic solvent* in the presence of a suitable metalorganic initiator, and let us assume that no electrophilic function (other than the unsaturation of the monomer) can react with the metalorganic sites. In that case neither transfer nor termination are likely to occur, and the polymerization scheme should reduce to only two reactions:

Initiation

$$R^{\ominus}Na^{\oplus} + CH_2{=}\underset{\displaystyle R}{\underset{|}{CH}} \longrightarrow R\text{—}CH_2\text{—}\underset{\displaystyle R}{\underset{|}{CH}}{}^{\ominus}Na^{\oplus}$$

that can be symbolized as:

$$I^{\ominus} + M \longrightarrow I\text{—}M_1^{\ominus} \ (\text{or } M_1^{\ominus})$$

Propagation

$$R\,M_n^{\ominus}Na^{\oplus} + M \longrightarrow R\,M_{n+1}^{\ominus}Na^{\oplus}$$

symbolized as:

$$I\,M_n^{\ominus} + M \longrightarrow I\,M^{\ominus}{}_{n+1} \ (\text{or } M_{n+1}^{\ominus})$$

The sites retain their activity for times that are long as compared to the duration of the polymerization process. This is the most striking difference with free radical processes, in which the life-time of the radicals is very small as compared to the time necessary to reach reasonable conversions. Consequently anionic polymerizations will easily proceed to complete conversion of the monomer, and the polymer chains that are formed still bear a metalorganic site at chain end. If initiation is fast enough, the number of polymer chains present in the reaction medium is equal to the number of metalorganic initiator molecules introduced. Consequently the number average degree of polymerization is given by the mole ratio of monomer converted to initiator introduced.

$$P_n = \frac{[\Delta M]}{[I]}$$

Further addition of monomer will result in an increase of the length of the existing chains.

Szwarc and his coworkers took a major part in research on the mechanisms, kinetics and applications of such *"living"* anionic polymerizations.

5.3.2 Kinetics and Molecular Weight Distribution of "Living" Polymers

Calling: $I^{\ominus}$ the initiator (the counter-ion is left out), M the monomer and $M_i^{\ominus}$ a "living" chain molecule of polymerization degree *i* (fitted at chain end with a carbanionic site), the initiation step can be schematized as:

$$I^{\ominus} + M \longrightarrow M_1^{\ominus}$$

The rate of consumption of the initiator is expressed by:

$$-\frac{d[I^{\ominus}]}{dt} = k_i[I^{\ominus}][M] \tag{5.21}$$

The successive propagation steps can be written as:

$$M_1^{\ominus} + M \longrightarrow M_2^{\ominus}$$
$$M_2^{\ominus} + M \longrightarrow M_3^{\ominus}$$
$$- - - - - - - - - - -$$
$$M_{i-1}^{\ominus} + M \longrightarrow M_i^{\ominus}$$

whereby k_p is supposed to be independent of chain length (as in free radical polymerization).

The *monomer consumption* is governed by the following rate equation*):

$$-\frac{d[M]}{dt} = k_p[M] \sum_{i=1}^{\infty} [M_i^{\ominus}] = k_p[M^{\ominus}][M] \tag{5.22}$$

in which $[M^{\ominus}]$ is the molar concentration of "living" chains of all sizes; if initiation is fast, $[M^{\ominus}]$ can be set equal to [I]. The reaction rate is first order with respect to monomer concentration. Upon integration one gets (Fig. 5.1):

*) Note that the total anion concentration $\sum_{i=1}^{\infty} [M_i^{\ominus}]$ is written for simplicity $[M^{\ominus}]$.

$$\ln \frac{[M]_0}{[M]} = k_p [M^{\ominus}] t \tag{5.23}$$

The distribution of molecular weights in a polymerization system involving rapid initiation, and neither termination nor transfer, is expected to be narrow. *Poisson-type distributions* are obtained upon distributing randomly p objects among q boxes, provided $p \gg q$. This is exactly the case of p monomer molecules being distributed among q growing chains, the number of which is constant throughout the process.

A more elaborate demonstration of the occurrence of a *Poisson-type* distribution arises from kinetics, by considering the concentration of species i (which contains i monomer units, the last one bearing the carbanionic site), as a function of time. The instantaneous variation of $[M_i^{\ominus}]$ can be expressed as:

For $i > 1$

$$\frac{d[M_i^{\ominus}]}{dt} = k_p [M]([M_{i-1}^{\ominus}] - [M_i^{\ominus}]) \tag{5.24}$$

and for $i = 1$

$$\frac{d[M_1^{\ominus}]}{dt} = -k_p [M][M_1^{\ominus}] \tag{5.25}$$

On the basis of equation (5.22) $d\bar{v}$ can be defined as the average number of monomer molecules added per living chain during time interval dt.

$$d\bar{v} = k_p [M]\, dt = \frac{d[M]}{[M^{\ominus}]} \tag{5.26}$$

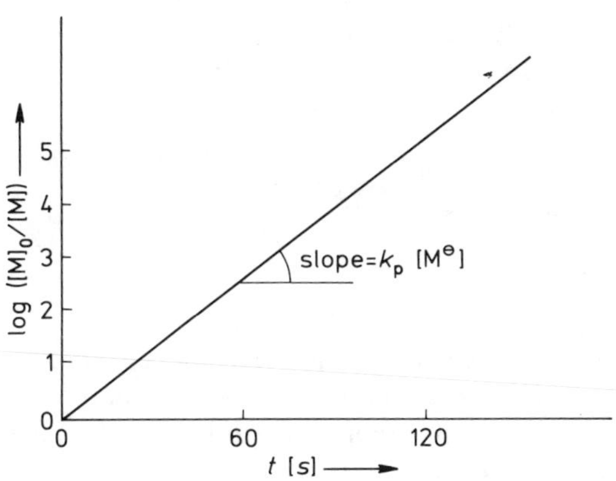

Fig. 5.1. Schematic first order plot of the anionic polymerization of styrene in tetrahydrofurane (with homoionic salt added to depress electrolytic dissociation); $[M^{\ominus}] = 5 \cdot 10^{-4}$ [*mol/l*]

As a consequence, $\bar{v}$ is the average degree of polymerization, more precisely the average number of growth steps which have taken place during the whole process, per "living" chain*). We can thus set:

*) In fact $\overline{P}_n = 1 + \bar{v}$, since the growth reaction starts once initiation has already introduced *one* monomer unit.

$$d[M_1^{\ominus}] = -[M_1^{\ominus}]\,d\bar{\nu}$$
$$d[M_2^{\ominus}] = ([M_1^{\ominus}] - [M_2^{\ominus}])\,d\bar{\nu}$$
$$\text{— — — — — — — — — — — — — —}$$
$$d[M_i^{\ominus}] = ([M_{i-1}^{\ominus}] - [M_i^{\ominus}])\,d\bar{\nu}$$

which leads upon integration to:

$$[M_1^{\ominus}] = [M^{\ominus}]e^{-\bar{\nu}}$$
$$[M_2^{\ominus}] = \nu[M^{\ominus}]e^{-\bar{\nu}}$$
$$\text{— — — — — — — — — — — —}$$
$$[M_i^{\ominus}] = [M^{\ominus}]\frac{\bar{\nu}^{i-1}\,e^{-\bar{\nu}}}{(i-1)!} \qquad (5.27)$$

$[M_i^{\ominus}]$ and $[M^{\ominus}]$ refer to numbers of molecules (molar concentrations) and not to concentrations by weight. The mole fraction x_i of molecules having reached a degree of polymerization i is given by:

$$x_i = \frac{[M_i^{\ominus}]}{[M^{\ominus}]} = \frac{\bar{\nu}^{i-1}\cdot e^{-\bar{\nu}}}{(i-1)!} \qquad (5.28)$$

This expression characterizes Poisson's distribution. To calculate the weight fraction w_i of the "living" species of polymerization degree i, it is assumed that the weight of the initiator fragment I is the same as that of one monomer unit (m_0); w_i is expressed as:

$$w_i = \frac{m_0(i+1)x_i}{m_0\sum_{i=1}^{\infty}(i+1)x_i} = \frac{(i+1)x_i}{\bar{\nu}+1}$$

which can also be written as:

$$w_i = \frac{\bar{\nu}}{\bar{\nu}+1}\cdot e^{-\bar{\nu}}\,\frac{i\,\bar{\nu}^{(i-2)}}{(i-1)!} \qquad (5.29)$$

In a Poisson distribution the number and weight distribution functions are very close to each other (Fig. 5.2), contrary to the case of the most probable distribution. It is interesting to compare Fig. 5.2 with Fig. 2.1, which both refer to the same number average degree of polymerization of $\overline{P_n} = 50$.

The number average molecular weight is given by:

$$\overline{M}_n = (\bar{\nu}+1)\,m_0,$$

the weight average molecular weight results from equation (5.29):

$$\overline{M}_w = m_0\frac{\bar{\nu}^2+3\bar{\nu}+1}{\bar{\nu}+1} = \left[1+\bar{\nu}+\frac{\bar{\nu}}{\bar{\nu}+1}\right]m_0$$

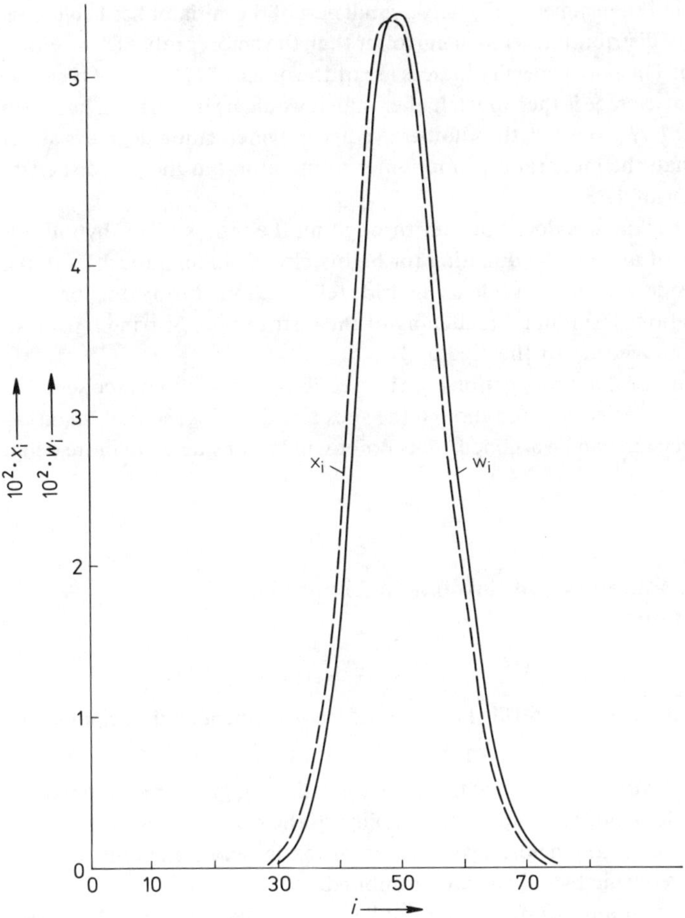

Fig. 5.2. Poisson-type distribution of polymerization degrees, calculated for $\overline{P}_n = 50$; (x_i = mole fraction; w_i = weight fraction)

and the polydispersity index can be expressed as:

$$\frac{\overline{M}_w}{\overline{M}_n} = \frac{\bar{\nu}^2 + 3\bar{\nu} + 1}{(\bar{\nu} + 1)^2} = 1 + \frac{\bar{\nu}}{(\bar{\nu} + 1)^2} \tag{5.30}$$

From equation (5.30) it follows that polymers with high average degrees of polymerization should exhibit low polydispersities, i.e. polydispersity indices close to 1.

Let us consider now the hypothesis made in this calculation:

● k_p is independent of the degree of polymerization. The same assumption is made in all chain polymerizations. Indeed the reactivity of the anionic sites remains constant as the chains grow.

● $k_i > k_p$: This may not always be the case. But the effect of slow initiation on molecular weight distribution is limited. Even in cases in which initiation is so slow that the monomer is entirely consumed while only a small part of the initiator has been used, the molecular weight distribution is much narrower than the "most probable" or Flory-Schulz distribution: The polydispersity indices are of the order of 1.2 to 1.4. Of course the molecular weight average is then much higher than it would be if initiation had been fast. For example, if $k_i/k_p = 10^{-6}$, the number average polymerization degree is about 100 times higher than the mole ratio of monomer to initiator, but the polydispersity index is of the order of 1.38.

● The total number of sites does not vary throughout the process. This hypothesis implies the absence of any proton donating (or highly electrophilic) impurities in the medium, and of side reactions, such as hydride elimination during the process. Accidental deactivations result in a broadening of the distribution. Stirring must also ensure adequate homogeneity of the medium.

● It was also assumed that propagation is an irreversible process, more precisely that depropagation can be neglected, even though the sites remain living, so that quantitative monomer conversion can be attained. This point will be considered in more detail in chapter 7.

5.3.3 Monomers, Initiators and Solvents in Living Anionic Polymerizations

5.3.3.1 Monomers

Various conditions have to be fulfilled for a monomer to undergo anionic polymerization:

a) It should be fitted with an electron withdrawing substituent, giving rise to polarization of the double bond, and capable of stabilizing the carbanion to be formed.
b) It should not contain any acidic hydrogen atom, as in acids, alcohols, amines, phenols. Even acetylenic hydrogens are prohibited.
c) It should not contain any strongly electrophilic function capable of yielding side reactions with carbanions or oxanions.
d) The carbanion arising from the monomer should be able to attack the monomer, thus allowing the polymerization to proceed.

5.3.3.2 Initiators: Metal Organic and Radical Anionic

There are two classes of anionic initiators: Those which are able to add onto a monomer, and those which transfer an electron to the monomer.

a) *Metal organic initiators* that add onto the monomers are quite numerous and diversified. Butyl lithium is one example.

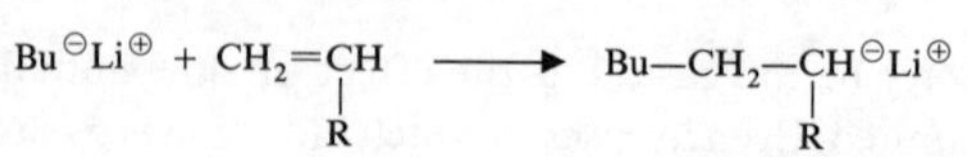

Table 5.1. Monomers polymerizable anionically

---

1. Vinylic and acrylic monomers

Styrene and substituted styrenes such as:
α-methylstyrene, p-methyl styrene, p-methoxy styrene, vinylmesitylene, p-dimethylaminostyrene, p-chlorostyrene, divinylbenzenes

Vinyl aromatics such as:
vinyl naphthalene, vinylbiphenyl, isopropenylnaphthalene ...

Conjugated dienes such as:
butadiene, isoprene, piperylene, phenylbutadiene ...

Vinyl pyridines

Alkyl methacrylates such as:
methyl, ethyl, t-butyl, octadecyl, dimethylaminoethyl ...

Acrylonitrile and methacrylonitrile

2. Heterocyclic monomers

3 and 4 membered heterocycles
oxirane (ethylene oxide) and substituted oxiranes
Thiirane (ethylene sulfide) and substituted thiiranes
Thietane and substituted derivatives

Lactones
propiolactone, caprolactone, glycolide

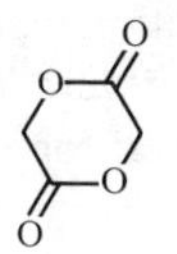
glycolide

Lactames
pyrrolidone, caprolactame NH—CO

Miscellaneous

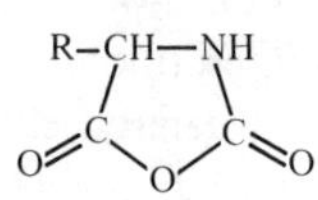
oxazoline diones

oxazolidine diones (Leuchs anhydrides)

Hexamethylcyclotrisiloxane (D3)

D3 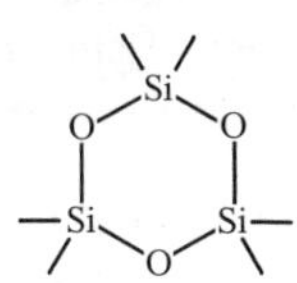

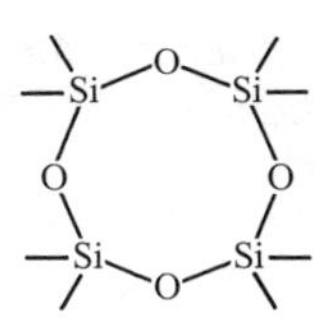 D4

Octamethylcyclotetrasiloxanes (D4)

Phosphoranes

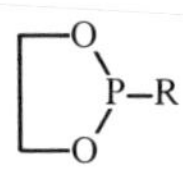
phosphoranes

The reactivity of a metalorganic compound as a nucleophile depends upon the substituents carried by the carbanionic site and upon the counterion. Thus, if we consider the series:

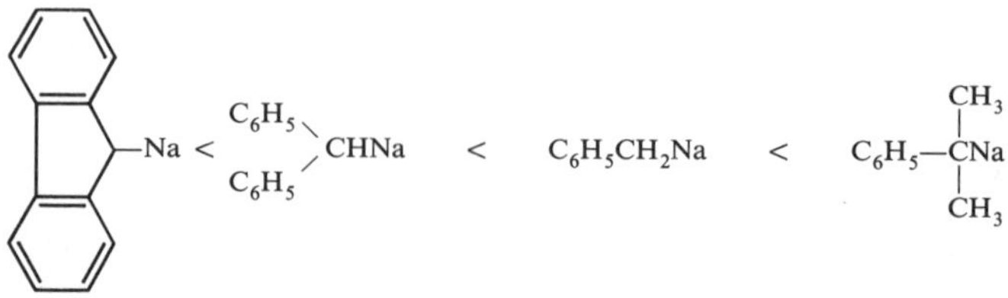

fluorenyl—Na diphenylmethyl—Na benzyl—Na cumyl—Na

in the same solvent, say THF, the greater the stability of the carbanionic site (because of electron delocalization) the lesser its reactivity. Cumyl-Na will readily initiate the polymerization of a monomer such as styrene (in which the electron withdrawing effect of the phenyl ring is weak); fluorenyl-sodium is of much lower reactivity and it will not attack styrene, though it can still initiate the polymerization of alkyl methacrylates, where the ester function exerts a strong inductive effect on the double bond. With this latter monomer cumyl-sodium would give rise to side reactions on the ester carbonyl, though of course initiation would occur, too.

Similarly, sec-butyl lithium is a more reactive initiator than n-butyllithium.

The consequence is that for each monomer the most appropriate initiator should be selected, sufficiently active to react rapidly and quantitatively with the monomer (to initiate the polymerization), but unable to give rise to side reactions. The most highly nucleophilic initiator is not always the best choice (see Scheme below).

Metal amides, alkoholates, thiolates or even silanolates are potential initiators for the anionic polymerization of heterocyclic monomers or of unsaturated monomers fitted with two highly electron withdrawing substituents (such as cyanoacrylates or vinylidene cyanide). For heterocyclic monomers, the best choice is an initiator exhibiting the same kind of active site as that arising from the cyclic monomer upon ring-opening: Alkoholates for oxiranes, thiolates for thiiranes, silanolates for cyclosiloxanes.

b) *Radical-ionic initiators* result from the *electron transfer reaction* of alkali metal onto an aromatic hydrocarbon in an aprotic polar (ion solvating) solvent:

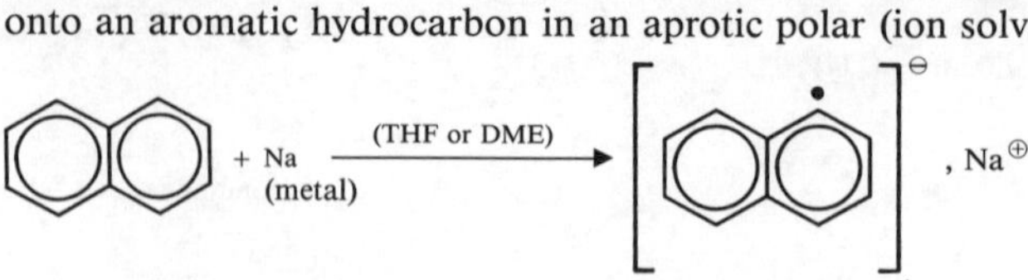

sodium naphthalene*)

The resulting species is kind of an electron reservoir, and it will transfer the additional electron back to any species exhibiting a greater electron affinity than the aromatic hydrocarbon chosen. Biphenyl and naphthalene are best suited. Homologues such as phenanthrene, naphthacene, and pyrene exhibit electron affinities that are too high (they even yield dianionic species) so that back-transfer to the monomer will be slow and difficult, or even impossible.

---

*) Systematic name: sodium dihydronaphthylide.

Scheme: Examples of systems in which nucleophilicity of initiator matches electron affinity of monomer

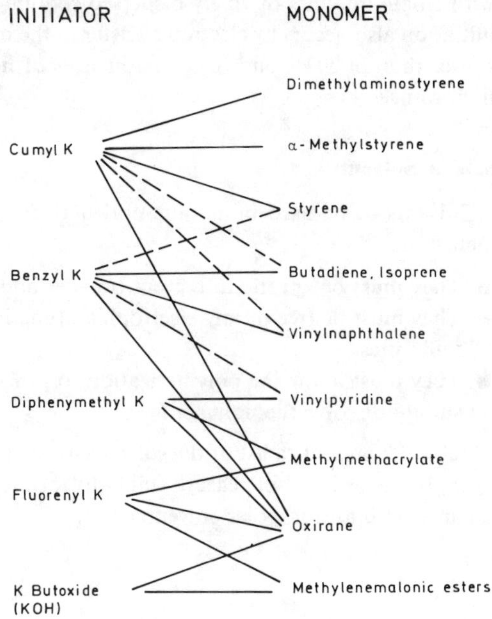

Full lines: No difficulty involved
Dotted lines: Side reactions may occur
No connection: Initiation impossible

Once electron transfer to the monomer has occurred, the radical-ionic species formed will undergo fast dimerization by coupling of the radicals, whereby a dianionic dimer is formed. This species can grow from both anionic ends. This is why a radical anionic species always acts as a bifunctional initiator.

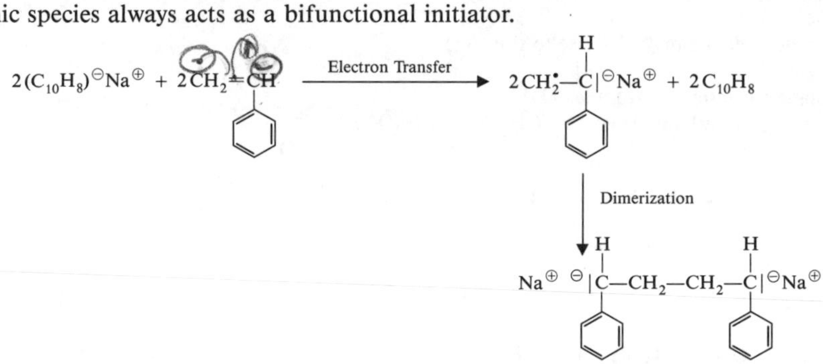

Direct electron transfer from an alkali metal, like sodium, to the monomer can also be used to initiate the polymerization. However initiation is heterogeneous, and it is slow as compared with the rate of propagation in a polar solvent. As one noteworthy exception, it can be used with α-methyl styrene at a temperature and a concentration at which propagation does not proceed. A dianionic oligomer is obtained, generally referred to as "living α-methyl styrene tetramer", which can in turn be used as an efficient difunctional anionic initiator for the polymerization of styrene and other monomers.

Finally it should be mentioned that styrene and dienes can be polymerized in the bulk with small amounts of finely dispersed sodium or lithium as initiator. In that case, initiation also occurs by electron transfer to the monomer, but the growth rate is much slower than in THF, and the consequences of heterogeneous initiation are of minor importance.

5.3.3.3 Solvents

Solvents to be used in anionic polymerizations must meet the following requirements:

- They must be aprotic to prevent transfer and termination.
- They must be free of any electrophilic function which may react with the carbanionic sites.
- They must allow the polymerization to proceed in homogeneous phase, i. e. they should dissolve the polymer.

Table 5.2 presents a list of the solvents most commonly used in anionic polymerizations, in the order of increasing ability to solvate metalorganic sites. This list includes polar as well as non-polar solvents.

Table 5.2. Aprotic solvents which can be used in anionic polymerization processes

alkanes (pentane, hexane, heptane, etc.)
cyclohexane
benzene (toluene may give rise to transfer)
dioxane
oxolane (or tetrahydrofurane, THF)
1,2-dimethoxyethane, DME $CH_3OCH_2CH_2OCH_3$

dimethylformamide, DMF (is not of general use) 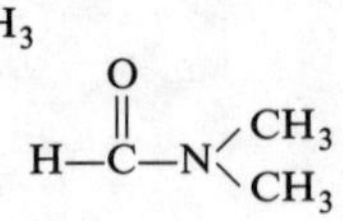

hexamethylphosphoramide, HMPA $O{=}P(N(CH_3)_2)_3$

Polar additives which can be used to enhance rates of polymerization in a non polar solvent:

glymes: $CH_3OCH_2CH_2OCH_2CH_2OCH_3$,
$CH_3OCH_2CH_2OCH_2CH_2OCH_2CH_2OCH_3$

crown ethers
cryptands

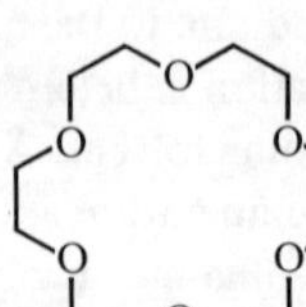

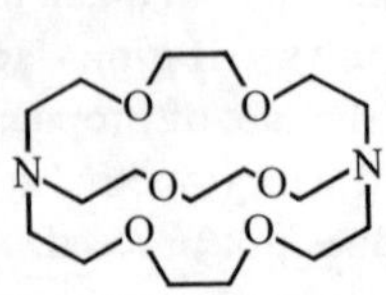

The reason why the choice of solvent is of major importance is illustrated in Table 5.3 by an example. Polymerization of dienes in polar solvents leads to unacceptably high 1.2 and 3.4 unit contents, while 1.4 placement (preferably 1.4-cis, or mixed 1.4-cis and 1.4-trans) is essential for applications as elastomers.

Table 5.3 also shows that the choice of the counter-ion is important. In diene polymerizations, only lithium will yield useful products.

Table 5.3. Homogeneous anionic polymerization of isoprene and of butadiene

Isoprene $H_2C_1{=}C_2(CH_3){-}CH_3{=}C_4H_2$

counterion	solvent	1,4-cis	1,4-trans	1,2	3,4
		mol-% of units inserted			
$Li^{\oplus}$	pentane	~94	—	—	~6
	pentane/THF (90:10)	~0	~25	~10	~65
	THF	—	—	~75	~15
	diethyl ether	—	~50	~5	~45
$Na^{\oplus}$	pentane	small	~40	~7	~50
$K^{\oplus}$	pentane	small	~55	~7	~35

1,3-Butadiene

counterion	solvent	1,4-cis	1,4-trans	1,2
		mol-% of units inserted		
$Li^{\oplus}$	pentane	~35	~50	~15
	THF	~0	~5	~95
$Na^{\oplus}$	pentane	~7	~25	~67
$K^{\oplus}$	pentane	~12	~38	~50

There are also very strong implications of kinetics for the choice of the polymerization solvent. This arises from the fact that the reactivity of the carbanionic sites is strongly affected by aggregation and by solvation phenomena, as schematized in Table 5.4. Sodium polystyryl is taken as an example, but the scheme is applicable to many systems.

Table 5.4. Effect of solvent polarity on the species present in anionic polymerization

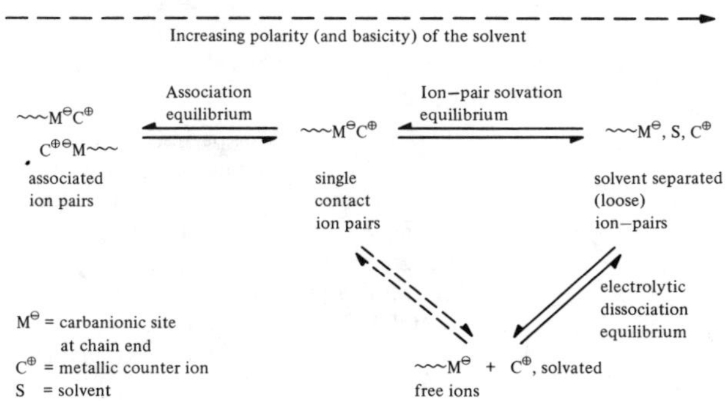

In non-polar solvents the ion pairs are predominantly associated, joining two (or more) living polymer chains via their metal organic sites. The aggregates are "dormant", i. e. they do not contribute to propagation. Their dissociation constant usually ranges from 10^{-7} to 10^{-4} [mol l^{-1}].

In polar solvents no tendency to form aggregates is observed. Electrolytic dissociation may become appreciable if the dielectric constant of the solvent is high enough. The reactivity of the sites as free ions is far higher than that of the corresponding ion pairs. It is therefore important to determine the proportion of sites that are dissociated into free ions. Moreover ion pairs can be tight or solvent separated, and this has also consequences on their reactivity. The complexity of these relationships is illustrated in Fig. 5.3 and 5.4 for sodium polystyryl in tetrahydrofurane as a function of temperature and of added glyme. The latter additive favors cation solvation and thus formation of more reactive solvent separated ion pairs.

The kinetic features of ionic polymerizations carried out in apolar or in polar solvents will be considered in the next sections.

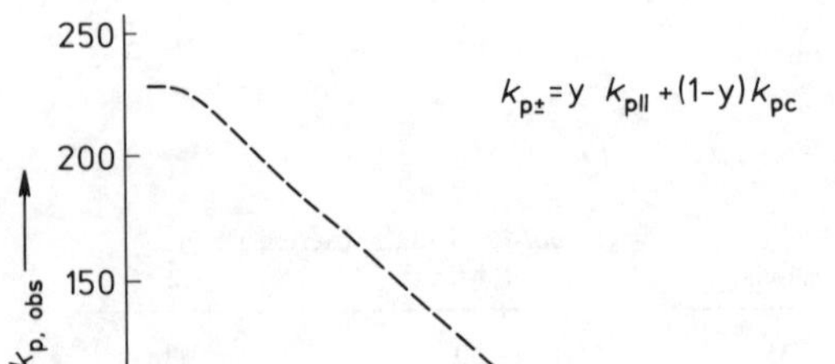

Fig. 5.3. Observed rate constant for anionic styrene polymerization in tetrahydrofurane versus temperature ($Na^{\oplus}$ = counter ion). The ionization is depressed by homoionic salt addition. The lower the temperature the higher the proportion of solvent separated ion-pairs

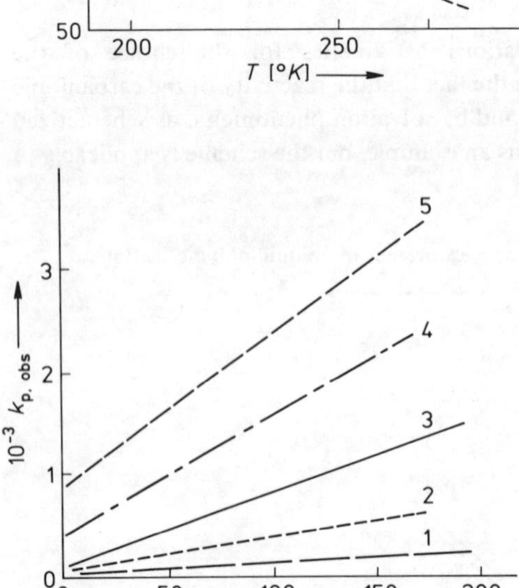

Fig. 5.4. Plot of experimental rate constants for anionic polymerization of styrene in tetrahydropyrane with added glyme, as a function of the reciprocal square root of the active site concentration. Curve 1: No glyme added 2 and 3: increasing amounts of glyme 4; 4 and 5: increasing amounts of glyme 5 added

5.3.4 Specific Features of Anionic Polymerizations in Apolar Solvents

In apolar solvents metal organic compounds usually aggregate to "inverse micelles". This applies to initiators, such as butyllithium, as well as to metal organic sites at chain ends. The degree of aggregation is still somewhat controversial, but it evidently influences the kinetics of the polymerization process. The mass action law is applicable to the association equilibrium. It is generally assumed that the unassociated fraction of the metal organic compound (or site) alone contributes to initiation and to chain growth.

In the case of butyllithium, for instance, the aggregation degree is close to 6, and the rate of initiation in a hydrocarbon solvent is therefore of the 1/6th order with respect to initiator concentration.

$$(\mathrm{BuLi})_6 \rightleftharpoons 6\,\mathrm{BuLi_f} \qquad (\text{or } \mathrm{BuLi} + (\mathrm{BuLi})_5)$$

$$\mathrm{K_e} = \frac{[\mathrm{BuLi_f}]^6}{[(\mathrm{BuLi})_6]} \qquad \text{or } [\mathrm{BuLi_f}] = (\mathrm{K_e}\,[(\mathrm{BuLi})_6])^{1/6}$$

Similarly, if the rate of propagation is found to be of the 1/2 order in concentration of active sites, one can conclude that the aggregation degree of the active sites is close to 2.

$$R_\mathrm{p} = \frac{-\mathrm{d}\,[\mathrm{M}]}{\mathrm{d}t} = k_\mathrm{p}\mathrm{K_e^{1/2}}\,[\mathrm{M}][\mathrm{M^-}]^{1/2} \qquad (5.31)$$

Integration of equation (5.31) leads to:

$$\ln\frac{[\mathrm{M}]_0}{[\mathrm{M}]} = k_\mathrm{p}\mathrm{K_e^{1/2}}\,[\mathrm{M^-}]^{1/2}\,t \qquad (5.32)$$

In spite of the aggregation phenomena occuring between active sites, all of them do participate in the propagation reaction. The statements concerning molecular weight (proportional to the mole ratio of [M] to [I]) and polydispersity index (Poisson-type distribution) remain entirely valid. This arises from the fact that the rates of aggregation and of dissociation are very high.

5.3.5 Specific Features of Anionic Polymerizations Carried out in Polar Solvents

The attraction force between two ions of opposite signs is given by *Coulomb's law*

$$F = \frac{Q_+ \cdot Q_-\, e^2}{\varepsilon \cdot l^2} \qquad (5.33)$$

where e is the charge of an electron, ε is the dielectric constant of the solvent, l is the distance between the ions and Q_+ and Q_- are their charges. Because ε is small in most organic solvents, F is quite high and it is usually impossible to detect electrolytic dissociation by conductimetry. There are however solvents in which organometallic sites are dissociated to a small extent. The mass action law applies, but the dissociation constant K_D is small.

5.3.5.1 Ions and Ion Pairs

Let us consider the propagation step of a polymerization and assume partial dissociation is involved. If the rate constants of propagation on ion pairs and on free ions are different, the situation can be schematized as below:

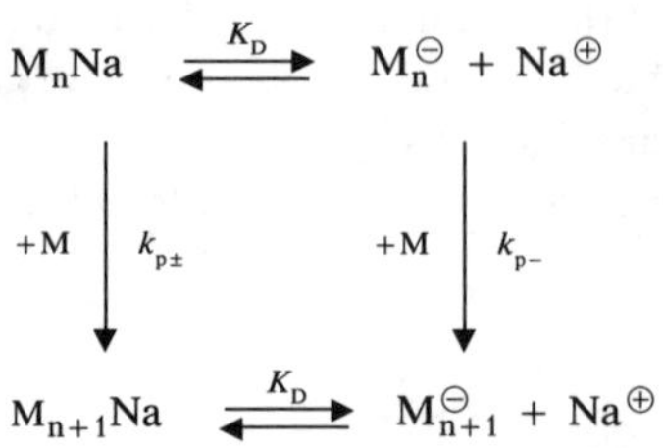

Calling $[M^\ominus]$ the concentration of active sites of any kind, whether dissociated or not, the mass action law can be written as:

$$K_D = \frac{\alpha^2 [M^\ominus]}{1 - \alpha} \tag{5.34}$$

assuming that no other ions are present but those originating from the dissociation of a fraction α of the active sites. If the degree of dissociation α is small, it can be approximated by:

$$\alpha = \left(\frac{K_D}{[M^\ominus]}\right)^{1/2}$$

From the first order plots of polymerization $\left(\ln\frac{[M_0]}{[M]} \text{ vs. } t\right)$ one can determine the experimental rate constant $k_{p_{obs}}$:

$$k_{p_{obs}} = \frac{1}{[M^\ominus]\,t} \ln\frac{[M]_0}{[M]} \tag{5.35}$$

$k_{p_{obs}}$ can be expressed as a function of the rate constants of propagation $k_{p\pm}$ and k_{p-} for ion pairs and for the free ions

$$k_{p_{obs}} = \alpha k_{p-} + (1 - \alpha) k_{p\pm} \tag{5.36}$$

Introducing the value of α, the overall rate of polymerization can be expressed as

$$-\frac{d[M]}{dt} = \left\{k_{p\pm} + \left(\frac{K_D}{[M^\ominus]}\right)^{1/2} (k_{p-} - k_{p\pm})\right\} [M^\ominus][M] \tag{5.37}$$

A plot of the experimental rate constant $k_{p_{obs}}$ versus $\left(\frac{K_D}{[M^\ominus]}\right)^{1/2}$ yields both $k_{p\pm}$ (from the ordinate intercept) and $(k_{p-} - k_{p\pm})$ from the slope of the line, provided K_D is known (Fig. 5.5).

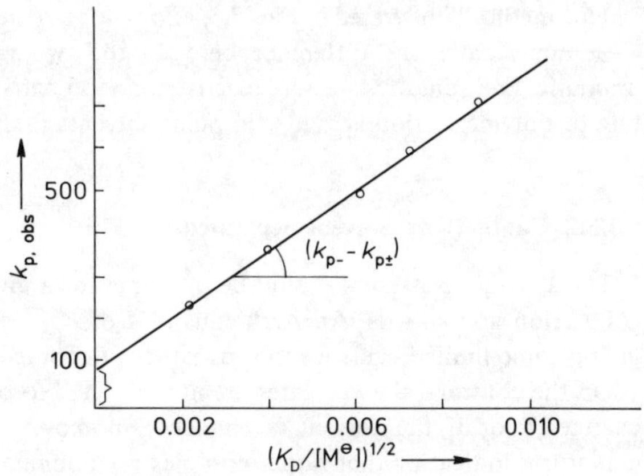

Fig. 5.5. Plot of the observed rate constant versus the dissociation degree of the active sites in the presence of added homoionic electrolyte. (Intercept $= k_{p\pm}$; slope $= (k_{p-} - k_{p\pm})$). Compare with Fig. 5.4

Another method can be used to obtain more reliable values of $k_{p\pm}$; it is based on depression of the ionic dissociation by means of a homoionic electrolyte. If K_ϕ is the dissociation constant of the electrolyte added (supposed $K_\phi > K_D$), c_ϕ its concentration and α_ϕ its degree of dissociation, calling $[M_d^{\ominus}]$ the concentration of free ions originating from the active sites, one can write, using Na-tetraphenylborate ($NaB\phi_4$) as homoionic electrolyte:

$$B\phi_4Na \rightleftharpoons B\phi_4^{\ominus} + Na^{\oplus}$$
$$RM_nNa \rightleftharpoons RM_n^{\ominus} + Na^{\oplus}$$

$$\left.\begin{aligned} K_\phi &= \frac{[B\phi_4^{\ominus}][Na^{\oplus}]}{[NaB\phi_4]} \\ K_D &= \frac{[M_d^{\ominus}][Na^{\oplus}]}{[M^{\ominus}]} \end{aligned}\right\} \tag{5.38}$$

Tetraphenylborates are often used for that purpose, since these salts are stable versus carbanions and soluble; their dissociation constant is rather high as compared to that of most organometallic sites occurring in polymerization reactions. The degree of dissociation α is given by:

$$\alpha = \frac{[M_d^{\ominus}]}{[M^{\ominus}]} = \frac{K_D}{[Na^{\oplus}]} \tag{5.39}$$

Since all $Na^{\oplus}$ ions come assumingly from $NaB\phi_4$, which is more extensively dissociated than the growing carbanionic sites, one can write:

$$[Na^{\oplus}] = K_\phi^{1/2}\,[NaB\phi_4]^{1/2} = \alpha_\phi c_\phi$$

It follows for the rate constants:

$$k_{p_{obs}} = k_{p\pm} + \frac{K_D}{\alpha_\phi c_\phi}(k_{p-} - k_{p\pm}) \tag{5.40}$$

This method, illustrated in Fig. 5.5, allows for a better accuracy on the value of $k_{p\pm}$.

In many cases the difference between the two rate constants $k_{p\pm}$ and k_{p-} is dramatic. Free ions are far more reactive than ion pairs. That is chiefly why the overall rate of polymerization is higher in polar solvents than in non-polar solvents.

5.3.5.2 Contact and Solvent Separated Ion Pairs

The k_{p-} rate constants should be the same for a given carbanion regardless of the counterion and so it is. An Arrhenius plot of k_{p-} versus $1/T$ yields a straight line, giving some limited evidence that no other phenomenon is involved.

On the contrary, the $k_{p\pm}$ rates are not expected to be the same if the counterion is changed, nor if the solvent is changed; moreover, Arrhenius plots often exhibit curvatures indicating that rather complex phenomena are involved.

Ion-pairs were shown to exist in at least two states:

- *Unsolvated contact ion-pairs or intimate (tight) ion-pairs.*
  Here the distance between anion and cation is the smallest possible; these contact ion-pairs are least reactive, and their dissociation into free ions is low.
- *Solvent-separated or non-intimate (loose) ion-pairs.*
  Here solvent molecules are located around the cation or between anion and cation, thus increasing the reactivity and the ability to ionic dissociation; in some cases even a shift of the spectral absorption (case of fluorenyl Li) has been observed.

Calling y the fraction of solvent separated ion-pairs, one may therefore write:

$$k_{p\pm} = y k_{p\|} + (1 - y) k_{pc} \tag{5.41}$$

where $k_{p\|}$ is the rate constant of propagation of the solvent separated ion-pairs and k_{pc} that of the contact ion-pairs. Obviously y depends on both the basicity of the solvent, i. e. on its ability to solvate the cations, even when they are associated with an anion, and upon the cation itself: its size and its polarizability are involved.

It should be reemphasized that the reactions involved in the solvation/desolvation equilibria or in the ionic dissociation equilibria are so fast that they have practically no influence on the molecular weight distributions of the obtained polymers.

To enhance solvation and/or dissociation into free ions of the active sites, *cation solvating additives* (such as *cryptands* or *crown ethers*) can be introduced into the reaction medium (see Table 5.2). The accelerating effect of such additions is impressive, especially in solvent media of low dielectric constant.

5.3.6 Applications of Anionic "Living" Polymerizations

The basic characteristics of anionic polymerizations in aprotic solvents is the lack of spontaneous transfer and termination reactions. Once total conversion of the monomer has been attained, the polymer chains remain "living". The active sites at the chain end can be used in a number of ways, as it will be shown below.

5.3.6.1 Protonation

To get rid of these active sites, induced deactivation by protonation is the easiest way. Water, alcohols, phenols, acids, etc. can be used for that purpose:

$$\sim\!\!\sim\!\text{CH}_2\text{—}\underset{|\atop\text{R}}{\text{CH}}\,|^{\ominus}\text{Na}^{\oplus} + \text{HOCH}_3 \longrightarrow \sim\!\!\sim\!\text{CH}_2\text{—}\underset{|\atop\text{R}}{\text{CH}_2} + \text{CH}_3\text{ONa}$$

The deactivation reaction (or "killing") is very fast in most cases.

5.3.6.2 ω-Functional and αω-difunctional Polymers

The organometallic sites at chain ends can also be used for the synthesis of ω-functional polymers, as they exhibit a high reactivity with respect to all kinds of electrophilic functions. The interest of such functionalization processes is still enhanced by the fact that the molecular weight can be chosen at will (from the monomer to initiator mole ratio) and by the narrow molecular weight distribution of the polymers arising from such processes.

Here are a few examples showing how the nucleophilic behavior of the terminal carbanionic sites can be utilized to synthesize *ω*-functional model polymers, or for purpose of chain extension, and for the synthesis of *a,ω*-difunctional polymers, if the living polymer is dianionic (resulting, for instance, from an electron transferring initiator).

- Polymers with end standing *carboxylic* acid functions can be obtained by nucleophilic attack of the carbanionic site onto CO_2 (in excess) to yield:

$$\sim\!\!\sim\!\text{CH}_2\text{—}\underset{|\atop\text{R}}{\text{CH}}\,|^{\ominus} + \text{CO}_2 \longrightarrow \sim\!\!\sim\!\text{CH}_2\text{—}\underset{|\atop\text{R}}{\text{CH}}\text{—C}\!\begin{smallmatrix}\nearrow\text{O}\\ \searrow\text{O}^{\ominus}\end{smallmatrix}$$

An alternative way is to react it with a cyclic anhydride, such as succinic anhydride:

In both cases the reaction is fast and quantitative.

- Polymers with end standing *hydroxyl* groups can be obtained upon reaction of the living polymer with oxirane (ethylene oxide) under conditions such as to avoid polymerization of the latter monomer:

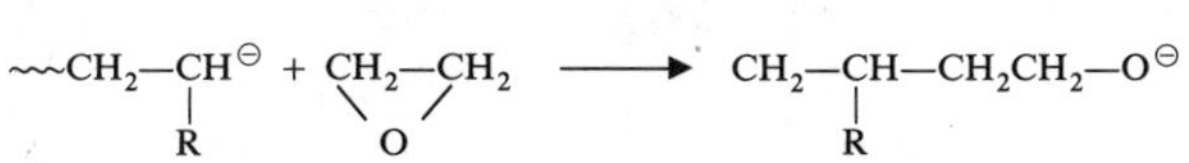

Here again the reaction is quantitative, and protonation of the alkoxide sites is required to get the alcohol functions at chain end.

- The end standing carbanionic sites can also react with functions such as:

— Esters	to yield ω-ketone functions
— Nitriles	to yield the same (after hydrolysis)
— Isocyanates	to yield amide functions
— Acid chlorides	to yield ω-ketone functions
— Benzylic or allylic halides	to yield ω-benzylic or ω-allylic polymers
— Lactones	to yield ω-hydroxyl groups

• With alkyl halides the reaction may involve both elimination and substitution, which means that the yield in alkyl groups at chain end is generally less than unity.

• The case of phosgene is interesting because it gives the possibility of getting acid chloride functions at chain end, provided $COCl_2$ is in large excess:

$$\sim\!\sim CH_2-\underset{\displaystyle R}{\underset{|}{CH}}\,|^{\ominus} + \underset{\text{(exess)}}{COCl_2} \longrightarrow \sim\!\sim CH_2-\underset{\displaystyle R}{\underset{|}{CH}}-C\begin{matrix}\nearrow O\\ \searrow Cl\end{matrix}$$

Otherwise coupling takes place, whereby an unfunctionalized chain having twice the molecular weight of the living precursor is formed.

• Coupling reactions result from the use of a difunctional reagent, such as xylylene dibromide, or diethyl succinate, (or phosgene, already mentioned), in stoichiometric amount, to deactivate a living polymer. If the polymer chain itself is also dicarbanionic such a reaction will induce "chain extension". This is in fact a step growth reaction involving one polymeric constituent and one micromolecular constituent. Large molecular weight increases can be obtained in this way.

5.3.6.3 Block Copolymers

The terminal metal organic sites of a polymer made anionically can also be used to synthesize *block copolymers,* as discussed in chapter 10. The principle of this synthesis is to use the terminal carbanionic sites to initiate the polymerization of another suitable monomer. This second monomer should exhibit a higher electrophilicity than the first, in order to provide fast and quantitative initiation.

5.3.6.4 Graft Copolymers

The anionic polymerization method can also be used to synthesize graft copolymers, upon reacting a monocarbanionic polymer with a polymer backbone fitted with electrophilic functions. An example of such a process is the reaction of "living" polystyrene onto a polymethylmethacrylate backbone:

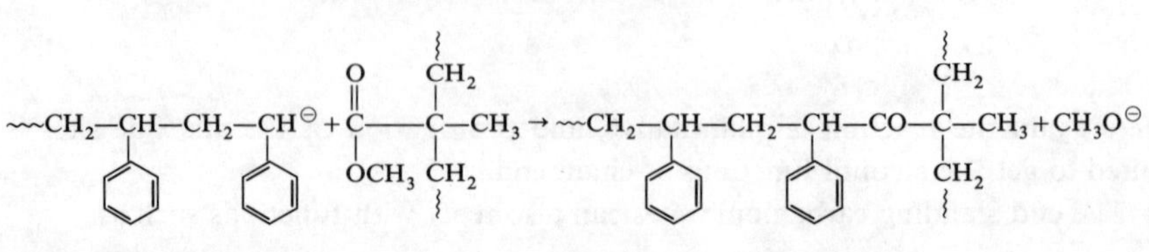

5.3.6.5 Networks

Anionic polymerizations have also found applications in network synthesis by so-called *end-linking methods.* Well defined α,ω-dicarbanionic polymer chains are reacted either with a small amount of a di-unsaturated monomer (the polymerization of which induces crosslinking) or with a plurifunctional electrophilic reagent in stoichiometric amount. Upon crosslinking the chains of the "living" precursor polymer become the elastically effective network chains.

5.4 Stereoelective Polymerizations

Anionic ring opening polymerizations involving optically active monomers such as alkylene oxide, do *not* affect the asymmetric carbon atom. No racemization takes place during the process, owing to the fact that the bond which opens is always that opposite to the asymmetric carbon atom.

Systems have been found in which the rate of addition of the two enantiomers onto the growing chain is not the same. As a consequence, the monomer mixture is enriched in the least reactive of the enantiomers, and the polymer formed contains an enhanced proportion of units originating from the other enantiomer. Starting from a racemic mixture of monomer molecules, one can expect to get a partly isotactic polymer, and the remaining monomer should exhibit some optical activity. This process is referred to as *stereoelective polymerization.*

Let us consider the general case of a chiral monomer, the concentrations of the enantiomers being $[R]$ and $[S]$. The optical activity of the initial mixture is:

$$\alpha = \alpha_D \frac{[R]}{[R] + [S]} - \alpha_D \frac{[S]}{[R] + [S]} = \alpha_D \frac{[R] - [S]}{[R] + [S]} \tag{5.42}$$

The overall degree of conversion is defined by:

$$\pi = \frac{[R]_0 + [S]_0 - [R] - [S]}{[R]_0 + [S]_0} = 1 - \frac{[R] + [S]}{[R]_0 + [S]_0} \tag{5.43}$$

where $[R]_0$ and $[S]_0$ are the concentrations at time zero. The degree of stereoelection ρ is defined as the ratio of the rate constants of chain growth for the two kinds of species. Calling C^* the concentration of active sites and assuming that the sites are undifferentiated, but add monomers R and S at different rates, the corresponding equations are:

$$-\frac{d[R]}{dt} = k_r[C^*][R] \qquad -\frac{d[S]}{dt} = k_s[C^\bullet][S]$$

or: $$\frac{d[R]}{d[S]} = \frac{k_r}{k_s}\frac{[R]}{[S]} \tag{5.44}$$

Introducing $\rho = k_r/k_s$ one gets:

$$\frac{[R]}{[R]_0} = \left(\frac{[S]}{[S]_0}\right)^{\rho} \tag{5.45}$$

The ratios $[R]/[R_0]$ and $[S]/[S_0]$ can now be expressed as a function of the degree of conversion π and of α/α_D which characterizes the optical activity of the remaining monomer at time t. From equation (5.42) it follows:

$$1 + \frac{\alpha}{\alpha_D} = \frac{2[R]}{[R]+[S]} \qquad \text{and} \qquad 1 - \frac{\alpha}{\alpha_D} = \frac{2[S]}{[R]+[S]}$$

Introducing the degree of conversion, from eq. (5.43) the relations searched for can be written as follows:

$$\frac{[R]}{[R_0]} = \left(1 + \frac{\alpha}{\alpha_D}\right)(1-\pi)\frac{[R]_0+[S]_0}{2[R]_0}$$

$$\frac{[S]}{[S_0]} = \left(1 - \frac{\alpha}{\alpha_D}\right)(1-\pi)\frac{[R]_0+[S]_0}{2[S]_0}$$

Thus

$$\left(1 + \frac{\alpha}{\alpha_D}\right)(1-\pi)\frac{[R]_0+[S]_0}{2[R]_0} = \left(1 - \frac{\alpha}{\alpha_D}\right)^{\rho}(1-\pi)^{\rho}\left(\frac{[R]_0+[S]_0}{2[S]_0}\right)^{\rho} \tag{5.46}$$

If the initial mixture was racemic ($[R]_0 = [S]_0$) this expression reduces to:

$$(1-\pi)^{\rho-1} = \frac{1+\alpha/\alpha_D}{(1-\alpha/\alpha_D)^{\rho}}$$

In most cases the degree of stereoelection, characterized by the value of ρ is not very high. There are, however, examples of more efficient stereoelection. They involve rather complex chiral initiators.

Problem 5.1

Calculate the distribution of degrees of polymerization in a polymer sample obtained anionically in an aprotic solvent under conditions such as no transfer or termination would occur. Assume initiation to be instantaneous and the initial ratio of monomer to initiator to be equal to 51 (which implies an average of 50 growth steps per chain).

For this calculation it is necessary to make use of *Stirling's* approximation:

$$\ln N! = N \ln N - N + 1/2 \ln 2 \pi N$$

Draw the distribution curve and compare it with the most probable distribution discussed in problem 3.1.

Problem 5.2

a) The rate of polymerization of a specific monomer M is given by the following equation:

$$R_p = -\frac{d[M]}{dt} = k_p[M^*][M]$$

where [M*] is the concentration of the active sites (supposed constant). If $[M]_0$ is the initial monomer concentration, what time is required to have [M] reduced to 1/10 of its initial value (conversion $\pi = 0.9$). Plot the variaton of conversion π versus time.

b) Carry out the same calculation and plot for a polymerization obeying a second order rate law.

$$R_p = -\frac{d[M]}{dt} = k_p[M^*][M]^2$$

What is the most striking difference between these two cases?
Numerical application:

$$k_p = 100\ [l \cdot mol^{-1} \cdot s^{-1}];\ [M]_0 = 1\ [mol \cdot l^{-1}];\ [M^*] = 10^{-4}\ [mol \cdot l^{-1}]$$

Problem 5.3

A solution of 104 *g* of styrene in dry dioxane (total volume 999 *ml*) is treated with 1 *ml* of a 1 *molar* solution of cumyl sodium, at 25 °*C*. The conversion degrees during the polymerization are measured at various times:

t	π
30 *s*	14%
60 *s*	26%
120 *s*	45%
240 *s*	70%
300 *s*	78%
600 *s*	95%

1) Demonstrate that the rate of polymerizaton R_p is first order with respect to monomer concentration [M].

2) Calculate the rate constant of propagation k_p.
3) Determine the number average molecular weight of the polymer formed after 10 minutes of reaction, assuming that no spontaneous transfer or termination has occured.

   It is assumed that initiation is instantaneous.

Problem 5.4

The anionic polymerization of styrene in THF initiated with cumyl potassium was carried out at various concentrations of "living" sites, but constant temperature. The following propagation rate constants were determined:

$$[M^{\ominus}] = 0.32 \times 10^{-3}\ [mol \cdot l^{-1}] \qquad k_p = 1\,130\ [l \cdot mol^{-1} \cdot s^{-1}]$$
$$[M^{\ominus}] = 2.1 \times 10^{-3}\ [mol \cdot l^{-1}] \qquad k_p = 510\ [l \cdot mol^{-1} \cdot s^{-1}]$$
$$[M^{\ominus}] = 8.4 \times 10^{-3}\ [mol \cdot l^{-1}] \qquad k_p = 280\ [l \cdot mol^{-1} \cdot s^{-1}]$$
$$[M^{\ominus}] = 34 \times 10^{-3}\ [mol \cdot l^{-1}] \qquad k_p = 160\ [l \cdot mol^{-1} \cdot s^{-1}]$$

Knowing that at that given temperature the electrolytic dissociation constant K_D is 0.8×10^{-7} for polystyryl potassium, calculate the values of the propagation constants for the free ions and for the ion pairs, k_{p-} and $k_{p\pm}$.

A second run used the most dilute solution with 2.7×10^{-2} $[mol \cdot l^{-1}]$ of potassium tetraphenylborate added, the dissociation constant of which is $K_\phi = 3 \times 10^{-4}$ at the same temperature. Calculate the new value of k_p that should be observed under these conditions.

Problem 5.5

A stereoelective polymerization of propylene sulfide is carried out with an optically active catalyst system. Starting from a racemic mixture of the monomer, it was found that when half of the monomer has been consumed the remaining monomer exhibits an optical activity which is one third of that measured for pure *R*-monomer (α is counted positive for the *R*-enantiomer). Calculate the degree of stereoelection of that catalyst system.

Problem 5.6

Single ended polystyrene macroanions are to be fitted with terminal acid chloride function using a difunctional terminator: terephthaloyl dichoride. Derive a relation between x, the fraction of all macroanions initially present which do not couple but carry an end standing acid chloride function, versus z, the mole ratio of difunctional terminator to macroanions initially present.

Problem 5.7

One mole of styrene is to be initiated by one millimole of sodium naphthalene. When conversion is complete the macroanions are killed with CO_2.

The injected styrene monomer contains 0.1 millimole of water, (i. e. 0.0018 g/104 g styrene, or 18 ppm). If each molecule of water kills one of the existing anions, taken at random, what weight fraction of the final polymer mixture consists of molecules having one — not two — carboxylic end groups.

How does the number average molecular weight of the polystyrene change as the proportion of water present in the styrene increases?

(It is assumed that the reaction of water with carbanion is much faster then the polymerization growth reaction.)

Chapter 6
Cationic Polymerizations

6.1 Introduction

Cationic polymerizations have been known for a long time, and they have been studied systematically in terms of mechanisms and kinetics over the past two decades. They are used in industry for the synthesis of butyl rubber (a copolymer of isobutene and isoprene), for various ring opening polymerizations (oxolane, trioxane), and for adhesive materials derived from polymerization of coumarone and the closely related indene.

The active sites are either *carbenium ions (i) or oxonium ions (ii)*, in some rare cases *ammonium ions*, with adequate counterions A^-.

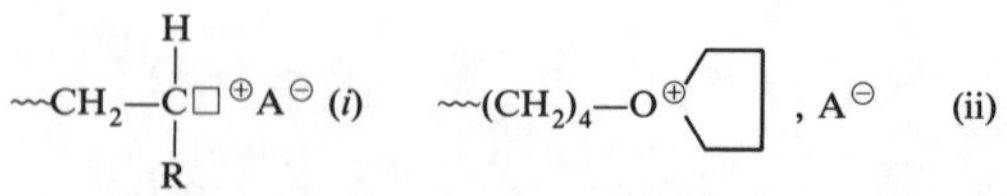

□ means a vacant orbital on the carbenium ion.

There are a number of differences between vinylic and heterocyclic monomers with respect to their behavior in cationic polymerization. These two cases will be examined separately in the present chapter.

The monomers fitted for cationic polymerization include all those with *electron donating* substituents, in which the polarization of the double bond makes them sensitive to electrophilic attack by a carbenium site.

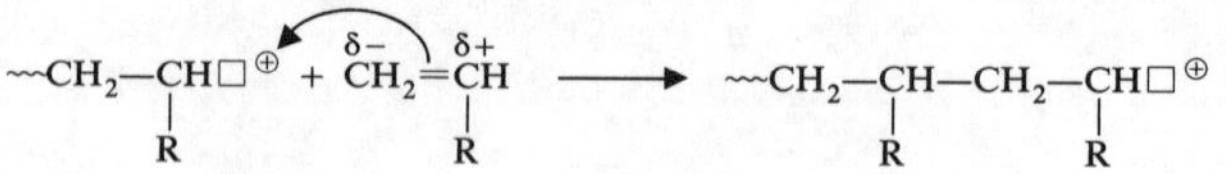

Monomers able to undergo cationic polymerization are listed in Table 6.1. Solvents used in cationic polymerizations should be stable towards acids and unable to react with electrophiles. Preferred are halogenated solvents (such as methylene chloride, methyl chloride, ethylene chloride, carbon tetrachloride), nitrocompounds such as nitromethane or nitrobenzene, and also dioxane. As in most anionic polymerizations the use of solvents is mandatory: Polymerization of undiluted vinyl monomers would be uncontrollable. Some "solvents" dissolve monomer but not polymer. Slurries are formed, of low viscosity, which is advantageous.

Table 6.1 Monomers polymerizable cationically

1. Unsaturated monomers

   Isobutene, alkenes

   Alkylvinyl ethers $CH_2{=}C(R)(R)$ $CH_2{=}C(R)(OR')$ $CH_2{=}C(OR)(OR)$

   Vinylacetals

   Indene x = CH_2

   Coumarone x = O

   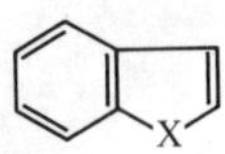

   Styrene and substituted styrenes

   α-methyl styrene

   Dienes

   N-vinyl carbazole

   β-pinene

2. Cyclic Monomers

   Oxiranes (epoxides) n = 2

   oxetane n = 3

   oxolane (THF) n = 4

   oxepane n = 6

   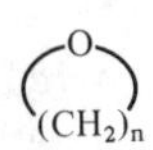

   Dioxolane p = 2

   Dioxepane p = 4

   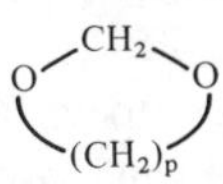

   N-substituted aziridines

   lactams

   cyclosiloxanes (D4)

   phosphoranes

   oxazolines

   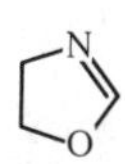

6.2 Cationic Polymerization of Vinylic Monomers

6.2.1 Initiators and Initiating Reactions

Carbenium salts are extremely reactive (far more than carbanions) and they can participate in a number of side reactions. It is of great importance that the anion involved exhibits a low nucleophilicity to prevent it from either recombining with the carbenium ion or extracting a proton from a growing chain: Both of these reactions would induce loss of the active site.

The initiators able to induce polymerization of vinylic monomers can be classified into three categories:

(a) Protonic acids: Their anions should be bulky (for efficient charge delocalization) and symmetrical (to avoid anion splitting).

Sulfuric acid (H_2SO_4), perchloric acid ($HClO_4$), fluorosulfonic acid (FSO_3H), trifluoromethylsulfonic acid (CF_3SO_3H), and even in some cases trifluoro acetic acid (CF_3COOH) have been employed. The initiation proceeds by protonation:

$$H^{\oplus}A^{\ominus} + CH_2{=}\underset{\displaystyle R}{\underset{|}{CH}} \longrightarrow CH_3{-}\underset{\displaystyle R}{\underset{|}{CH}}{}^{\oplus}A^{\ominus}$$

(b) Lewis acids such as $SnCl_4$, $AlCl_3$, BF_3, when reacted with some nucleophilic compound, can also initiate the polymerization of vinylic monomers. These so-called "initiator-coinitiator systems" give complexes that are the real cationic initiators as shown in the examples below:

$$BF_3 + H_2O \rightleftarrows [H^{\oplus}BF_3OH^{\ominus}]$$

$$AlCl_3 + RCl \rightleftarrows [R^{\oplus}AlCl_4^{\ominus}]$$

$$SbF_5 + HF \rightleftarrows [H^{\oplus}SbF_6^{\ominus}]$$

$$H^{\oplus}A^{\ominus} + M \rightarrow HM^{\oplus}A^{\ominus}$$

$$R^{\oplus}A^{\ominus} + M \rightarrow RM^{\oplus}A^{\ominus}$$

There has been a long debate as to whether a *pure* Lewis acid is capable of initiating polymerization or not. The problem is not easy to settle as these species can react with so many compounds. In some instances even self ionization equilibria have been observed, as is the case of $SbCl_5$:

$$2\,SbCl_5 \rightleftarrows SbCl_4^{\oplus}, SbCl_6^{\ominus}$$

This could explain the initiation in the absence of any "coinitiator" added on purpose, as it has been postulated in some cases.

(c) Existing carbenium salts can also be used to initiate cationic polymerizations. The stability of such compounds is enhanced (and their electrophilicity is lowered) because of substituents allowing charge delocalization, as in the case of triphenylmethyl (trityl) salts, or because of resonance stabilization, as in the case of oxocarbenium salts*).

$$\phi_3CCl + SbCl_5 \longrightarrow \phi_3C^{\oplus}, SbCl_6^{\ominus} \text{ (trityl hexachloroantimonate)}$$

$$\phi{-}COF + SbF_5 \longrightarrow \phi CO^{\oplus}, SbF_6^{\ominus} \text{ (benzoyl hexafluoroantimonate)}$$

6.2.2 Propagation Reaction

The growth reaction proceeds, as mentioned above, by nucleophilic attack of the monomer onto the cationic site. However, owing to the very high reactivity of the carbenium sites, side reactions occur and, contrary to anionic polymerization, there is no relation between the molecular weight of the polymer formed and the monomer to initiator mole ratio.

---

*) $R{-}C{\equiv}O^{\oplus} \leftrightarrow R{-}\overset{\oplus}{C}{=}O.$

Some side reactions occur upon propagation, as hydride shifts can take place if carbenium ions of enhanced stability can result. Such *intramolecular hydride shifts* can be exemplified by the polymerization of 3-methyl butene:

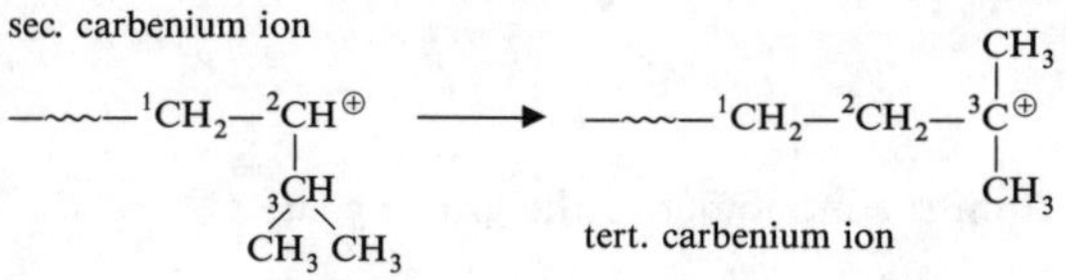

If this isomerization takes place after each growth step the resulting polymer comprises 3 carbon atoms per monomer unit (instead of 2). The driving force for this hydride shift is the enhanced stability of tertiary carbenium ions with respect to secondary ones. However this is a rather slow process. It occurs to a large extent only when the polymerization is carried out at low temperature, because then the rate of propagation is so greatly reduced that the probability of hydride shift becomes important. If this is not the case, the resulting polymer is in fact a random copolymer of units that have undergone hydride shift, and of such that have not.

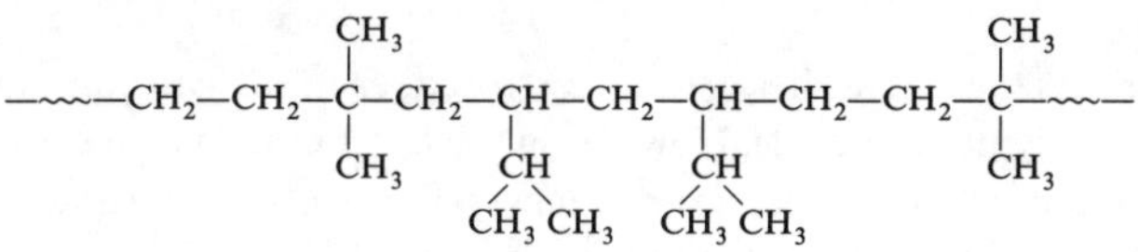

6.2.3 Transfer and Termination Reactions

Several reactions may take place to stop chain growth, either by transfer, or by termination.

(a) Spontaneous termination occurs in many cases by reaction between the growing cation and its counterion:

* *Anion-cation recombination* provides for an irreversible loss of sites, upon formation of a terminal ester function:

* *Anion splitting* originates from the attack of the carbenium ion onto the weakest part of the counter ion. This is why symmetrical anions such as BF_4^-, SbF_6^-, with extended charge delocalization are preferred, as they are less likely to undergo anion splitting:

(b) Transfer without termination of the kinetic chain can involve:

∗ *Proton transfer* to the monomer according to:

∗ or *hydride ion transfer* from the monomer to the growing site:

∗ or *proton expulsion* from the growing site (which is basically the same process):

In these three examples the active sites are not destroyed; each site can consequently lead to more than *one* polymer molecule. However in the latter case (which is the most frequent one), renewed initiation is necessary to have polymerization resume. This is why steady-state conditions can be applied to the concentration of carbenium sites in such systems, even though the sites are not lost upon transfer, as it will be seen.

(c) Transfer implying termination of the kinetic chain

Substances such as water, ethers, tertiary amines, and esters may react with the carbenium ions at chain end to yield cationic species of lower reactivity, often unable to give rise to further chain growth.

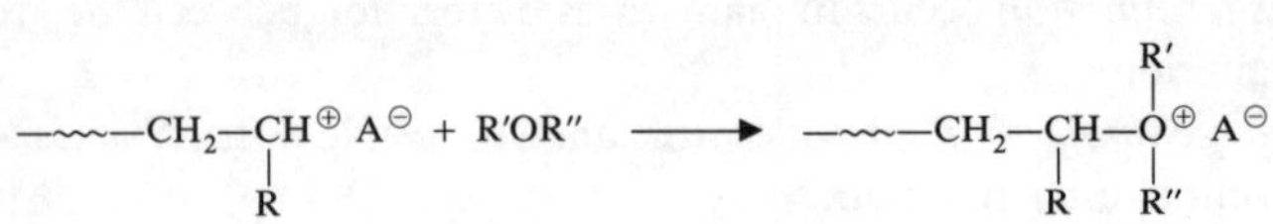

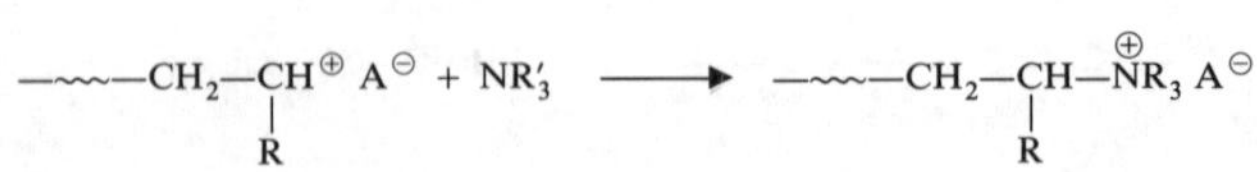

A systematic search for all possible reactions occurring upon cationic polymerization of vinylic monomers by means of a "catalyst-cocatalyst" system has been performed by Kennedy and Gillham using the initiating system t-butyl chloride-trimethylaluminum. The monomer chosen, 2,2,4-trimethylpentene, is a non-polymerizable dimer of isobutene. The reaction mixtures were submitted to gas chromatographic analysis to determine the nature and the yields of the substances formed. Ion splitting, alkylation, deprotonation reactions have thus been clearly identified.

6.2.4 Kinetic Expressions for Cationic Polymerizations

In contrast to the relative simplicity of radical polymerizations and of anionic polymerizations in aprotic solvents, a variety of scenarios can be postulated for cationic polymerizations, and the kinetic schemes evolving from the postulates are both complicated and hard to verify experimentally. Three points should be mentioned at the outset:

(a) Rates of polymerization are greater by one or more orders of magnitude than in anionic and radical polymerizations, so fast that keeping the temperature constant may be difficult.

(b) Often chlorinated solvents are used, which offer control of temperature in "runaway" reactions by vaporization at low temperatures, and which cause the polymer being formed to precipitate as porous, non-coherent aggregates that form a slurry. Though advantageous by maintaining low viscosity, the formation of the second phase immediately invalidates kinetic expressions derived for *single phase* homogeneous reactions.

(c) Transfer reactions, especially of growing chains to monomer, are very important, in marked contrast to anionic polymerization in aprotic media. Several cases have to be considered, depending upon the nature of termination or transfer processes occurring.

If initiation proceeds by means of a "catalyst cocatalyst complex" involving an equilibrium, [C] being the catalyst concentration (BF_3, $AlCl_3$, etc.) and [HA] the "cocatalyst" concentration, for example HOH or HCl, the equilibrium can be written as:

$$\mathrm{C + AH} \rightleftarrows \{\mathrm{H}^{\oplus}\,\mathrm{CA}^{\ominus}\} \qquad K = \frac{[\mathrm{H}^{\oplus}\,\mathrm{CA}^{\ominus}]}{[\mathrm{C}][\mathrm{AH}]} \tag{6.1}$$

Initiation then originates from the attack of the ion pair complex onto the monomer (protonation or alkylation):

$$\{\mathrm{H}^{\oplus}, \mathrm{CA}^{\ominus}\} + \mathrm{M} \xrightarrow{k_i} \mathrm{HM}^{\oplus}, \mathrm{CA}^{\ominus}$$

(where $\mathrm{HM}^{\oplus} = \mathrm{M}_1^{\oplus}$ is a single monomer unit carrying the carbenium ion) whereby the rate of initiation is expressed as:

$$R_i = k_i[\mathrm{H}^{\oplus}\mathrm{CA}^{\ominus}][\mathrm{M}] = k_i K[\mathrm{M}][\mathrm{C}][\mathrm{AH}] \tag{6.2}$$

Propagation involves growth steps which can be written as:

$$\mathrm{HM}_{n-2} - \mathrm{M}^{\oplus}, \mathrm{CA}^{\ominus} + \mathrm{M} \longrightarrow \mathrm{HM}_{n-1} - \mathrm{M}^{\oplus}, \mathrm{CA}^{\ominus}$$

where $\mathrm{HM}_{n-1}\,\mathrm{M}^{\oplus}$ will be called for simplicity $\mathrm{M}_n^{\oplus}$. The rate of monomer consumption is first order in monomer concentration and first order in active site concentration, as in most radical and anionic polymerizations:

$$R_p = -\frac{d[M]}{dt} = k_p[M^{\oplus}][M] \tag{6.3}$$

$[M^{\oplus}]$ stands for the molar concentration of active sites of any length. Let us now consider the cases discussed above:

(a) If termination occurs solely by combination of the carbenium ion with its counterion (or by any other monomolecular process*) like alkylation), one can write:

$$M_n^{\oplus} A^{\ominus} \xrightarrow{k_t} M_nA \qquad R_t = k_t[M^{\oplus}] \tag{6.4}$$

By assuming steady-state conditions over small time intervals one gets:

$$k_t[M^{\oplus}] = k_i K[M][AH][C] \tag{6.5}$$

and the rate of polymerization is expressed as:

$$R_p = -\frac{d[M]}{dt} = \frac{k_p k_i K}{k_t}[AH][C][M]^2 \tag{6.6}$$

This equation shows that R_p is second order with respect to monomer concentration, as it is for anionic polymerizations in a protic solvent.

In this case, the number average degree of polymerization is expressed by the average number of growth steps occurring during the lifetime of a carbenium site:

$$\bar{P}_n = \frac{k_p[M]}{k_t} \tag{6.7}$$

(b) If the solvent (or any other compound present) is solely responsible for transfer, one can still assume steady-state conditions if the species formed is unreactive with monomer. This can be equated as:

$$k_{tr}[S][M^{\oplus}] = k_i K[C][AH][M] \tag{6.8}$$

where [S] stands for the molar concentration of the compound responsible for the transfer reaction.

The rate of polymerization is expressed accordingly:

$$R_p = k_p \frac{k_i K}{k_{tr}} \frac{[C][AH][M]^2}{[S]} \tag{6.9}$$

The number average degree of polymerization is given by:

$$\bar{P}_n = \frac{k_p[M]}{k_{tr}[S]} \tag{6.10}$$

(c) If the initial complex H^+CA^- is reformed as a result of proton abstraction or of transfer, the same equation applies as in case *(a)*.

---

*) Cation and anion are part of the same chemical species. Therefore any reaction involving attack of the cation by its anion is a monomolecular process.

In some instances these expressions have been confirmed experimentally. However in most cases several transfers or terminations coexist, and so none of the above equations is satisfactory. It becomes futile to try to account for the phenomena observed, as the number of parameters to take into consideration (the various rate constants) becomes too high.

Furthermore, equations (6.1) to (6.10) are based upon the assumption that the rate determining step in the initiation process is the protonation of the monomer (or its alkylation by $R^{\oplus}, CA^{\ominus}$). This is what is generally observed. There are cases, however, in which the rate determining step is the reaction between initiator and coinitiator; in other words, the actual formation of the initiating complex is the slowest reaction:

$$AH + C \longrightarrow H^{\oplus}, CA^{\ominus}$$

If this is the case the rate of initiation should be written as:

$$R_i = k_i'[C][AH] \tag{6.11}$$

and the expressions for the polymerization rate have to be modified. In cases *(a)* and *(c)* the rate equation should be written:

$$R_p = k_p \frac{k_i'}{k_t} [C][AH][M] \tag{6.12}$$

and in case *(b)*:

$$R_p = k_p \frac{k_i'}{k_{tr}} \frac{[C][AH][M]}{[S]} \tag{6.13}$$

Under these conditions the rate of propagation is always first order in monomer concentration.

As far as *molecular weights* are concerned, it should be remembered that the probability for transfers of any kind to occur increases with increasing reaction temperature. The ratio of the rate constant of transfer to monomer to the rate constant of propagation (which is called monomer transfer constant C_M) decreases by a factor 100 when the temperature decreases from 30 to −60 °C in the polymerization of isobutene. This allows the molecules to grow far more at low temperature, provided the solvent and the constituents of the initiating system are not themselves giving rise to transfer reactions. For this reason initiating systems containing carboxylic acids, phenyl ethers, anhydrides, and even diethyl ether should be avoided if the process aims at formation of high molecular weight compounds.

If in equation (6.6) we replace the group $\frac{k_p k_i}{k_t}$ by its Arrhenius form:

$$\frac{k_p \cdot k_i}{k_t} = \frac{A_p A_i}{A_t} \exp[(-E_p - E_i + E_t)/RT] \tag{6.14}$$

The sum $E_R = E_p + E_i - E_t$ is of the order of −5 to +10 *kcal/mol.*

This means that the effect of temperature on the overall rate of polymerization is generally *smaller* than it is for free radical polymerization. Similar considerations lead also to the value of the overall "activation energy" for the molecular weights E_M which is *always* negative. The molecular weight is therefore expected to increase as the reaction temperature decreases. But this effect is greatly amplified by the lower probability of occurrence of transfer reactions, as stated above.

As mentioned before, if hydride shift is involved in the process of propagation, the probability for such isomerizations of the active site increases as the temperature decreases. This is especially the case with monomers such as 4-methylbutene or of vinylcyclohexane.

6.2.5 Polyisobutene and Butyl Rubber

The chief large scale application of cationic polymerization of an unsaturated monomer is that of isobutene and of its copolymers, to yield butyl rubbers.

As the molecular weight is chiefly determined by the occurrence of transfer reactions, it is essential to carry out these polymerizations at very low temperatures (from −100 to −30 °C), the highest molecular weights being obtained at the lowest temperatures.

An important aspect of cationic polymerizations arises from the very high propagation rates, even at low temperature. The corresponding heat release rates are difficult to control by heat transfer surfaces (refrigeration) unless the viscosity of the reaction mixture is kept very low. This can be achieved by causing slurry formation, upon using a solvent of the monomer which does not disolve the polymer formed. This will be further discussed in chapter 13.

In some respect large scale cationic polymerization of isobutene is similar to the free radical "mass" polymerization process for vinyl chloride: Two phases are formed; The solvent can be used as a heat release regulator upon vaporization; The biphasic mixture can be stirred easily and efficiently.

6.2.6 Inifer Processes

Inifer systems, first developed by Kennedy, have been widely applied to functionalizations of cationic polymers at chain end. The name arises from the fact that the initiating system — constituted of an alkyl chloride and boron trichloride — is also able to give yield to transfer reactions. In the presence of an adequate monomer, such as isobutene, the reaction scheme can be written as:

Initiator formation:

$$RCl + BCl_3 \rightleftarrows \{R^{\oplus}, BCl_4^{\ominus}\}$$

Initiation by alkylation:

$$\{R^{\oplus}, BCl_4^{\ominus}\} + CH_2{=}\overset{|}{\underset{|}{C}} \longrightarrow R{-}CH_2{-}\overset{|}{\underset{|}{C}}{}^{\oplus}, BCl_4^{\ominus}$$

Propagation:

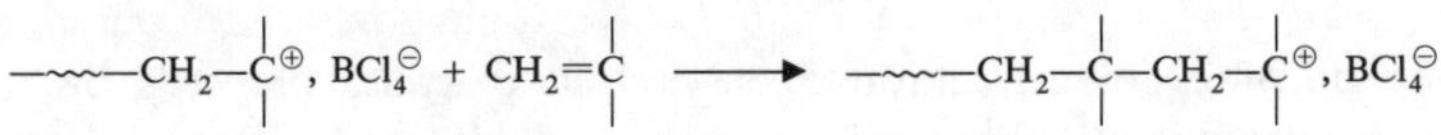

Transfer to the alkyl halide:

Ion splitting:

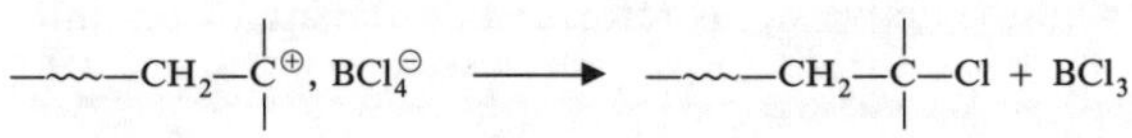

The interesting feature of this process arises from the fact that ion splitting and transfer to the alkyl halide give exactly the same product.

Transfer to the monomer may occur, but the probability of transfer to the alkyl halide exceeds by far that of transfer to monomer. Approximate molecular weight control is rendered possible by an adequate choice of the mole ratio of monomer to alkyl chloride. The range of molecular weights accessible is however limited to low values: 10^3 to 10^4.

Inifer systems provide the molecules with halogenated functions at chain end, the alkyl group sitting at the other end. Difunctional initiator systems have also been developed and they yield polymer molecules carrying C-Cl bonds at both chain ends.

In some instances the tertiary chloride sites are themselves capable of acting — together with BCl_3 — as a macromolecular alkylating agent for another monomer, to yield block copolymers, as e. g. poly(α-methylstyrene-b-isobutene).

Further transformation of the tertiary C-Cl bond can be performed by chemical reactions. An example is given below, showing how formation of OH functions at chain ends can be obtained in three steps: dehydrohalogenation, borohydration, oxidative hydrolysis.

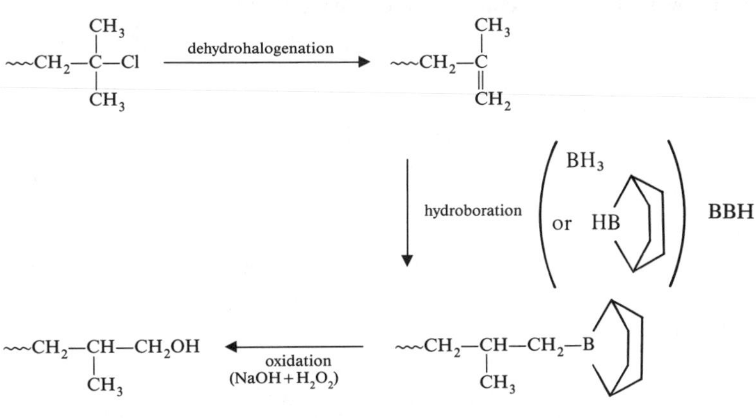

6.3 Cationic Polymerization of Heterocycles

The ring opening polymerization of several heterocycles proceeds by cationic initiation via *oxonium sites* that are formed upon alkylation (or protonation) of the monomer:

The mechanism of chain growth involves nucleophilic attack of the oxygen of an incoming monomer onto a carbon atom in α-position with respect to the oxonium site, whereby the cycle opens and the active site is reformed on the attacking unit:

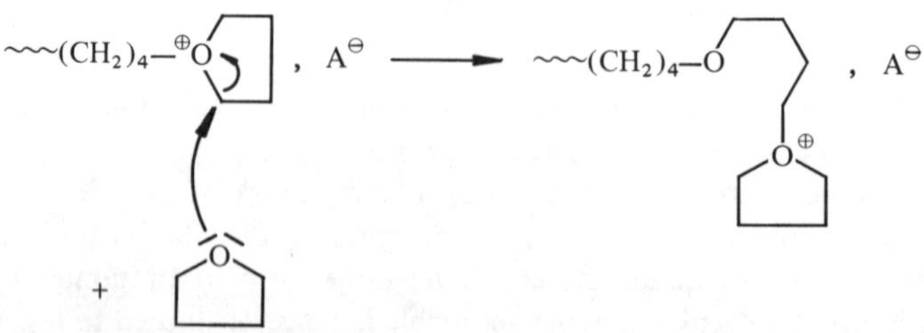

All the five-, six-, and seven-membered rings are quite stable, and the change in free energy involved in ring opening polymerization is small. As a consequence, the depropagation reaction cannot be neglected and conversion is limited to values lower than unity. The equilibrium monomer concentration is strongly temperature dependent, as it will be shown in chapter 7.

6.3.1 Initiation

The initiators are the same as in cationic polymerization of vinylic monomers. Initiation occurs either by protonation or by alkylation, though in some cases hydride ion transfer was found to occur. Here are a few examples:

Protonation:

Acylation:

Alkylation:

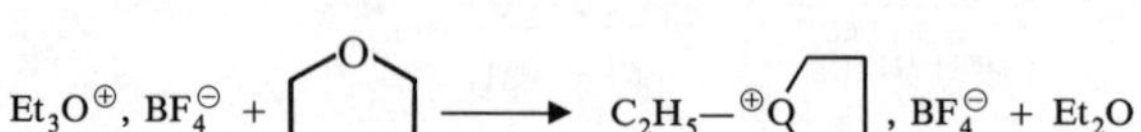

Hydride transfer:

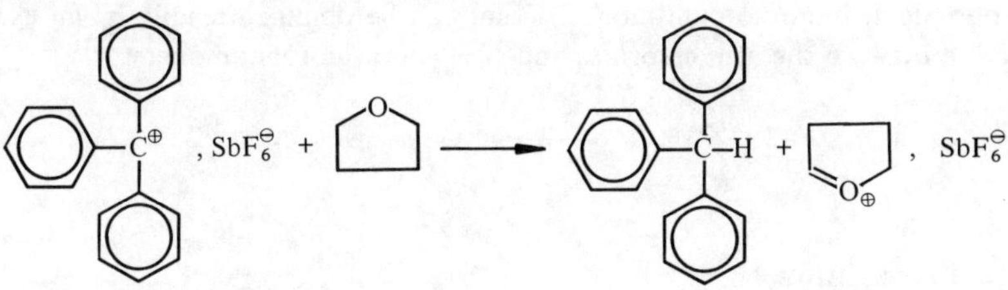

followed by:

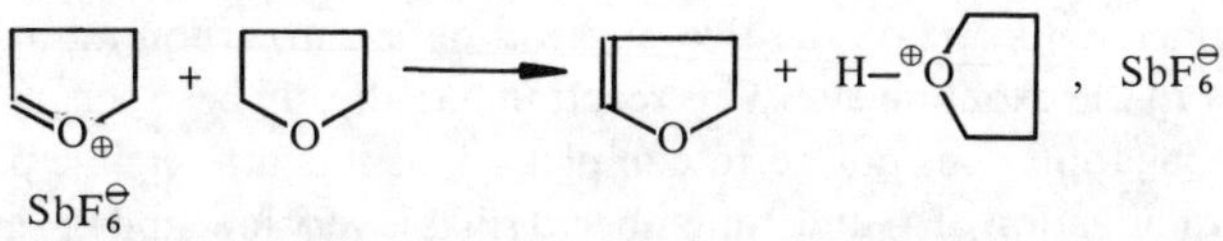

Whenever initiation involves alkylation it implies that the initiator fragment stays at chain end of the polymer. In other words, initiation then occurs by addition onto the monomer, contrary to what happens in the two other cases.

It should be noted that diphenyl methylium salts give rise to alkylation whereas trityl salts yield hydride shift with oxolane (THF) and similar monomers.

The special case of *"triflic initiators"* (derived from trifluoromethyl sulfonic acid) deserves some comments. In this case the coupling reaction between cation and anion to form an ester is reversible. Therefore the active sites do not get lost upon recombination between the oxonium ion and its counterion. Let us take the example of methyltriflate as the initiator. Initiation occurs by alkylation:

whereupon chain growth can take place:

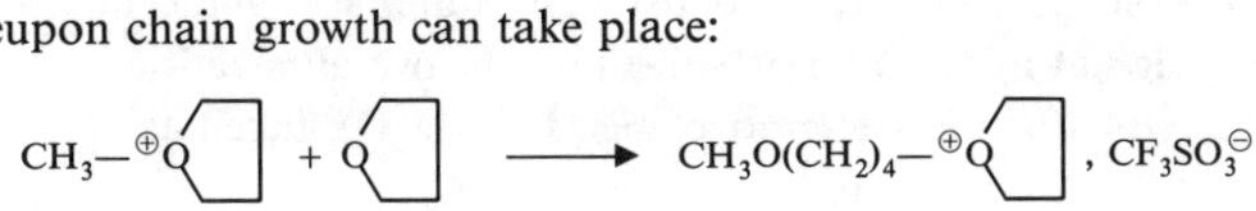

As indicated, the ion pairs are in equilibrium with the ester form:

If instead of the ester, *triflic anhydride* is used as the initiator, the same reaction can occur at both chain ends, demonstrating that the anhydride behaves as a bifunctional initiator:

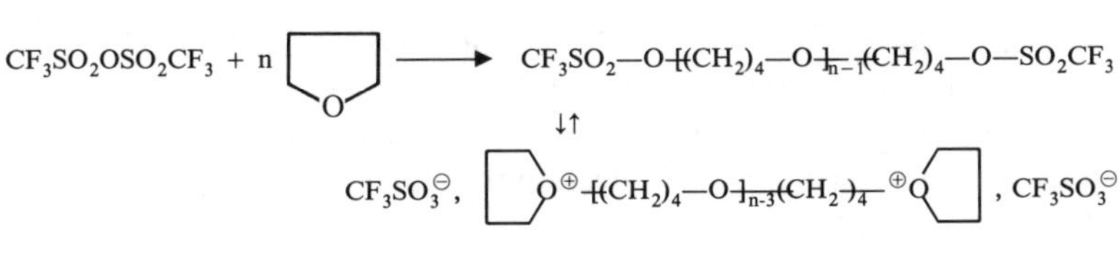

Other bifunctional initiators include dioxocarbenium salts such as adipoyl (or terephthaloyl) hexafluoroantimonates, that can be obtained readily by an exchange reaction between the acid chlorides and silver hexafluoroantimonate:

$$ClCO(CH_2)_4COCl + 2\,AgSbF_6 \longrightarrow 2\,AgCl\downarrow + \overset{\oplus}{O}C(CH_2)_4CO^{\oplus},\ 2\,SbF_6^{\ominus}$$

6.3.2 Propagation

As already stated propagation involves nucleophilic attack by the oxygen doublet of the incoming monomer onto one of the two electron deficient carbon atoms in α-position with respect to the oxonium site. This reaction has already been schematized.

In most cases propagation does not go to completion, as it is a reversible process. The enthalpy of polymerization of 5- and 7-membered rings is very low, and the entropy change involved is also small, so that the free energy of polymerization is close to zero at room temperature. Depropagation has to be taken into consideration for many of these heterocyclic monomers. The following two points should be kept in mind:

- The higher the temperature, the faster the reaction, but the lower the limiting degree of conversion.
- The reversibility of the propagation reaction induces broadening of the molecular weight distribution.

6.3.3 Transfer and Termination

If such catalysts as [BF_3, H_2O] or [$AlCl_3$, HCl] are used to initiate the polymerization of cyclic ethers, the polymers formed usually exhibit low molecular weights because of transfers originating from anion splitting. The overall reaction scheme can be written as shown below for the system oxetane/(BF_3, H_2O), taken as an example:

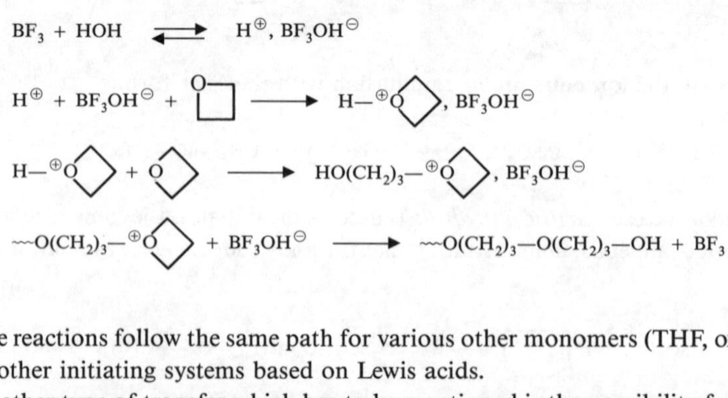

The reactions follow the same path for various other monomers (THF, oxirane) and with other initiating systems based on Lewis acids.

Another type of transfer which has to be mentioned is the possibility for a growing oxonium-ion to react with an ether function to yield an open chain tertiary oxonium-ion of lesser reactivity:

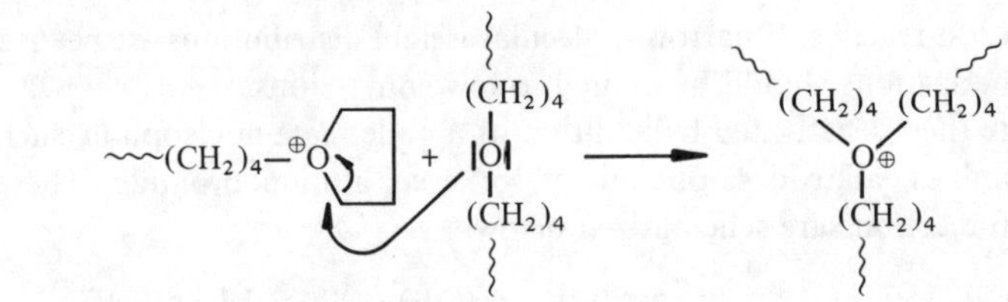

The probability for such reactions to occur increases with the amount of polymer formed (i. e. with conversion) as in all transfer reactions involving the polymer. Upon deactivation of these oxonium salts one of the three equivalent $>O^{\oplus}$—C bonds has to be cleaved. Such a process leads obviously to a *redistribution* of chain lengths by a random process, and it contributes to an *increase* in the polydispersity index. Here again, the limit of the $\overline{P}_w/\overline{P}_n$ ratio is two, as in all kinds of "reshuffling" reactions, leading to Flory's most probable distribution.

This reaction can also occur intramolecularly, and it is then called *back-biting*. The consequence of back-biting should be formation of *cyclic macromolecules*.

6.3.4 "Living" Cationic Ring Opening Polymerizations

A few heterocyclic monomers (oxolane, t-butylaziridine) have been polymerized under conditions such as to have the number of sites remaining constant throughout the process. If transfer processes are prevented as well, these polymerizations are truly "living": The number average polymerization degree of the polymer formed is given by the mole ratio of monomer to initiator. Active oxonium sites are retained at chain ends. Upon addition of further monomer, the existing chains are extended correspondingly.

Such "living" polymerizations require initiators, fitted with a stable counter-ion (that cannot undergo ion-splitting), and able to react rapidly, quantitatively and unambiguously (by addition) with the monomer. Most of these initiators involve preexisting carbenium ions, which are good alkylating agents for heterocycles. Examples of such efficient initiators are given below:

$C_6H_5—\overset{+}{C}=O,\ SbF_6^-$ $\quad$ $CH_3—CH_2—\overset{+}{C}=O,\ PF_6^-$

oxocarbenium salts

$(CH_3)_2CH^+,\ SbF_6^-$ $\quad$ $(C_6H_5)_2\overset{+}{C}H,\ BF_4^-$

dialkyl methylium salts

$CF_3—SO_2—O—CH_3$ — triflic ester

$CF_3—SO_2—O—SO_2—CF_3$ — triflic anhydride

Under these conditions the growing sites are long-living, provided the reaction medium does not contain any transfer agents such as alcohols, ethers, amines, etc.

The reaction mechanism then involves only two steps: Initiation and propagation. The polymers are living. However, as conversion increases, the probability of transfer to the polymer (already mentioned) increases. The resulting open-chain tertiary oxonium sites are far less reactive. If narrow molecular weight distributions are needed these cationic polymerizations should be limited to low conversions.

The active sites are then deactivated by addition of an adequate nucleophile, such as water, tertiary amines, alkoxides, phenolates, or even lithium bromide. These induced deactivation reactions are schematized below:

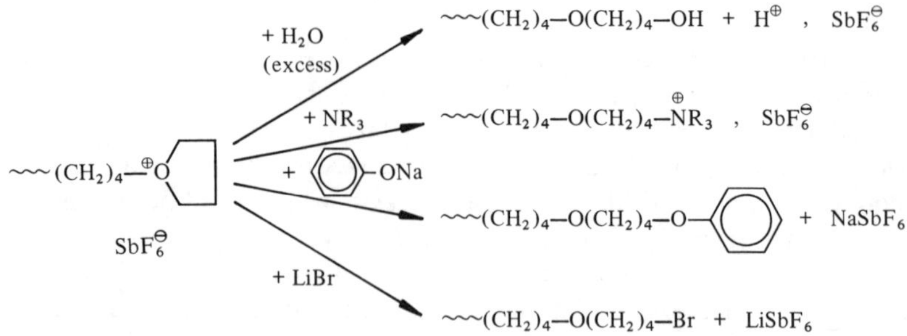

6.3.5 Some Specific Cases

The polymerization of *oxolane* (THF) in the bulk or in a solvent such as CH_2Cl_2 has been studied intensively and, provided adequate initiators are chosen, "living" polymer species are obtained. However, THF-polymerization never goes to completion because of the *"ceiling" temperature* and because of progressive inactivation of the sites, owing to transfer to ether functions of the polymer already formed. Both reversibility and chain redistribution contribute to increase the width of the molecular weight distribution, which is initially Poisson-type and eventually reaches Flory's most probable distribution at equilibrium.

Since the *oxirane* and *oxetane* rings are more strained than that of THF, the polymerization of these monomers is more favored than that of THF. Equilibrium monomer concentrations are lower. But the chief characteristics of THF-polymerization are found here too, though some side reactions are known to occur.

Dioxolane is a rather special case owing to the presence of a methylene group between two oxygen atoms. The active site can therefore involve two possible structures:

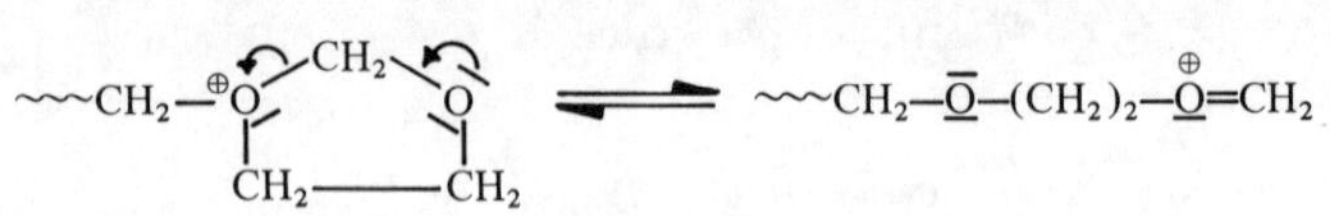

the open chain form being more reactive. Initiation often occurs by hydride-ion abstraction, e. g.:

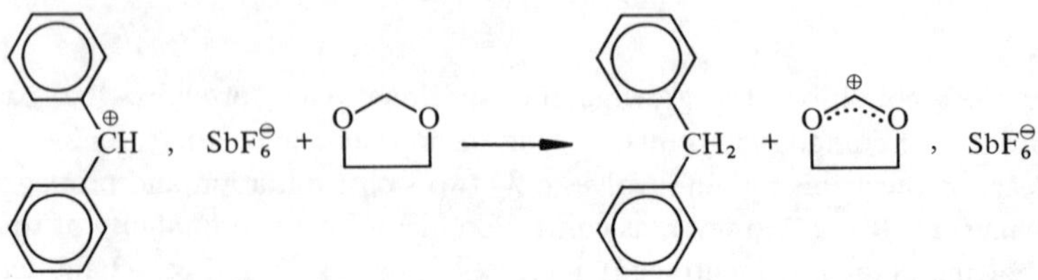

The dioxolenium cation is sensitive to the attack by nucleophiles, whereby strong nucleophiles will be fixed at the carbon, and weaker ones at the oxygen:

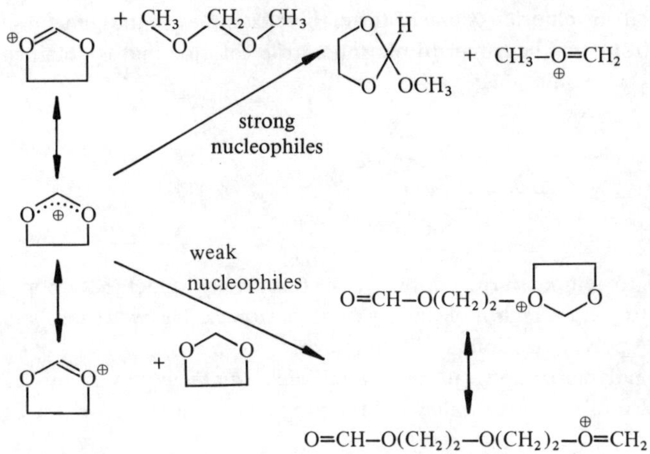

The latter type of reaction is the propagation process; it should be observed that the contribution of the open chain active site is predominant, because it involves a stabilized methylene oxonium group.

N-substituted aziridines readily undergo cationic polymerization, but with most N-alkyl substituents isomerizations leading to deactivation or transfer prevent these processes from being "living". However the bulky and symmetric tert.-butyl group does not comprise hydrogen in α-position to the nitrogen; therefore the life-time of the active sites is extended very much and the "living" polymers obtained can be functionalized by adequate deactivators in a close to quantitative manner. The narrow molecular weight distributions also testify to the "living" character of poly(tert.-butylaziridine) obtained with efficient initiators, such as triflic derivatives:

Initiation

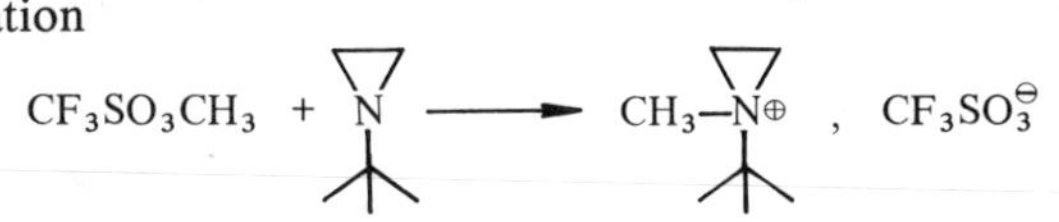

Propagation

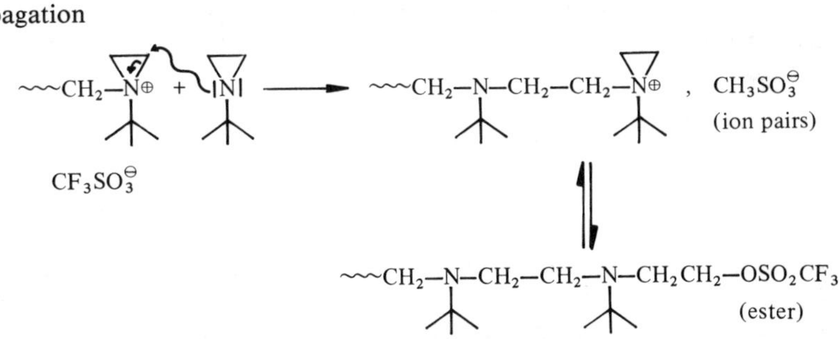

6.3.6 Other Ring Opening Polymerizations

6.3.6.1 Polyphosphonitrilic Chloride (Polyphosphazenes)

Polyphosphonitrilic chloride is one of the earliest examples of inorganic polymers, as it originates from a cyclic trimer of phosphonitrile chloride that is obtained upon reaction of PCl_5 onto ammonia:

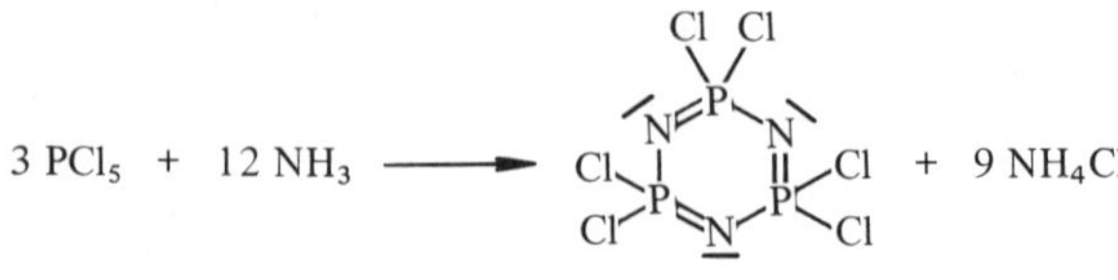

When heated to temperatures above 250 °C this trimer undergoes spontaneous polymerization to yield a high molecular weight elastomer. The exact mechanism of this ring opening polymerization is still unknown.

However, the polymerization can also be initiated *anionically* by means of weak bases, such as alcohols or metal salts, and the propagating site may well be of $\sim\!\!-\overset{Cl}{\underset{Cl}{P}}=N^{\ominus}$ type. This may lead to the conclusion that the no-catalyst polymerization occurring at elevated temperature is of *zwitterionic character*, at least at the beginning:

$$^{\oplus}PCl_2{=}N{+}PCl_2{=}N{+}_n\, PCl_2{=}N^{\ominus}$$

Branching reactions are likely to occur, that is why the materials obtained are *"self vulcanized"*.

Chemical transformation can be carried out on the polyphosphazene chains by substituting the reactive PCl-bonds (that are quite sensitive to hydrolysis) by much less reactive POR-bonds upon reaction with alcohols in the presence of some HCl scavenger.

6.3.6.2 Polydimethylsiloxanes (PDMS)

Early access to polydimethylsiloxanes resulted from controlled hydrolysis of dichlorodimethylsilane:

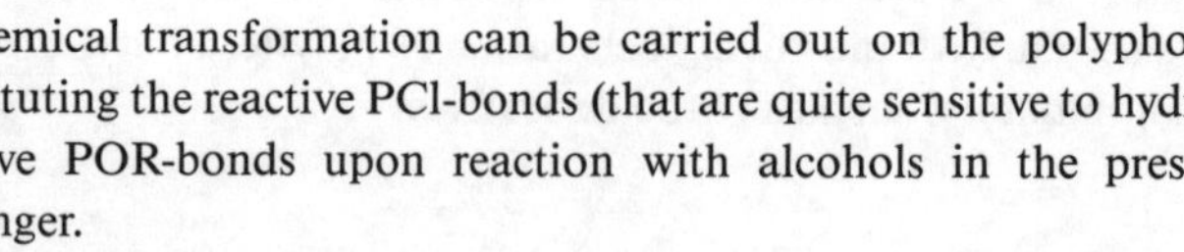

Upon reaction of ⪢SiOH with ⪢Si—Cl a —Si—O—Si— (siloxane) link is formed very rapidly. This reaction can be considered as a step growth polymerization.

However it has been established later that this reaction can be carried out under conditions such as to favor the formation of *cyclic siloxanes*, the six membered *D3* (hexamethyl cyclotrisiloxane) and the eight membered *D4* (octamethyl cyclotetrasiloxane) being the most likely to form. These are now the starting compounds for the PDMS synthesis by ring opening polymerization. *D3* can be polymerized anionically with initiators such as butyl lithium or silanolates, whereby under certain conditions the polymers obtained can be considered to be "living". *Cryptands* added to the reaction medium proved very effective and increased the rate of propagation. *Block copolymers* of styrene and dimethyl siloxane have also been made, by initiating the polymerization of *D3* using the lithium organic sites of a living polystyrene.

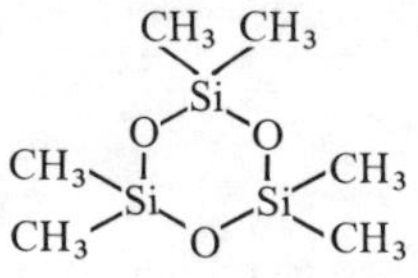

D3

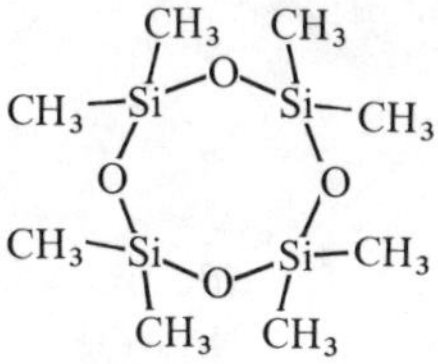

D4

D4 is of far greater importance as it is formed in much larger proportion in the controlled hydrolysis of dichloro dimethyl silane. It was assumed for a long time that *D4* was insensitive to the attack by anionic sites, but it was shown recently that indeed *D4* can be polymerized anionically in the presence of *activating cryptands*. Nevertheless the preferred method of PDMS synthesis is the cationic ring opening polymerization of *D4*. The active sites may be either *oxonium ions* (as in the case of THF polymerization) or *silicenium ions*. The catalyst systems preferred in industry are *acid clays*. The interesting feature is that a control of the average degree of polymerization (and of the functions located at chain end) can be gained by introducing into the reaction mixture a known amount of a disiloxane as a transfer agent. The success of this method might incline one to think that propagation proceeds by ring expansion until the transfer process opens the ring and stops the chain growth process. The most efficient functionalization agents used are hexamethyl disiloxane and its functional derivatives, dihydrogenotetramethyl disiloxane (M'_2) and divinyltetramethyldisiloxane:

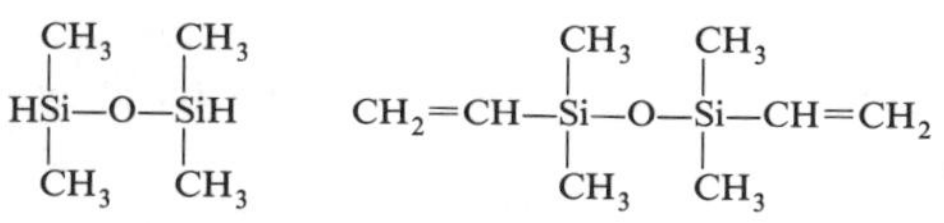

The PDMS chains formed are then fitted at both ends with hydrogenosilane or with vinylsilane functions, which are capable of giving hydrosilylation reactions. This is a way commonly used to make block copolymers with polydimethylsiloxane blocks, and also PDMS model-networks.

The interest of PDMS arises from its low glass transition temperature and from the *large flexibility* of its chain. Low molecular weight PDMS are used as *lubricating* agents, high molecular weight homologues as potential *elastomers.* To have them exhibit elastic properties some vulcanization has to be achieved, generally by free radical processes. To enhance their ability to undergo crosslinking, the chain can be fitted with some dangling unsaturations, arising from a small proportion of unsaturated *D4**) added to the regular *D4* in the synthesis of the polymer. It should be noted at this point that finely dispersed silica — generally used as a filler in silicone rubbers — also contributes to the chain bonding process. The exact mechanism of this *"physical vulcanization"* is still controversial, but it is obvious that the polysiloxane oxygen acts as hydrogen bond acceptor for the very strong silanol (—SiOH) donors on the surface of the silica, causing very strong adsorption of parts of the polysiloxane chains to accessible SiOH-sites on the particles.

Problem 6.1

An equimolar mixture of boron trifluoride and water is used to initiate the cationic polymerization of isobutene in methylene chloride solution. Assuming that transfer to solvent and to monomer is negligible (which is questionable), that ion-splitting is the most likely mechanism of termination, and that protonation of the monomer is the slowest reaction step, establish the expression of the monomer consumption rate.

Problem 6.2

Consider the cationic polymerization of isobutene initiated with the $TiCl_4$/water system, which is carried out in methylene chloride (admittedly unable to give rise to transfer reactions). Assuming the ratio k_p/k_t to be independent of temperature, and the transfer to monomer constants C_M to be equal to 21×10^{-4}, 6.6×10^{-4}, and 1.6×10^{-4} at -20, -50 and -78 °*C*, respectively, and knowing that at -78 °*C* the initial molecular weight of the polymer formed is 100 000, what would it be at -20 and at -50 °*C*? Take $[M]_0 = 1$ [$mol \cdot l^{-1}$].

Problem 6.3

Oxolane is polymerized cationically in the bulk at 0 °C, using benzoyl hexafluoroantimonate as the initiator, under conditions such as to prevent transfer and termination reactions.

---

*) Tetramethyltetravinyl cyclotetrasiloxane:

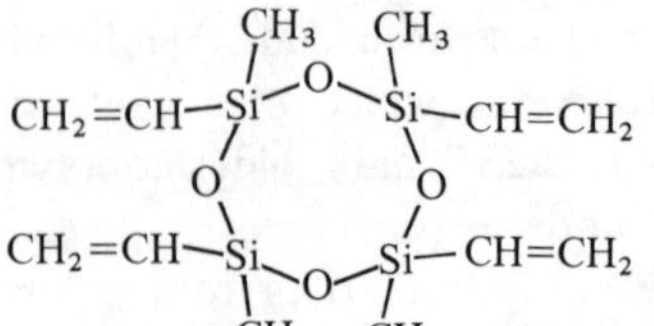

It is assumed that the propagation reaction is irreversible. What is the time needed to get a conversion degree π of 10%, when the amount of initiator used is such as to get a poly-THF of molecular weight $\overline{M}_n$ of 10000 at that precise conversion. Initiation is assumed to be fast and quantitative.

Data needed: Density of oxolane: 0.886; rate constant of propagation $k_p = 2 \cdot 10^{-3}$ $[l \cdot mol^{-1} \cdot s^{-1}]$ at 0 °C.

Problem 6.4

The reactivity ratios for isobutene (A) and isoprene (B) in CH_3Cl at — 103 °C, initiated by $AlCl_3$ are quoted as $r_a = 2.5$ and $r_b = 0.4$. Assume that the polymer formed initially is to have an isoprene content $F_b = 0.05$. What monomer ratio [A]/[B] should be used? What will be the integrated value $\langle F_b \rangle$ of the B-content of the polymer if the conversion is carried to $\pi = 0.80$?

Chapter 7
Special Features of Ionic and Pseudo-Ionic Polymerizations

7.1 Propagation-Depropagation Equilibria

7.1.1 General

It was already mentioned (chapters 5 and 6) that in various ionic polymerizations conversion is limited by a propagation-depropagation equilibrium. The monomer concentration remaining, when the equilibrium is attained, is strongly temperature dependent. Dainton and Ivin gave a thermodynamic explanation of the phenomena involved: The free energy change ΔG due to polymerization in such systems is small. Since ΔG is concentration dependent, there exists a concentration at which — for a given temperature — ΔG becomes equal to zero. Further net conversion of monomer to polymer becomes then impossible. Since the phenomenon originates from thermodynamics, it should be independent of the nature of the active site. The reversibility of the growth step is solely related to the gain of free energy which arises from the transformation of a monomer into its polymer.

As ionic "living" polymerizations involve long lasting active sites at chain end, they have been of great importance in the detailed investigation of the propagation-depropagation equilibria. Two cases have been studied very thoroughly: The cationic ring-opening polymerization of oxolane (THF), and the anionic polymerization of α-methylstyrene. A few other cases are listed in Table 7.1.

The occurrence of termination reactions makes similar investigations in free radical polymerizations very difficult, because once termination has occurred, depropagation as well as further growth are rendered impossible. Nevertheless, the equilibrium exists, and it was first detected in alternating free radical copolymerization of ethylene and SO_2.

Qualitatively it can be stated that the monomers concerned cannot be polymerized to quantitative yields in the temperature range in which the active sites would be efficient. Conversely, any site created on a polymer chain in that temperature range would result in unzipping (depropagation) with monomer release. This does not prevent such polymers from being used in manifold applications: In poly(methyl methacrylate), for instance, unzipping becomes appreciable only at temperatures far above the glass transition temperature of the polymer. In some cases, special care has to be taken to avoid active sites to be formed, as unzipping would be induced. In polyoxymethylene, for instance, the macromolecules are fitted with terminal ester functions to prevent depropagation.

Table 7.1

Monomers (or systems) that undergo reversible polymerizations*)

carbonyl monomers

$CH_2{=}O$	$CH_3{-}CH{=}O$	$CCl_3{-}CH{=}O$
formaldehyde	acetaldehyde	chloral

cyclic monomers

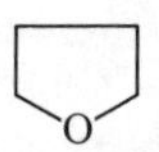

oxolane (THF)

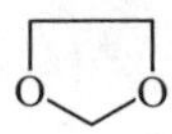

dioxolane

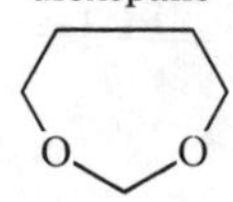

dioxepane

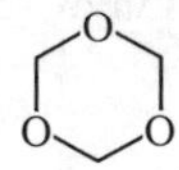

trioxane

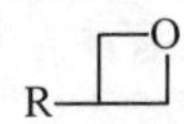

oxetanes

vinyl monomers

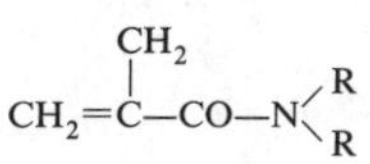

some substituted methacrylamides

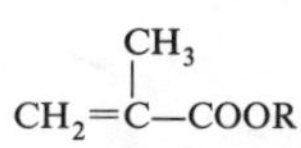

some methacrylic esters

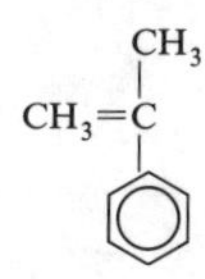

α-methylstyrene

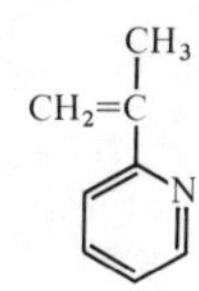

isopropenylpyridine

*) All polymerizations quoted come to equilibrium at a conversion π less than identically l, as will become apparent in equation (7.10) below. For most monomer-polymer systems π is very close to unity, i. e. 0.99 or more. Thus, in a strict sense all polymerizations are reversible. This list cites some of those which reverse (by temperature increase and having an active chain end) to give back monomer.

7.1.2 Kinetics and Thermodynamics of Reversible Propagation

Let us consider a chain with an active site (*) at chain end and assume that propagation is reversible (the i-th growth step being shown here):

$$\overset{i-1}{—\sim\sim—}M^* + M \underset{k_u}{\overset{k_p}{\rightleftarrows}} \overset{i}{—\sim\sim—}M^*$$

k_p and k_u are the rate constants of propagation and of depropagation (unzipping). At equilibrium, both reactions occur at the same rate, and we can write *approximately**):

$$-\frac{d[M]}{dt} = (k_p[M] - k_u)[M^*] = 0 \tag{7.1}$$

which implies that the equilibrium monomer concentration is given by:

$$[M]_e \cong \frac{k_u}{k_p} \tag{7.2}$$

*) See eq. (7.12).

This equilibrium monomer concentration has to be accounted for, when studying the overall kinetics of a polymerization. The rate equation can be expressed as:

$$-\frac{d[M]}{dt} = k_p[M^*]\cdot([M] - [M]_e) \quad (7.3)$$

which upon integration yields the modified first-order rate equation:

$$\ln\frac{[M]_0 - [M]_e}{[M] - [M]_e} = k_p[M^*]\,t \quad (7.4)$$

This equation was found to apply quite satisfactorily to living polymerizations, in which the concentration of active sites [M*] is known to stay constant over the entire process (Fig. 7.1).

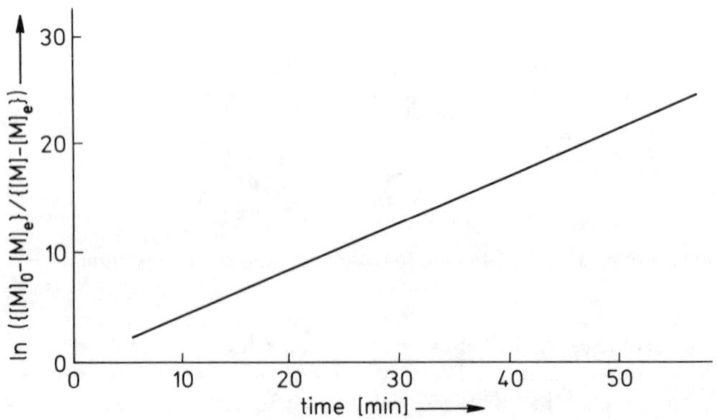

Fig. 7.1. Bulk polymerization of tetrahydrofurane; first order plot ($[M]_0$ = 12.3 $[mol \cdot l^{-1}]$; $[M]_e$ = 3 $[mol \cdot l^{-1}]$ at 30 °C; initiator: $\phi_2\,CH^+$, SbF_6^-; (I) = $2.8 \cdot 10^{-3}$ $[mol \cdot l^{-1}]$)

To examine the variation of $[M]_e$ as a function of temperature, it should be recalled that any rate constant can be expressed in the Arrhenius form as:

$$k_p = A_p \exp\left(-\frac{E_p}{RT}\right) \qquad k_u = A_u \exp\left(-\frac{E_u}{RT}\right) \quad (7.5)$$

where E_p and E_u are the activation energies of the propagation and of the depropagation reactions, and A_p and A_u are the corresponding frequency factors. The difference between the energies of the two opposite reactions is obviously the enthalpy change upon polymerization ΔH. On the other hand, the overall variation of the entropy during the polymerization is given by:

$$\Delta S = R\ln\frac{A_p[M_s]}{A_u} + R\ln\frac{[M]}{[M_s]} \quad (7.6)$$

wherein $[M_s]$ is the monomer concentration in the standard state, often taken as 1 $mol \cdot litre^{-1}$; the use of this factor is to keep the logarithmic concentration term dimensionless. This can also be expressed as:

$$\Delta S = \Delta S^0 + R \ln([\mathrm{M}]/[\mathrm{M_s}]) \tag{7.7}$$

where ΔS^0 is the molar standard entropy of reaction. Starting from equation (7.2) and inserting equations (7.5), one gets at the equilibrium:

$$\ln \frac{A_p [\mathrm{M}]_e}{A_u} = \frac{E_p - E_u}{RT} \tag{7.8}$$

and thus:

$$R \ln \frac{A_p}{A_u} [\mathrm{M_s}] + R \ln \frac{[\mathrm{M}]_e}{[\mathrm{M_s}]} = \frac{\Delta H}{T_c} \tag{7.9}$$

which expresses the relation between temperature and the equilibrium monomer concentration. Equation (7.9) can be set in more explicit presentations as:

$$T_c = \frac{\Delta H}{\Delta S^0 + R \ln([\mathrm{M}]_e/[\mathrm{M_s}])} \tag{7.10}$$

or as:

$$\ln([\mathrm{M}]_e/[\mathrm{M}]_s) = \frac{\Delta H}{RT_c} - \frac{\Delta S^0}{R} \tag{7.11}$$

The so-called *"ceiling-temperature"* T_c is the temperature above which further net conversion of monomer to polymer cannot take place beyond the equilibrium monomer concentration $[\mathrm{M}]_e$. Therefore a ceiling temperature always is referred to a stated monomer concentration. Generally, either one molar or bulk molar concentrations are taken as references for the definition of T_c.

ΔH being negative, the equilibrium monomer concentration increases as the temperature increases and a plot of $\ln[\mathrm{M}]_e$ versus $\frac{1}{T}$ yields a straight line with a negative slope ($\Delta H/R$) and an ordinate intercept equal to $-\Delta S^0/R$. For a given system no polymerization can ever occur when the value of $[\mathrm{M}]_e$ is higher than the molar concentration of the monomer in the bulk (Fig. 7.2).

7.1.3 Influence of the Reversibility of Chain Growth on the Molecular Weight Distribution

A polymerization reaction that does not involve spontaneous termination or transfer reactions leads to narrow *(Poisson-type)* molecular weight distributions, as it was shown in Fig. 5.2 of chapter 5. If it is assumed that the propagation steps ae reversible and that the polymerization is allowed to reach its equilibrium conversion degree, at the temperature chosen, the molecular weight distribution drastically changes. Let us consider that both rate constants k_p and k_u are independent of the chain length.

At equilibrium the monomer concentration and that of any "living" species containing i monomer units have become constant, which implies that:

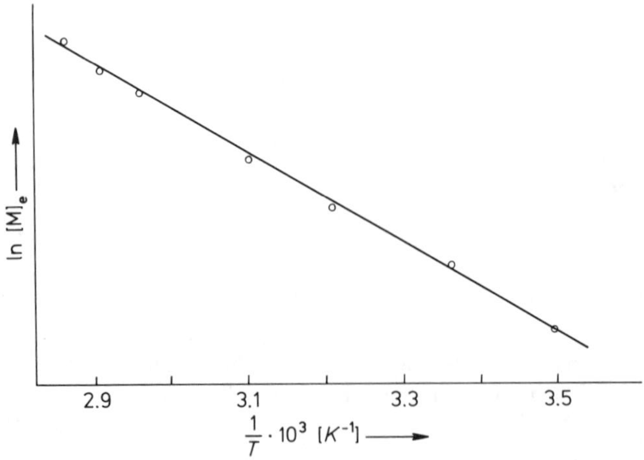

Fig. 7.2. Plot of the equilibrium monomer concentration in bulk polymerization of tetrahydrofurane versus $1/T$

$$\frac{d[M]}{dt} = 0 \quad \text{and} \quad \frac{d[M_1^*]}{dt} = 0$$

Equation (7.1) can be written in a more precise fashion:

$$-\frac{d[M]}{dt} = k_p[M]\left[\sum_{i=1}^{\infty}[M_i^*]\right] - k_u \sum_{i=2}^{\infty}[M_i^*] \tag{7.12}$$

because depropagation cannot occur on species M_1^* which results from the initiation process. The equilibrium monomer concentration can be calculated from Eq. (7.12) by setting $d[M]/dt = 0$*).

$$[M]_e = \frac{k_u}{k_p}\left[1 - \frac{[M_1^*]_e}{[M^*]_e}\right] \tag{7.13}$$

where [M*] stands for the total number of active sites of all degrees of polymerization:

$$[M^*] = \sum_{i=1}^{\infty}[M_i^*]$$

The equilibrium concentration of species $[M_1^*]$ can be calculated in the same manner, by setting $d[M_1^*]/dt = 0$:

$$\frac{d[M_1^*]}{dt} = -k_p[M_1^*][M] + k_u[M_2^*] = 0 \tag{7.14}$$

The first term of this expression refers to formation of species M_2^* upon addition of one monomer unit to M_1^*, and the second term concerns depropagation from M_2^* leading to M_1^*. At equilibrium, both terms are equal, which leads to:

*) Note that comparison with eq. (7.2) shows that $[M_1^*]_e/[M^*]_e \ll 1$.

$$[M_2^*]_e = \frac{k_p}{k_u}[M]_e[M_1^*]_e \tag{7.15}$$

For species i the equation comprises 4 terms, two of them refering to propagation and the two others to depropagation:

$$\frac{d[M_i^*]}{dt} = k_p[M]([M_{i-1}^*] - [M_i^*]) - k_u([M_i^*] - [M_{i+1}^\bullet]) = 0 \tag{7.16}$$

Since $k_p[M][M_{i-1}^*] = k_u[M_i^*]$ at equilibrium, as shown by recurrence from eq. (7.14), it is again possible to express $[M_{i+1}^*]$ as a function of $[M_i^*]$ by solving equation (7.16):

$$[M_{i+1}^*]_e = \frac{k_p}{k_u}[M]_e[M_i^*]_e$$

Introducing the monomer concentration as expressed in eq. (7.13), one gets:

$$[M_2^*]_e = [M_1^*]_e\left[1 - \frac{[M_1^*]_e}{[M^*]}\right]$$

$$[M_3^*]_e = [M_2^*]_e\left[1 - \frac{[M_{1e}^*]}{[M^*]}\right]$$

$$[M_i^*]_e = [M_{i-1}^*]_e\left[1 - \frac{[M_1^*]_e}{[M^*]}\right]$$

These equations can be put together in the following expressions for the molar (number) concentration of species M_i^* when the equilibrium has been reached.

$$[M_i^*]_e = [M_1^*]_e\left[1 - \frac{[M_1^*]_e}{[M^*]}\right]^{i-1} = [M_1^*]_e\left[\frac{k_p[M]_e}{k_u}\right]^{i-1} \tag{7.17}$$

For the mole fraction of "living" molecules exhibiting a polymerization degree *i* (at equilibrium) it can be written:

$$x_i = \frac{[M_i^*]_e}{[M^*]} = \frac{[M_1^*]_e}{[M^*]}\left(1 - \frac{[M_1^*]_e}{[M^*]}\right)^{i-1} \tag{7.18}$$

This expression is identical with Flory's most probable distribution (see Fig. 2.1):

$$x_i = p^{(i-1)} \cdot (1 - p)$$

with

$$p = \left(1 - \frac{[M_1^*]_e}{[M^*]}\right) = \frac{k_p}{k_u}[M]_e$$

Material balance of the monomer implies that:

$$[M]_0 = [M]_e + \sum_{i=1}^{\infty} i[M_i^*]_e \tag{7.19}$$

$[M]_0$ being the initial monomer concentration. The number average degree of polymerization is then given by:

$$\overline{P_n} = \frac{[M]_0 - [M]_e}{[M^*]} = \sum_{i=1}^{\infty} \frac{i\,[M_i^*]_e}{[M^*]} = (1-p)\sum_{i=1}^{\infty} i p^{(i-1)} = \frac{1}{1-p} \qquad (7.20)$$

Flory's most probable distribution implies that the polydispersity index tends towards two as p tends towards unity. The molecular weight distribution is thus far broader than it would be if the propagation step were irreversible*). Strategically this means that a living polymerization involving a propagation-depropagation equilibrium will have gradually its molecular weight distribution shift from Poisson's to the "most probable", at constant M_n. Thus to make block copolymers (see chapter 10) the second monomer should be added when conversion is not quite completed, if species of nearly constant molecular weight and composition are desired.

7.1.4 Consequences of the Reversibility of the Propagation Step

When a *"living"* polymer solution, already at equilibrium, is progressively heated to higher temperatures, depropagation tends to occur, and the polymer eventually vanishes entirely.

However, if the same operation is carried out with a *"dead"* polymer (after having "killed" the active sites) depropagation cannot occur unless new sites are created. New sites are easily achieved with poly(oxolane) (=poly-THF) as the presence of any protonating or alkylating agent suffices to create active oxonium sites on the chain from which depropagation will originate. In the case of poly(trioxane) (=poly-(methylene oxide) = "Delrin^{TM}") depropagation mainly starts from the chain ends, and for usage of this polymer it is required to have the chains *end-capped* with functions unable to yield active sites. This is why the poly(oxymethylenes), that are commercially available, are treated with acetic anhydride to provide terminal ester functions that prevent depropagation.

Vinyl polymers such as poly(α-methyl styrene) and, to a lesser extent, poly(methyl methacrylate) are subject to unzipping if a site has been created somewhere on the chain upon cleavage of a C—C bond. Irradiation with UV light (or with γ-rays or X-rays) can induce such chain scissions. Depending upon temperature, depropagation may occur, and monomer is recovered. Poly(ethylene sulfones) are specially sensitive to attack by various types of radiation, and the alternating copolymer is then rapidly degraded by unzipping. This property has been applied in manufacturing *printed circuits* in microelectronic components: A film of the polymer is irradiated (generally with an electron beam) through a mask, and the areas irradiated degrade readily, whereas those not irradiated are still coated with the film.

Depropagation is not to be compared with degradative processes, such as dehydrochlorination of poly(vinyl chloride), pyrolysis (yielding chemical fragments that are not

*) More elaborate equations are required if the initiation step is also reversible, whereupon initiator is not consumed entirely, either. The cationic *chloral* polymerization and the cationic ring opening polymerization of *lactams* are such cases.

monomers), or oxidation, which are chemically and thus thermodynamically unrelated.

Depropagation takes on a special significance when polymers are exposed to ionizing radiation (in nuclear reactors, high voltage equipment, space vehicles, etc.). Under such exposure, radicals are being introduced which can set up conditions for depropagation. What will happen depends then on whether the monomer released can diffuse away, or by remaining builds up concentration locally to restore equilibrium, and so on.

7.2 Polymerization Processes Involving Dipole-Dipole Interactions

A number of polymerization methods cannot be classified among the classical anionic or cationic processes. They are sometimes referred to as pseudo-ionic, more often as insertion polymerizations, because the incoming monomer is inserted into the active site by means of an SN_2 type reaction, as schematized below:

The term "insertion polymerization" is however quite unspecific, and provides no information on the actual polymerization mechanism. In this section, some of these polymerizations — often exhibiting "living" character — will be considered for their interest as such, and for their potential applications to the synthesis of well-defined polymer samples.

The basic concept underlying all these methods is that an electrophilic monomer will react with a nucleophilic site, or vice-versa. The site is not necessarily an ion pair, it can also be a dipolar group. Conversely, a monomer can be activated chemically and reacted as an ion onto a dipolar site. The main feature of any polymerization process is that after each growth step the site is restored at chain end.

7.2.1 "Living" Cationic Polymerization of Vinylic or Vinylenic Monomers

It was emphasized in chapter 6 that cationic polymerization of unsaturated monomers is severely impeded by transfer reactions, involving either the counter-ion or the monomer itself. The high reactivity of carbenium ions is responsible for these side reactions. As a consequence, no control of the molecular weight of the polymer can be exerted, even at very low temperatures, and no linear relation is observed between molecular weight and conversion.

It has been attempted in various ways to decrease the reactivity of the sites, in order to lower the probability of transfer. Higashimura was the first to find initiating systems involving sites of lower electrophilicity. They are composed of hydroiodic acid and iodine, and the initiation reaction can be written as:

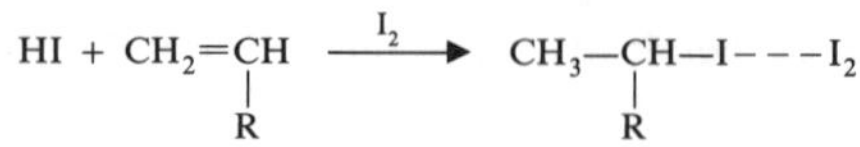

The site is not ionic. It is a carbon-iodine bond, activated by the presence of the electron withdrawing iodine. The propagation reaction involves insertion of the incoming monomer between the C—I bond:

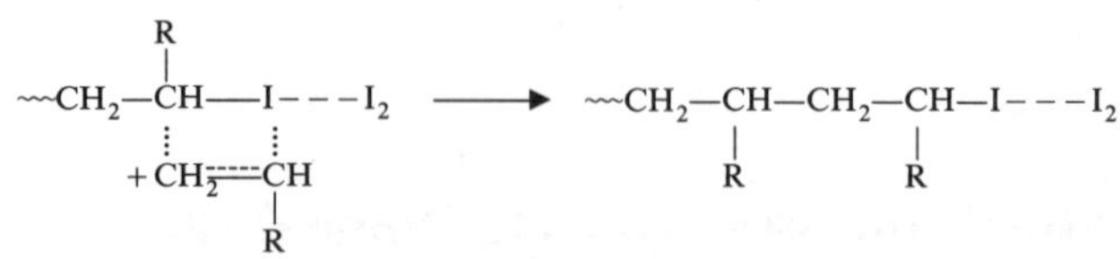

and the site is recovered at chain end after each insertion step.

The consequence of this reaction scheme is that:

- The molecular weight increases linearly with the degree of conversion.
- The final number average degree of polymerization is equal to the ratio of the molar concentrations of monomer and initiator.
- The polydispersity of the samples is low (<1.1).

These results imply that transfer and termination are negligible in such systems. Each site will lead to one macromolecule and the system can be quoted as "living". Functionalizations at chain end can be performed either by reacting the end-standing carbon-iodine bond with adequate nucleophiles, or by using a functional iodide to initiate the process. This method applies to a limited number of nucleophilic monomers, such as vinylethers.

Another such pseudo-cationic method, developed by Kennedy, applies to many monomers, including isobutene, which are of economic importance. The method involves active sites that are ester functions, activated by means of a Lewis acid.

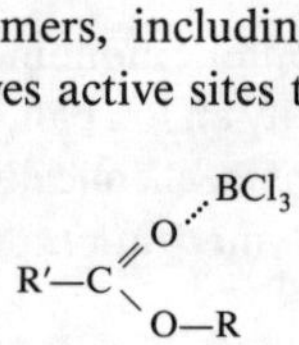

This activation results in the ability of the O—R bond of the ester to react with an electrophilic monomer, the site being restored at chain end after each monomer insertion:

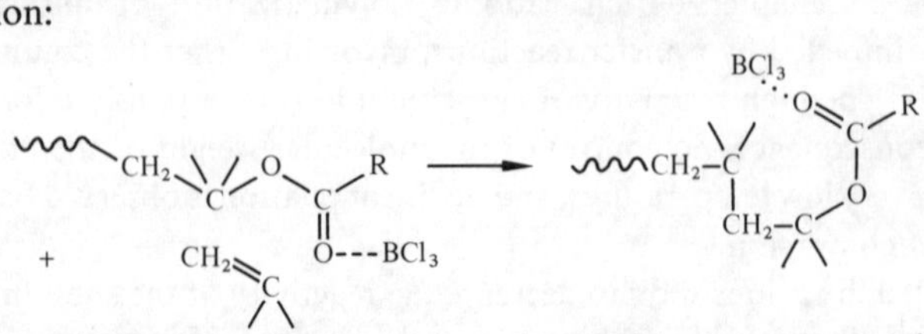

Here again, the polymerization goes to completion and the molecular weight increases linearly with conversion. Deactivation of the sites at chain end can be carried out in two ways: either by heating the mixture, or by adding a nucleophile, such as triethylamine, pyridine or methanol, but it produces only chloro-terminated polymer.

The initiating systems are constituted of an ester activated by BCl_3. A number of esters have been found efficient. Random copolymers of isobutene and some isoprene have also been made by this methods, yielding butyl rubber.

7.2.2 Group Transfer Polymerization (GTP)

GTP is another kind of insertion process recently developed by Webster and Sogah, particularly applicable to unsaturated monomers carrying electron withdrawing substituents, such as methacrylic or acrylic esters, the polymers of which, at moderate molecular weights (ca. 50000), serve as excellent components of automobile finishes, requiring far less solvent than used in conventional automotive lacquers.

The reaction is based upon Michael's addition. The initiating site is a silylketene acetal, which can be made from an ester and chlorotrimethylsilane, in the presence of a secondary lithium amide:

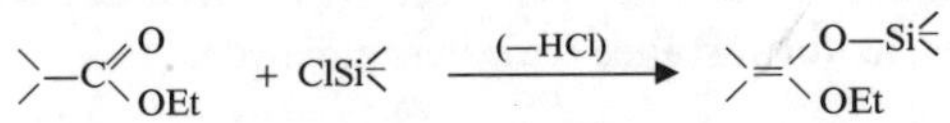

It can also be made by hydrosilylation from methylmethacrylate and trimethylsilane.

When a methacrylic ester reacts with the silylketene acetal, in the presence of a catalyst that activates the silylether bond, a dimer is formed. It still bears the silylketene acetal function, which stands available for further growth steps, illustrated by methyl methacrylate:

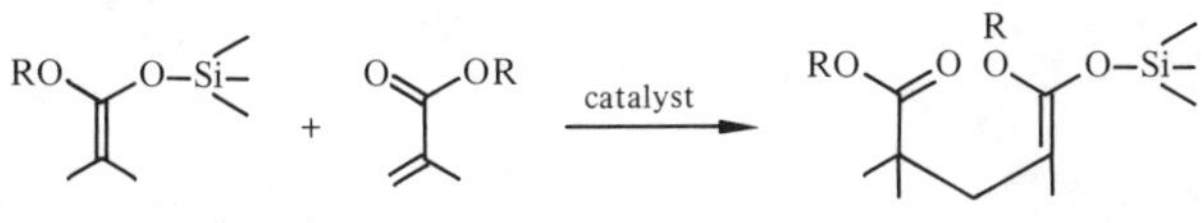

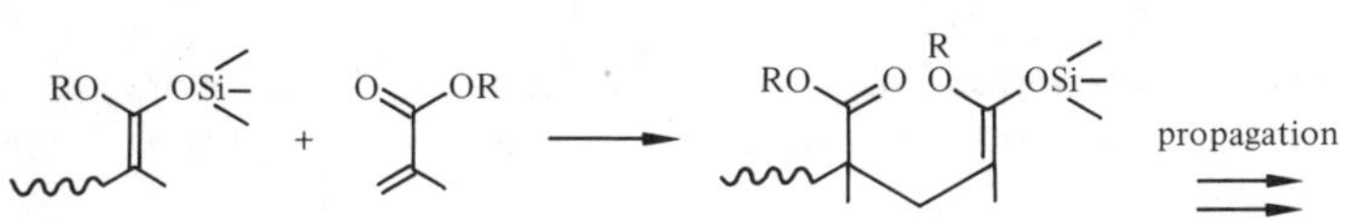

The rate constant for acrylate addition is so much greater than for methacrylic addition that, given an equimolar mixture of the two monomers, the chain adds almost exclusively the acrylate. To make a random copolymer, the ratio of MMA to MA has to be chosen very large. This is somewhat similar to the problem of making random styrene butadiene rubber via anionic solution polymerization, which implies the use of high styrene/butadiene mole ratios.

Allylic groups (e. g. in allyl methacrylate) are unreactive in GTP, as they are in anionic polymerization, and they can thus be introduced to small amounts into acrylate or methacrylate copolymers, if subsequent crosslinking (e. g. by air oxidation) is desired to produce networks via the allylic groups.

The activating catalysts needed for GTP include cyanide, fluoride and F_2H^- anions, the latter being the most efficient. In many cases tris[dimethyl-amino]sulfonium hydrodifluoride is used. This compound is made conveniently from SF_4 and dimethylamino-trimethylsilane according to:

$$SF_4 + 3\,Me_2N{-}SiMe_3 \longrightarrow (Me_2N)_3S^{\oplus}\ SiMe_3F_2^{\ominus}$$

$$\downarrow H_2O$$

$$(Me_2N)_3S^{\oplus}\ HF_2^{\ominus}$$

These polymerizations exhibit the characteristics of "living" processes: the polymerization degree of the polymer is determined by the monomer to initiator (silylketene acetal) mole ratio, and the molecular weight distribution within a sample is usually narrow. Once monomer conversion is completed, the silylketene acetal sites at chain ends remain active, and the addition of another suitable monomer may result in block copolymer formation. Functionalizations can be performed by induced deactivation of the silylketene acetal functions, as for instance with benzaldehyde:

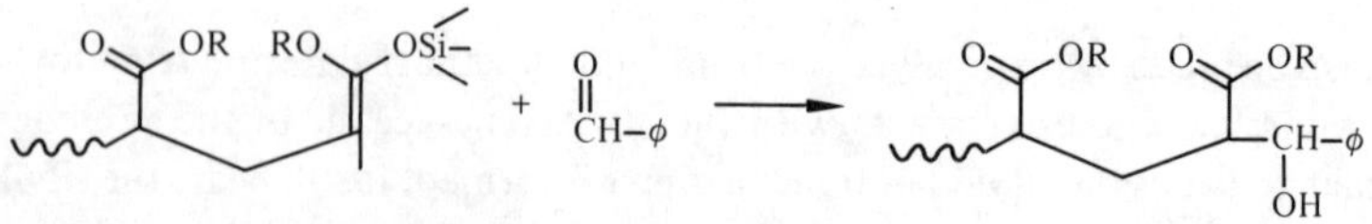

An alternate method of functionalization is to use a functional silylketene acetal as the initiator. Here is an example:

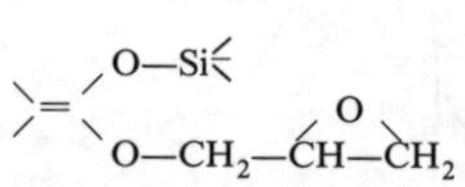

As in anionic polymerizations, monomers exhibiting labile hydrogens would deactivate the sites, and these proton donating sites should be blocked either by silylation or by means of dihydropyrane:

Ex. 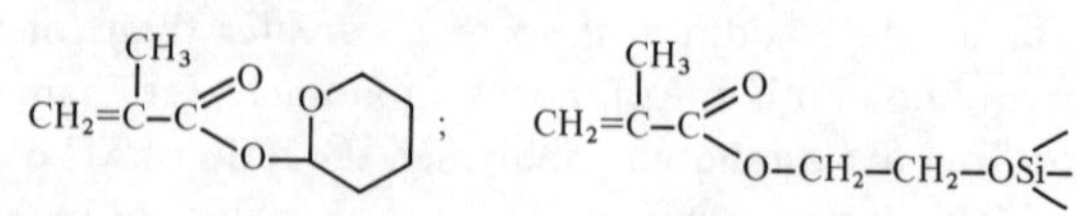

GTP exhibits thus many similarities with ionic polymerizations, although the mechanism is quite different. The propagating sites are not ion pairs, but rather strongly dipolar chemical functions.

In the "reversed" GTP, also referred to as Aldol-GTP, the same basic reaction (modified Michael's addition) is used to polymerize a silyl enol, the initiator being in that case an activated aldehyde function. The activation is provided by a Lewis acid such as $ZnBr_2$.

Initiation can be written as:

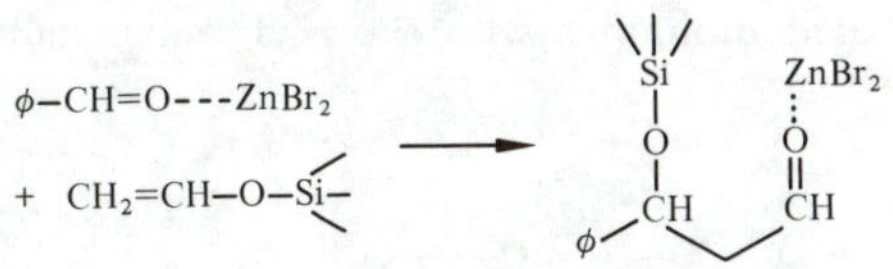

and propagation proceeds in the same way:

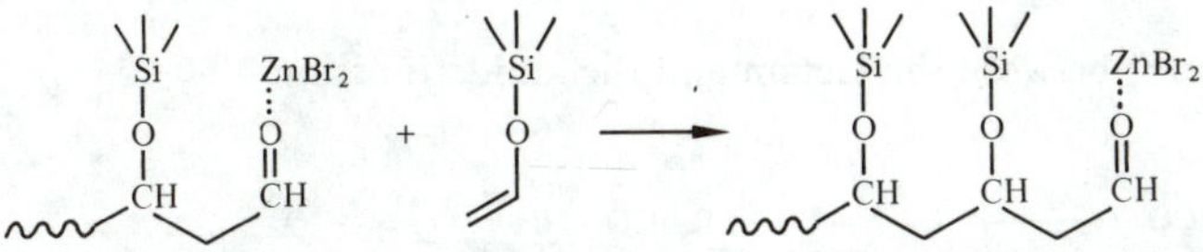

This polymerization method allows control of the molecular weight, and it yields samples of low polydispersity. Once the polymerization is completed, hydrolysis of the silylether functions is easy to perform, and it gives access to poly(vinylalcohol). This is the only direct method to make poly(vinylalcohol) samples of well-defined structure and chain length.

7.3 Activated Monomer Polymerizations

Some cyclic monomers undergo ring-opening polymerization according to an unusual mechanism. This is the case of lactams and of oxazolidine diones (Leuchs' anhydrides), the former being the precursor of some polyamides (nylon-6), the latters of polypeptides, often used as models for proteins. These so-called anionic "activated monomer" polymerizations have their counter-part in the cationic "activated monomer" polymerization of strained monomers such as oxiranes.

7.3.1 Anionic "Activated Monomer" Polymerizations

The basic feature of these polymerizations is that the cyclic monomer is deprotonated whereby it becomes a nucleophile. It then reacts with the dipolar function located at chain end of the growing polymer and gets incorporated. Protonation of the active site subsequently takes place at the expense of another monomer molecule, which is then ready to react in turn. The driving force of such a process is the larger acidity of the cyclic compound, as compared to the corresponding open chain function (lactam versus open chain amide). This explains why it is the incoming monomer that carries the charge, instead of the last unit of the polymer chain. The reaction steps of the "activated monomer" polymerization of a lactam are schematized below:

- Deprotonation of the monomer

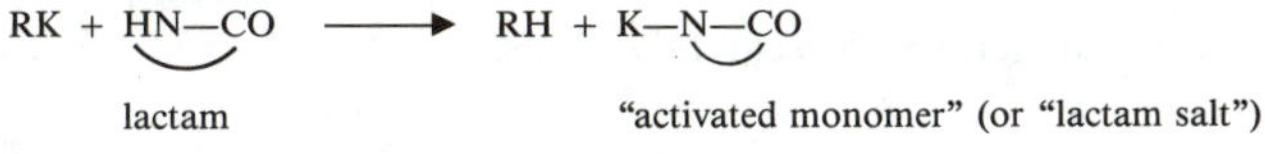

- Initiation: Reaction of the activated monomer with the acyl lactam added as "promoter"

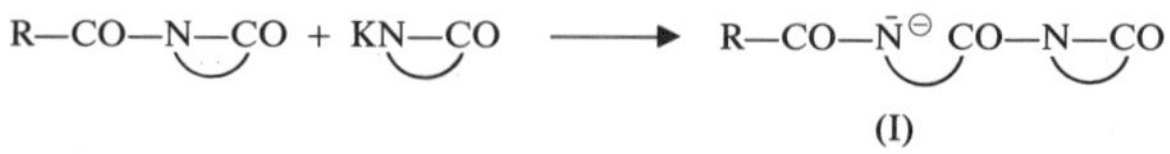

- Proton exchange between the lactam and the amide function

- Propagation proceeds exactly according to the same mechanism, and each step of growth is followed by a proton transfer reaction, which yields a new "activated monomer" unit to be added subsequently.

It should be pointed out that the presence of the promoter (acyl lactam) is not absolutely necessary, since it also can be formed slowly upon dimerization of the lactam.

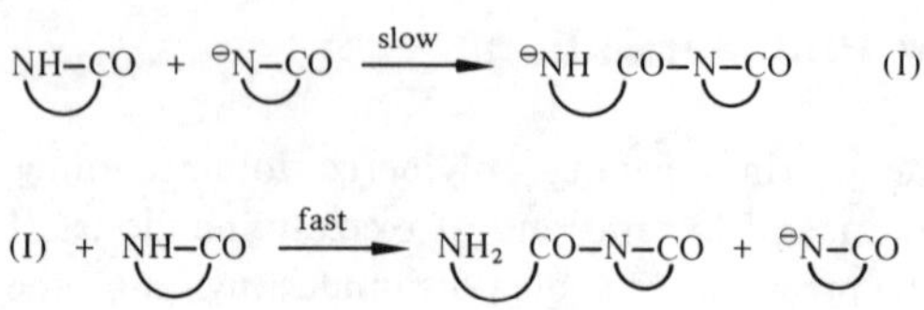

However, if fast initiation is wanted, to allow some control of the molecular weight of the polyamide to be formed, the addition of a known amount of acyllactam (the promoter) at the onset of the process is required.

Such polymerizations can take place in the bulk or in solution in aprotic solvents. The polymers obtained generally exhibit narrow molecular weight distributions — even though the monomer involves a proton donating function — because of that particular mechanism of chain growth.

A very similar mechanism is obeyed in the case of the oxazolidine diones, cyclic compounds that result from the reaction of phosgene with an amino acid:

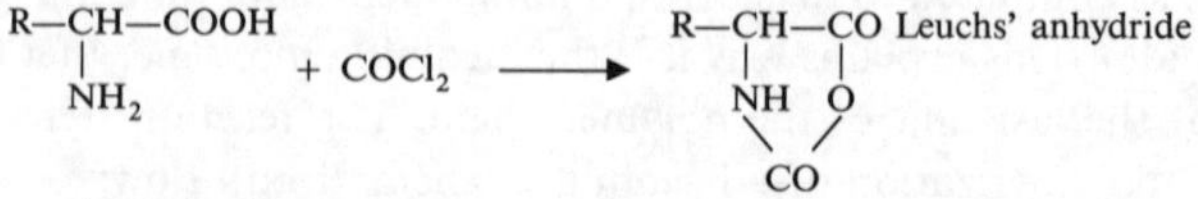

The only difference between Leuchs' anhydrides and lactams is that after each step of growth a carbamate function is formed, that undergoes decarboxylation first; subsequently the proton transfer reaction can take place, as in the mechanism described above:

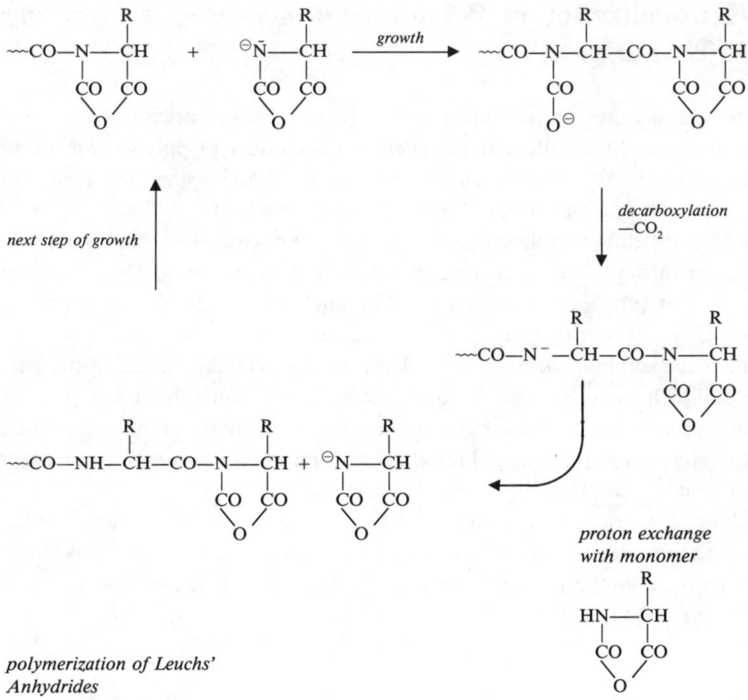

polymerization of Leuchs' Anhydrides

7.3.2 Cationic "Activated Monomer" Polymerizations

It has been shown above that the anionic polymerization of lactams and of oxazolidinediones follows an uncommon pathway. The monomer carries the negative charge (lactam salt) and each propagation step is a nucleophilic attack on the end-standing acyllactam function, followed by proton transfer.

This "activated monomer" polymerization has its equivalent in cationic ring-opening polymerization, as shown by Penczek: The polymerization of an oxirane can be carried out in the presence of some alcohol and of a trace of a strong protonic acid. Propagation then proceeds via the attack of the alcohol function sitting at chain end onto the protonated monomer: The oxirane cycle is opened, and the alcohol function is reformed.

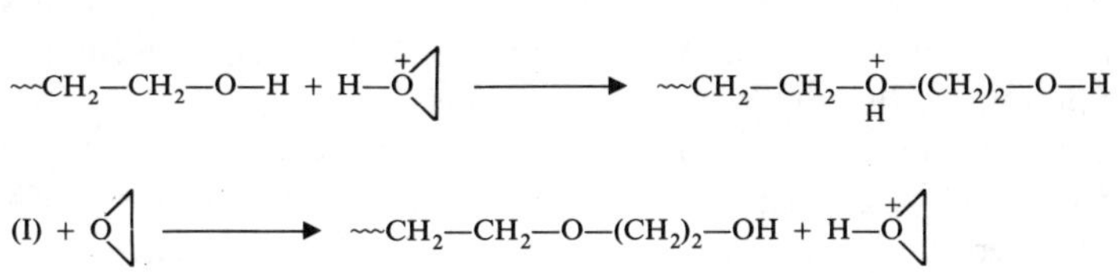

(I)

7.4 Rationalization of Polymerization Processes Involving Ions or Dipoles

Classical ionic polymerizations involve active sites which are ion pairs (free ions to a small extent). On the other hand, there are a number of polymerization methods involving no ionic sites, but instead strong dipoles as active sites. The mechanisms of these "insertion polymerizations" are quite diversified.

The scheme below, established and discussed by Saegusa, is an attempt to rationalize all polymerization processes involving either ionic or dipolar species. Their common feature is that propagation always involves nucleophilic attack of the site onto the monomer, or vice-versa.

Since a nucleophile can only react with an electrophile, an active site exhibiting nucleophilic character will be fit to attack an electrophilic monomer (e. g. carrying electron withdrawing substituents). Conversely, a nucleophilic monomer will react with an active site exhibiting electrophilic character. However, the site at chain end as well as the monomer can be either a dipole or an ion pair.

In classical anionic vinyl polymerizations the active site is a nucleophilic anion (carbanion or oxanion) fitted with a metal ion for electrical neutrality. It is able to react with a covalent monomer exhibiting an electrophilic character, due to its electron withdrawing substituent R.

$$\sim CH_2-\underset{\underset{R}{|}}{\bar{C}H}\,K^+ + CH_2{=}\underset{\underset{R}{|}}{CH} \longrightarrow \sim CH_2-\underset{\underset{R}{|}}{CH}-CH_2-\underset{\underset{R}{|}}{CH}^-\,K^+$$

In classical cationic ring-opening polymerizations propagation proceeds by nucleophilic attack of the cyclic covalent monomer onto the electron deficient carbon atom vicinal to the oxonium ion.

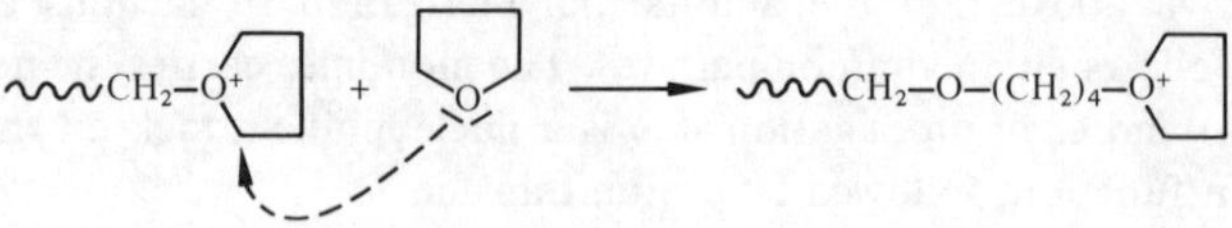

The so-called anionic "activated monomer" polymerization of lactams is an example of a growth reaction involving a dipolar site and an "ionic" monomer: the "lactam salt" (or deprotonated lactam) is a strong nucleophile, which reacts on the acyllactam function sitting at chain end. This latter function is restored at chain end. A new lactam salt is formed upon proton transfer (see 7.3.1).

The reverse case, i. e. cationic "activated monomer" polymerization has been mentioned above. It involves the attack of the nucleophilic alcohol function at chain end on the protonated (ionic) heterocyclic monomer.

Oxazoline is a monomer of particular interest in so far as, depending upon the type of initiator used, the mechanism of chain growth can be purely cationic, or involve dipole-dipole interactions.

Saegusa's Rationalization of ionic and dipolar polymerizations

Monomer / Active site	Nucleophilic		Electrophilic	
	covalent	ionic	covalent	ionic
Electrophile Ionic	Cationic polymerization Ex: THF + $CF_3SO_3CH_3$ isobutene + BF_3, H_2O	Zwitterionic Polymerization		
Covalent (Dipole)	Cationic Insertion Polymerization Ex: oxazoline + MeI isobutene + Ester (BCl_3)	Anionic "Activated Monomer" Polymerization Ex: Lactams + acyllactames		
Nucleophile Ionic			Anionic Polymerization Ex: Styrene + BuLi oxirane + KOR	Zwitterionic Polymerization
Covalent (Dipole)			Anionic Insertion Polymerization Ex: Ziegler-Natta Polym. GTP	Cationic "Actioned Monomer" Polymerization Ex: THF + H^+ + ROH

If it is reacted with an ionogenic compound such as methyl tosylate or methyl triflate*⁾, the polymerization is truly cationic:

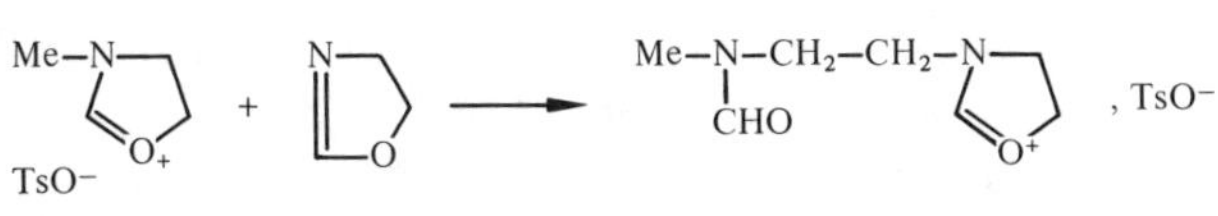

If it is reacted with a non-ionogenic alkyl halide, such as methyl iodide, the polymerization is slower and proceeds via a quite different mechanism, involving propagation reactions between dipoles of opposite signs:

*⁾ Tosylate = p-methylphenyl sulfonate
Triflate = trifluoromethyl sulfonate

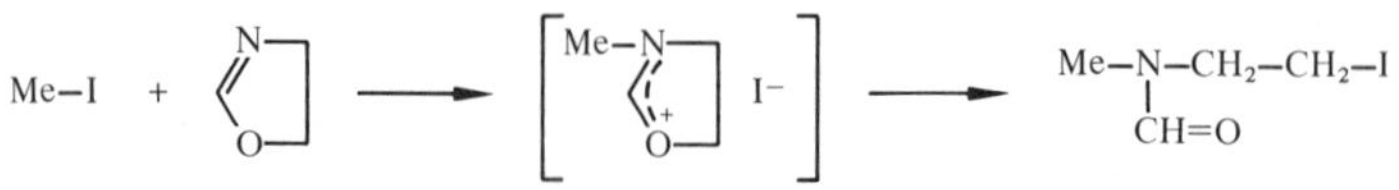

After each growth step the carbon-iodine bond is restored at chain end, ready for the next step.

The anionic counter-part of this type of polymerization involving dipoles is group transfer polymerization (GTP). An electrophilic covalent monomer is polymerized by means of active sites that are also dipolar and covalent: the silylketene functions. Again, the propagation occurs by insertion of the monomer, the silylketene site being restored at chain end.

Even the Ziegler-Natta polymerization of ethylene has been sometimes considered as a dipole-dipole insertion process: The π electrons of ethylene have a nucleophilic character, and tend to occupy the vacant orbital on the transition metal. Subsequently, the monomer is incorporated in the chain and the new vacant site can be saturated with a new monomer molecule. It thus can be considered as a coordinated dipole-dipole reaction on the transition metal complex.

In Saegusa's scheme there is still one case to be considered: The case of an ionic monomer reacting with an active site that is itself ionic. This could be exemplified by the so-called zwitterionic copolymerization. A number of such reactions have been described. A characteristic example of this kind is the oxazoline-propiolactone system. Oxazoline is a nucleophile and propiolactone is an electrophile. Upon reaction between these two species a zwitterionic dimer is formed:

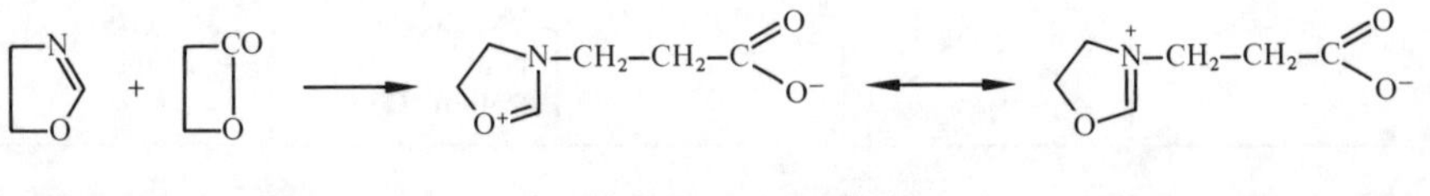

which finally leads to an alternating polymer, whose repeat unit is:

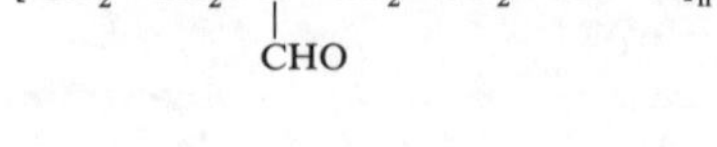

A characteristic feature of this polymerization is that polymeric zwitterions can react together, as in the case of step growth polymerization. A number of systems of this kind have been investigated. Usually polymerization starts upon heating a mixture of the two monomers. These spontaneous polymerizations are often referred to as "no-catalyst" polymerization.

Noteworthy is the fact that acrylic acid can play the role of the lactone in the alternating copolymerization quoted above. This is one of the few cases known in which a copolymerization process can involve a heterocyclic monomer and an unsaturated compound. The zwitterionic dimer is formed upon dipolar addition of acrylic acid to the heterocycle, followed by proton transfer:

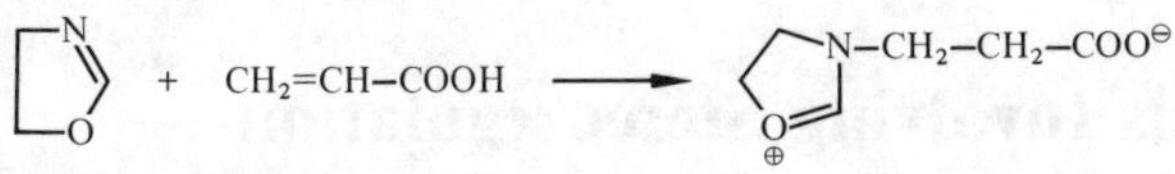

Other nucleophilic monomers include dihydrooxazine (I), azetidine (II), ethylenephosphonite (III). The most efficient electrophilic monomers are maleic anhydride, acrylic acid, acrylamide. Even hydroxyethyl methacrylate can yield zwitterionic alternating copolymerization with dihydro oxazine derivatives.

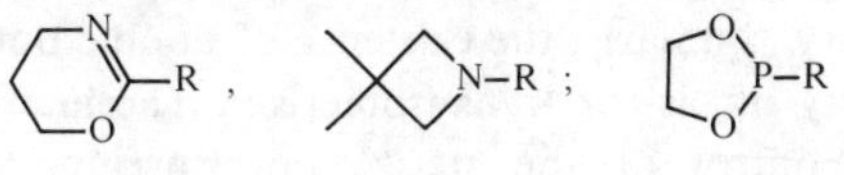

(I) (II) (III)

Problem 7.1

The thermodynamical parameters of the ring opening polymerization of oxolane (THF) are given. Calculate the equilibrium monomer concentrations at 0 °*C*, 20 °*C* and at 40 °*C*. ($\Delta H = -5.6$ [K *cal/mol*]; $\Delta S = -20.6$ [$cal \cdot mol^{-1} \cdot K^{-1}$]).

Problem 7.2

The cationic polymerization of oxolane involves neither transfer nor spontaneous termination. The propagation reaction is however reversible and the equilibrium monomer concentration is 2.3 [$mol \cdot l^{-1}$] at 20 °*C*. The polymerization is carried out in the bulk with benzoyl hexafluoroantimonate as initiator, the concentration of the latter being 3.10^{-3} [$mol \cdot l^{-1}$]. Initiation is supposedly instantaneous. The rate constant of propagation is $k_p = 30.10^{-3}$ [$l \cdot mol^{-1} \cdot s^{-1}$] at 20 °*C*, and the density of oxolane at 20 °*C* is 0.866 [$g \cdot ml^{-1}$].

a) Calculate the initial rate of polymerization.
b) What conversion π has been attained after 15 minutes of reaction?
c) Calculate the molecular weight of the polymer formed.
d) What can be said about the polydispersity of the polymer obtained under such conditions: $[M] \gg [M]_e$?
e) Calculate the time necessary to reach the propagation-depropagation equilibrium within 1%.
f) What would be the molecular weight and polydispersity ratio of polyoxolane obtained as in *e)*?

Chapter 8
Polymerizations Involving Stereoregulation

8.1 Introduction

It was mentioned in chapter 1 that every substituted carbon atom on a polyvinylic chain is a potential center of asymmetry. Similarly, the existence of double bonds in a polydiene chain grants the possibility of cis- or trans-isomerism. The interest of polymerization processes involving control of the steric conformation of the asymmetric carbon atoms — or of the cis/trans conformation of the double bonds — arises from the fact that the thermomechanical properties of stereoregular polymers (in most cases isotactic) are often quite different from those of atactic polymers. This can be illustrated by two examples:

● Atactic polypropylene is rather soluble in various solvents and its glass transition temperature is too low (70 °C) to allow any application. Isotactic polypropylene is highly crystalline, its melting point is close to 170 °C, and it is insoluble in all common solvents at temperatures up to 100 °C. Syndiotactic polypropylene is also crystalline, but its melting temperature is around 130 °C.

● Polybutadienes exhibit glass transitions at very low temperatures (~ −100 °C), but their behaviour strongly depends upon their chain conformations: All-trans polybutadiene melts at 141 °C and is consequently a tough material at room temperature, whereas the all-cis polymer melts at +1 °C, and is a viscous liquid, well fitted for applications as elastomers after it has been vulcanized by building chemical bridges between individual chains.

Conversely, in many cases, especially for common polymers such as polystyrene, poly(vinyl chloride), poly(methyl methacrylate), polyvinylacetate, stereoregulation has no major interest, as the atactic amorphous polymer already exhibits reasonably good properties and is suitable for applications.

Polymerizations involving stereocontrol do not yield fully isotactic (or all-cis) polymers. The chain contains a certain proportion of units that have escaped stereoplacement. However, the content of units of a given type often exceeds 90%, which is enough to allow crystallization and to enhance the mechanical behaviour of polymers such as polypropylene.

In free radical polymerizations the addition of a monomer to the radical site does not involve steric control. Differences in the rate of addition, arising from interactions between the substituent of the site and that of the incoming monomer, may slightly favour one type of addition (say R) with respect to the other (S), but this effect is of minor importance. As an example it can be stated that in PMMA or in PVC obtained by free radical polymerizations the racemic diads are slightly more numerous than the meso diads, implying some predominence of syndiotactic triads. Small differences in the activation energies corresponding to the formation of meso and of racemic diads have been observed. Consequently, the syndiotactic diad content is enhanced if the polymerization is carried out at low temperature.

Ionic polymerizations involving free ions (sp^2) are likewise unable to provide stereocontrol. They also usually yield atactic polymers, even though the meso diads are in slight excess.

The only efficient way to enhance stereospecific placement of the monomer units along the chain is to have the monomers oriented *prior* to their addition to the growing site. In other words, propagation should occur by insertion of the monomer between the chain and the active site rather than by mere addition to a loose chain end.

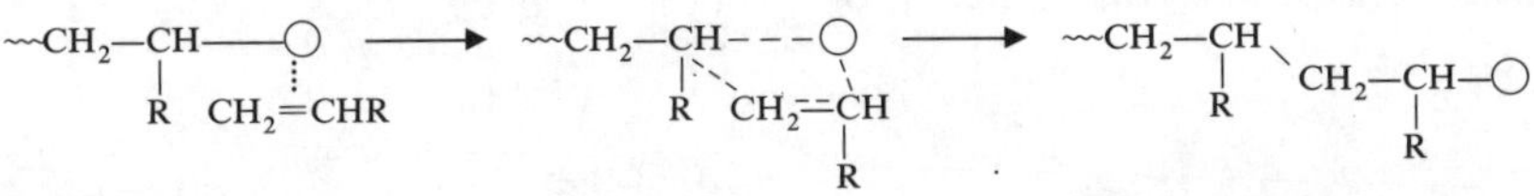

orientation of the incoming monomer with respect to the active site — 4-centered transition state — insertion of the monomer into the C-active center bond (implying restoration of the active site at the chain end)

The methods of polymerization involving stereocontrol have expanded tremendously after the pioneering work of Ziegler and Natta in the fifties. They are presently of common use in industry: most of the polyethylene and all of the polypropylene are made by such stereospecific polymerization processes. They are also commonly used to synthesize polydienes exhibiting high 1,4-cis unit contents, used in the rubber industry. It is interesting to note, however, that the mechanisms of such polymerization reactions are still controversial.

8.2 Ziegler-Natta Polymerizations

It has been the merit of Ziegler and Natta to develop efficient systems that provide coordination (and orientation) of the monomer with respect to the growing chain end, prior to its insertion. These initiating systems are in most cases heterogeneous, but still of high efficiency. Convenient stereoregulating processes to polymerize ethylene, propylene, and dienes have been developed, though the use of these methods extends far beyond these monomers, as shown on Table 8.1. Stereospecific polymerizations do not apply to highly polar monomers, though.

The catalytic systems developed by Ziegler and Natta usually comprise two constituents: an organometallic compound involving a metal cation of groups I to III (such as triethyl aluminum) and a transition metal compound arising from groups IV to VIII in the periodic chart of elements (such as titanium trichloride).

Triethyl aluminum is in itself a weak anionic initiator. When reacted with ethylene it usually yields unsaturated oligomers such as hexene, octene, decene. Monomer is inserted between the Al—C bond, but chain growth competes with transer which implies hydride shift, yielding the alkene and an Al—H bond. In turn, the latter is able to insert ethylene. Hydrogen is also a very active transfer agent in these reactions. The ratio of propagation to transfer can be varied widely, reaching from the dimer to low molecular weight linear polyethylene, depending upon the reaction conditions chosen.

Table 8.1 List of monomers which can be polymerized by Ziegler-Natta type initiators

ethylene	$H_2C=CH_2$
propene	$H_2C=CH-CH_3$
butene	$H_2C=CH-CH_2-CH_3$
styrene and some substituted styrenes	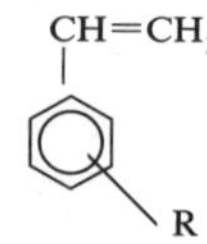
vinyl ethers	$CH_2=CH-OR$
butadiene, isoprene, and various other dienes	$H_2C=CH-CH=CH_2$ $H_2C=C(CH_3)-CH=CH_2$
cycloalkenes: C_3 and C_4	
methylene cycloalcanes and spiro | 2.2 | pentane	
acetylene, phenylacetylene	$CH\equiv CH$ 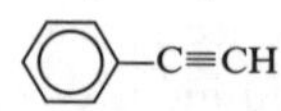

However, these polymerizations are slow and the polymers obtained are of no particular interest for applications.

Insertion (propagation)

$$R_2Al-Et + C_2H_4 \longrightarrow R_2Al-CH_2CH_2-Et$$

$$R_2Al-CH_2CH_2-Et + (n-1)C_2H_4 \longrightarrow R_2Al-(CH_2CH_2)_n-Et$$

Transfer reactions

$$R_2Al-(CH_2CH_2)_n-Et \longrightarrow R_2AlH + CH_2=CH-(CH_2CH_2)_{n-1}-Et$$

$$R_2Al-(CH_2CH_2)_n-Et + C_2H_4 \longrightarrow R_2AlCH_2CH_3 + CH_2=CH-(CH_2CH_2)_{n-1}-Et$$

Titanium halides are electron deficient and can be considered as weak Lewis acids. However, they usually do not induce cationic polymerizations.

Quite different are the binary initiating systems constituted of triethyl aluminum and a titanium halide. Upon mixing these two constituents in an inert solvent fine crystalline particles are formed, which were shown to efficiently initiate polymerizations of various monomers. The growth reaction often involves steric control of the monomer addition, although not always. Even under the most favorable conditions, however, the proportion of stereospecific growth seldom approaches unity: some atactic polymer is formed simultaneously in the reaction mixture, in addition to the highly crystalline stereoregular polymer. The former can be removed easily owing to its much higher solubility.

A large family of binary initiating systems capable of yielding stereospecific growth have been developed over the past 25 years.

- The metal organic constituents include aluminum, zinc, magnesium, beryllium, and lithium derivatives. Aluminum compounds are preferred because of their high activity. The $AlEt_3$ can be replaced in some instances by $AlEt_2Cl$ or by $AlEtCl_2$, which are less flammable and easier to handle. The organic moiety can also be varied: Ethyl, methyl, isopropyl, t-butyl, benzyl derivatives have been used.
- The transition metal compound also offers a wide choice: Titanium, vanadium, zirconium, molybdenum, chromium, and tungsten compounds have been tested. Halides (chlorides or iodides) are most commonly employed, but various other derivatives have been cited in the literature, such as alkoxides, acetylacetonates, or partially alkylated or cyclopentadienylated compounds.

Each individual system can be characterized by its efficiency — defined as the number of grams of polymer produced per gram of transition metal — and by its ability to sterically control the growth reaction, evaluated from its iso- and syndio-triad content, or in a more qualitative fashion by the degree of crystallinity of the polymer formed. Very high efficiencies have been attained with various systems, of the order of one ton polymer produced with one gram of Ti compound. In such cases removal of the initiator constituents from the polymer is unnecessary.

Factors influencing the efficiency of a binary initiating system and the degree of stereocontrol it can provide are quite numerous, namely:

- The nature of the transition metal, and its oxidation degree. Titanium and vanadium derivatives are quoted among the best, but many others are efficient. The transition metals should not be at their highest oxidation degree.

Ti^{3+} is far better than Ti^{4+}, although $TiCl_4$ is used frequently as the transition metal constituent: $TiCl_3$ then results from an in *situ* reduction according to:

$$TiCl_4 + AlEt_3 \longrightarrow TiCl_3 + Et_2AlCl + Et^\bullet$$

- The ligands of the transition metal compound also play a role. In most cases halides are used, or alkoxide derivatives. As a noteworthy exception one should quote the soluble catalyst systems constituted of dicyclopentadienyl titanium dichloride and a trialkylaluminum, which allow, in some instances, to make highly syndiotactic polymers in homogeneous phase.
- The crystal structure and the particle size of the transition metal halide also influence the process markedly. The hexagonal α-$TiCl_3$ and the cubic γ-$TiCl_3$ are far more efficient in terms of yield and of stereocontrol than the brown β-$TiCl_3$, which is formed spontaneously upon *in situ* reduction of $TiCl_4$ at low temperature. The efficiency of the catalyst system also increases with increasing specific surface of the particles. Grinding is essential to get good polymerization rates, as the number of active sites is greatly enhanced.
- The nature of the metalorganic compound also plays an important role. Although many such compounds have been tested, the aluminum compounds are usually preferred for their high stereoregulating activity. It should be pointed out that trialkyl aluminums exhibit a strong tendency to associate to dimers, especially if R is methyl

or ethyl. If R is either longer or bulkier, the efficiency of the catalyst system is lower, and so is the ability of the system to priviledge isotactic growth.

● The mole ratio of the two constituents also plays an essential role in the process. In most cases there is an optimum value to this ratio, and if the proportions are shifted away from this value, the yield and the stereospecificity of the polymerization are decreasing.

● The addition of small amounts of an electron donating compound, such as an amine or an ether, tends in many cases to increase both the efficiency and the stereospecificity of the polymerization.

8.3 Mechanisms Involved in Ziegler-Natta Polymerizations

Upon mixing the two constituents of the Ziegler-Natta system a complex is formed, involving electron deficient bonds. Its structure can be schematized as:

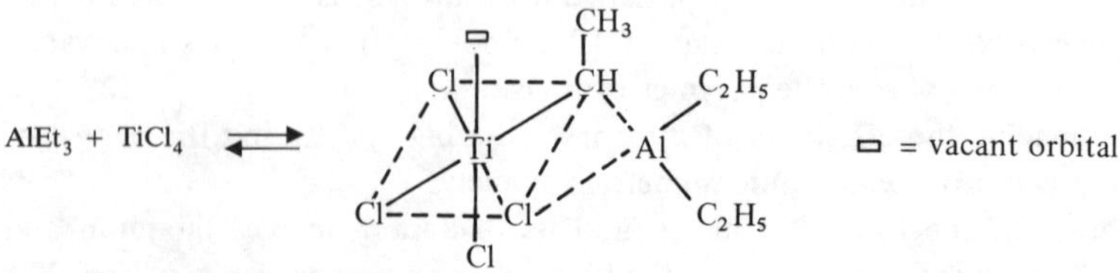

The *octahedral coordination* of the titanium is of great importance. In most cases this complex is insoluble, and it is dispersed as colloïdal particles in the reaction mixture. There are, however, cases in which the colored complex formed is soluble, as in the example below:

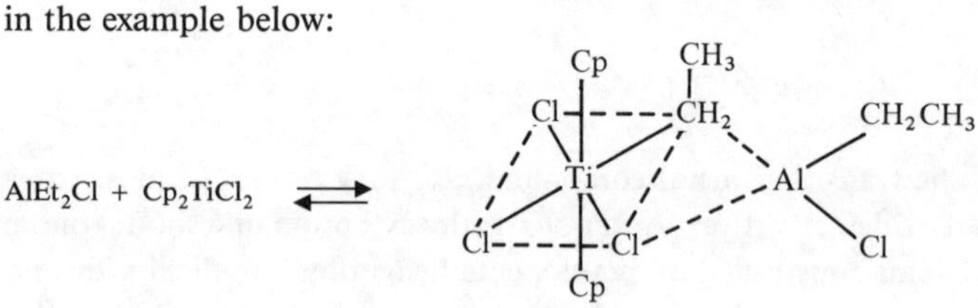

where Cp = cyclopentadienyl =

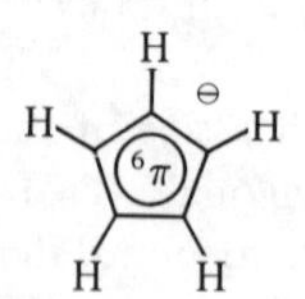

This complex formation often involves *alkyl migration.* It was established that activation of the polymerization initiator is due to alkylation of the transition metal (here titanium), as it is shown in the second example. The structures of these complexes have not been fully elucidated yet.

The metal organic compound acts as a reducing agent with respect to the transition metal, whereby radicals are formed. But these radicals do not play a role in further reactions.

$$TiCl_4 + AlR_3 \longrightarrow TiCl_3 + AlR_2Cl + R^{\bullet}$$

$$TiCl_4 + AlR_2Cl \longrightarrow TiCl_3 + AlRCl_2 + R^{\bullet}$$

It should be remembered that the coordination number of aluminium is 4 and that transition metals have coordination numbers ranging from 4 to 6. In the latter case they are octahedral in shape, and this seems to be favorable.

The feature that enables stereospecific growth in these Ziegler-Natta polymerizations is the ability of the transition metal ion (which is *electron deficient*) to coordinate the incoming monomer, and to be simultaneously connected with the end of the growing macromolecule by a σ-bond. Whether the transition metal is solely responsible for coordination with the monomer *and* for linking with the chain, or whether it is in fact the bimetallic complex that does it, has been subject to a long debate. Presently the "monometallic mechanism" is preferred to the "bimetallic model".

Among the drawbacks of the latter mechanism one should quote the impossibility to account for syndiotactic growth, that implies some kind of alternation in the placement (and coordination) of the incoming monomer with respect to the growing chain end.

The monometallic insertion mechanism of the monomer between transition metal and chain end can be schematized as:

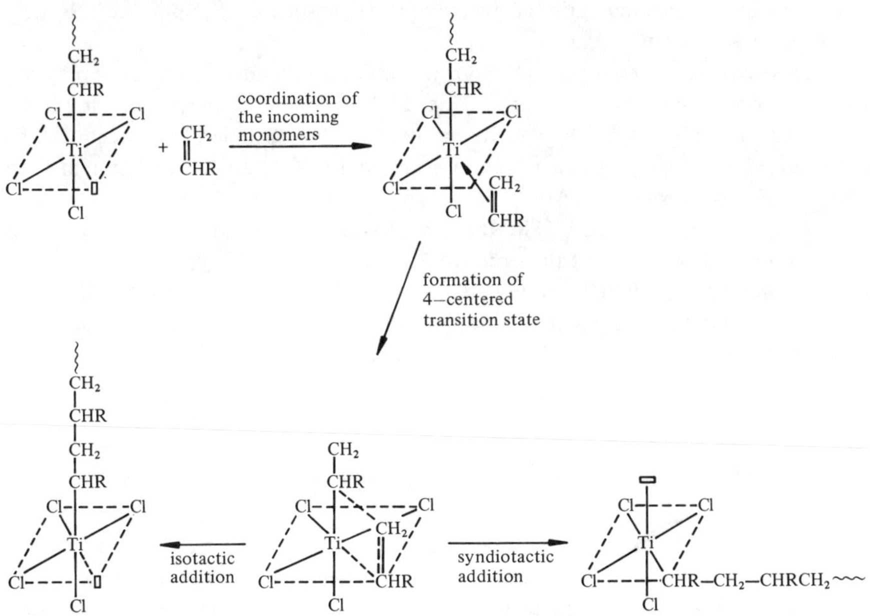

First, an electron deficient bond is formed between the monomer and Ti at a vacant orbital. Thereafter a *four centered* transition state is established, which implies cis-opening of the double bond when the monomer is finally inserted into the chain. Once that has occurred, two possibilities are offered:

Either the position of the growing chain with respect to the transition metal ion is the same as initially, and the vacant orbital is also restored at the same location; then the process can be repeated and the system is suited for *isotactic* growth of the polymer chain.

Or the chain is linked to the transition metal ion via the orbital that was previously vacant, and the orbital to which it was linked before has become vacant. The next monomer to be added will then be placed in the reverse steric position with respect to the chain, which provides *syndiotactic* growth.

The nature of the linkage between the catalyst site and the monomer has been subject to controversies. The current trend is to assume a coordinated anionic link between the chain end and the active site. This should be the case for ethylene, propylene, styrene, and dienes. This assumption accounts for the fact that metalorganics such as BuLi can be highly activated by addition of $TiCl_4$ or VCl_3, and that the efficiency of the systems decreases as one goes from ethylene to propylene and to butene (a monomer more sensitive to cationic polymerization).

However, monomers such as isobutylvinylether — known to polymerize cationically — can also undergo stereospecific polymerization by means of a Ziegler-Natta system, provided strong electron acceptors such as $TiCl_4$ or $AlRCl_2$ are present. In these cases, one might speak of coordinated cationic polymerization. It should be remembered that, owing to its various degrees of oxidation, titanium can play the role of a cation or of an anion, depending upon the nature of its substituents, i. e., upon their electron donating or electron-withdrawing character.

A given system can stereoregulate the growth for one monomer, yield atactic polymer for another monomer, and be inactive for a third one. Small changes in the initiating system can provide major changes in the process. For instance, butadiene can be polymerized to 1,4-cis-polymer using the system $AlEt_3/TiI_4$, to 1,4-trans-polymer (to over 90%) by means of $AlEt_3/VCl_4$, and to a polymer that is essentially 1,2-poly(butadiene) with $AlEt_3/Ti(OBu)_4$, whereby in all three cases the Al/transition metal mole ratio has to be of the order of 3.

Summarizing, it should be stated that Ziegler-Natta systems are generally quite efficient for initiating the polymerization of monomers of low polarity, and that they rather often induce stereospecific growth, though not always.

8.4 Kinetics of Ziegler-Natta Polymerizations

Polymerizations initiated by binary stereoregulating systems are quite difficult to investigate kinetically, because no general rule can apply, owing to the very large variety of individual systems. The processes are usually heterogeneous and the initiation mechanism is difficult to understand, as it depends upon the activation, by the metalorganic constituent, of well defined sites on the surface of the crystalline transition metal derivative. The number of effective sites created is in fact very small with respect to the actual mole amount of the transition metal compound (often less than 10^{-2} moles per mole Ti). Furthermore, whatever growth mechanism is considered, the rate of propagation is strongly influenced by the rate of coordination of the monomer onto the site, prior to the actual growth step.

The conversion curves are usually S-shaped, and from the plots of the polymerization rate versus time (Fig. 8.1) it appears that the number of sites slowly builds up, and that termination processes have to be considered:

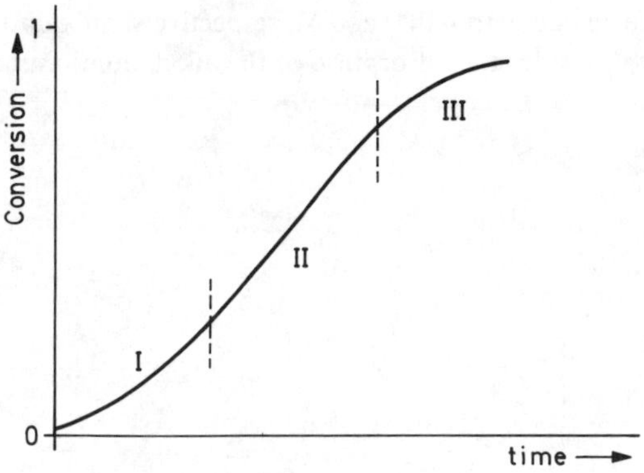

Fig. 8.1. Schematic representation of the conversion versus time in a Ziegler-Natta polymerization

I Induction period
II Steady state period
III Rate decrease as sites are destroyed

● The induction period corresponds to the formation of the active sites, at specific locations, upon reaction of the metalorganic constituent with the transition metal compound. It may also be due in part to the slow diffusion of the monomer to the sites where it should be adsorbed (or coordinated). It may happen that additional sites are formed during the process owing to fractures of the crystalline catalyst particles occuring as the polymer accumulates (see chapter 13 for further details).

● When a stationary state is observed (corresponding to a constant rate of polymerization) it means that the number of active sites is steady, and that the rate of diffusion of the monomer to the active sites remains constant.

● A decrease in the rate of polymerization implies that active sites are being destroyed, and/or that the rate of diffusion of the monomer to the active sites has decreased. Access to these sites tends to become more difficult as bulky polymer molecules are attached to them. This latter phenomenon is clearly evidenced by the effect of hydrogen, known to be a transfer agent in such processes: The reaction results in cutting the bonds between active sites and polymer molecules, whereby easier access is granted to the incoming monomer.

In the most simple cases, the polymerization rate can be considered to be governed by the kinetics of adsorption of the metalorganic compound onto the crystalline transition metal compound. However, if it is assumed that both the monomer and the metalorganic compound are adsorbed on the surface of the Ti compound, one has to apply the Langmuir-Hinshelwood model. Thus initiation as well as propagation supposedly result from reactions between previously adsorbed monomer and the sites which are formed upon activation of the crystalline transition metal salt by adsorbed metalorganic constituent.

These hypotheses do not imply that the "bimetallic mechanism" has to be preferred, as the role of the organometallics (AlR_3, etc.) is primarily to activate the transition metal compound.

The fractions θ_{Al} and θ_M of surface coated with AlR*) and M, respectively, are quite low, usually less than 1%. Thus if we consider the adsorption of the alkyl aluminium

and that of the monomer

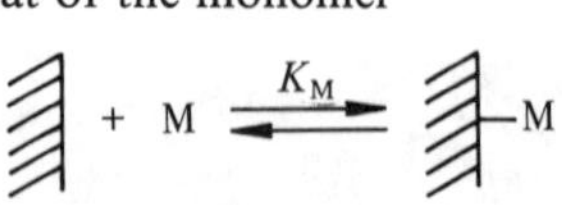

the values of θ_{Al} and θ_M are given by:

$$\theta_{Al} = \frac{K_{Al}[AlR]}{1 + K_{Al}[AlR] + K_M[M]}$$

$$\theta_M = \frac{K_M[M]}{1 + K_{Al}[AlR] + K_M[M]}$$

The initiation, growth, transfer and termination reactions can then be expressed as:

Initiation:

Alkylation of an octahedral titanium site, on the surface of the titanium halide crystal, upon reaction with a previously adsorbed metalorganic compound.

Propagation (steady-state growth):

Insertion of previously adsorbed (coordinated) monomer between the activated Ti site and the polymer chain attached to it.

Transfer to the monomer:

Hydride shift between the incoming (previously adsorbed) monomer and the polymer chain. It results in breaking the bond between the Ti and the polymer, and in the formation of a new Ti-alkyl bond: The site thus remains active.

Termination:

Inactivation of the site, which may result from a reaction between adsorbed monomer and the metalorganic activator. Other possible deactivation processes can be thought of.

---

*) The organometallics will by symbolized as AlR.

Other transfer processes are possible as well, namely intramolecular hydride transfer, and transfer to AlR. They should be taken into account, as well. However, the more reactions are included, the more complicated the rate expression becomes, and it is then fitted with so many parameters that an experimental check becomes worthless.

Calling [C*] the number of active sites per unit area of crystalline surface (in *mole/l*), θ_M being the fraction of surface that is coated with monomer (dimensionless), one can write for the rate of polymerization:

$$R_p = -\frac{d[M]}{dt} = k_p[C^*]\,\theta_M + k_{trM}\,[C^*]\,\theta_M \tag{8.1}$$

k_p and k_{trM} are expressed in $[s^{-1}]$.

Assuming steady state conditions, which means setting initiation and termination rates equal, one can write:

$$k_i\,\theta_{Al}\theta_M = k_t[C^*]\,\theta_M \tag{8.2}$$

and the concentration of active sites per unit area of the crystalline precipitate results

$$[C^*] = \frac{k_i\theta_{Al}}{k_t} \tag{8.3}$$

(k_i being expressed in $[mole/m^2 \cdot s]$ and k_t $[s^{-1}]$). For the rate constant of polymerization it follows:

$$R_p = -\frac{d[M]}{dt} = \frac{k_p + k_{trM}}{k_t} \cdot k_i \frac{K_{Al}K_M[AlR][M]}{(1 + K_{Al}[AlR] + K_M[M])^2} \tag{8.4}$$

This rate equation applies for the system defined above. If intramolecular hydride transfer occurs (whereby an additional term has to be inserted, that does not contain θ_M), the rate expression becomes much more complicated and involves a large number of parameters, that are not easily accessible experimentally.

In spite of the fact that the active sites cover generally less than 1% of the surface of the transition metal compound, the monomer consumption is fairly rapid. This testifies the high efficiency of the active sites in Ziegler-Natta polymerization.

In homogeneous Ziegler-Natta systems kinetic equations are also quite complicated, and still controversial. There are analogies, however, between the treatments proposed and the kinetic equations for ionic polymerizations involving transfer reactions which have been discussed in chapter 5.

8.5 Ziegler-Natta Ring Opening Polymerizations

Cyclic unsaturated monomers such as cyclobutene, methylene cyclobutane, and even spiropyrane do undergo polymerization with Ziegler-Natta initiating systems, whereby in most cases isomerization takes place upon chain growth, and generally 1,4-cis- or 1,4-trans-poly(isoprenes) are obtained:

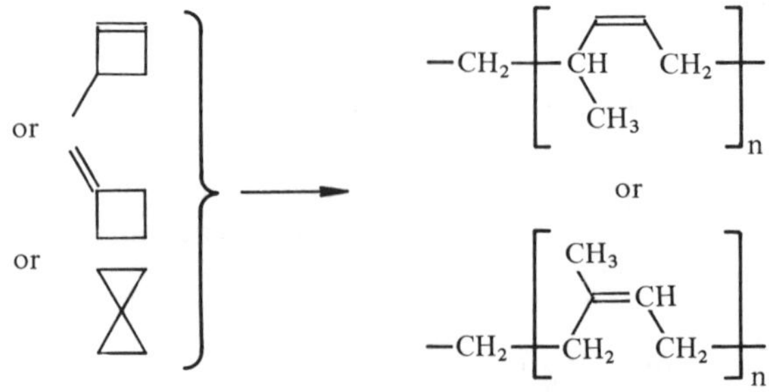

Cyclooctadiene is a different case, as both double bonds are involved and each monomer unit incorporated is bicyclic:

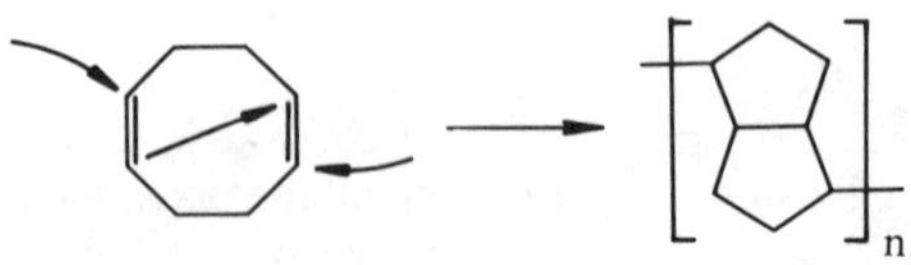

This case resembles the cyclopolymerization involving intrachain propagation in free radical polymerization of some unconjugated diene monomers.

8.6 Ziegler-Natta Copolymerization Reactions

An interesting feature arises from the investigation of the copolymerization of two monomers by means of Ziegler-Natta systems: In general, the reactivity ratios r_a and r_b obey the relation:

$$r_a \cdot r_b = 1$$

which implies that Ziegler-Natta copolymerization is close to "ideal": The active sites exhibit the same preference for monomers A and B, as if their reactivity were independent of the monomer unit added last to the chain. These observations were made with systems containing ethylene, propylene and butene. However, the reactivity ratios reflect the ability of the individual monomers to polymerize, and the instantaneous composition of the copolymer formed differs widely from that of the feed. Adjustment of the feed composition is necessary to provide adequate compositional homogeneity.

One specific case should be mentioned: Cis- and trans-butene-2 can be copolymerized with ethylene, provided the feed is very rich in butene (to compensate for its much lower reactivity). In that case, the degree of alternation is very high, and when isotactic placement of the butene units can be achieved, the polymer formed is nothing else but an isotactic head-to-head poly(propylene):

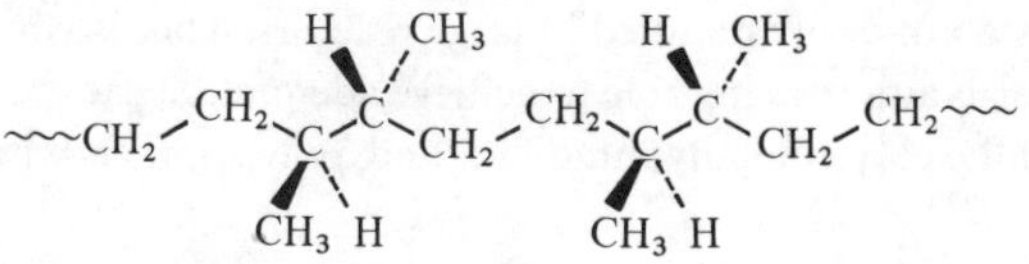

8.7 Applications of Ziegler-Natta Polymerizations

Although Ziegler-Natta initiating systems have been used in many different instances, the domains in which they have led to new products or to improved polymers are in limited number, but of decisive importance. Here are a few examples of industrial applications of stereospecific polymerizations.

Polyethylene can be obtained by free radical processes, under very high pressures and elevated temperatures. The polymerization then involves a large number of transfer steps, yielding many branches, short as well as long. This impedes the ability of the polymer to crystallize, and the resulting polyethylene exhibits a lower density, owing to the large amorphous parts of the material (LDPE). With Ziegler-Natta catalyst systems, ethylene can be polymerized smoothly, at low pressures and at temperatures close to ambient, to linear, highly crystalline, high density polyethylene (HDPE). Upon copolymerization of ethylene with small amounts of butene, it has been possible to introduce a known proportion of short side chains (ethyl), thus enabling to accurately control the ability of the polymer to crystallize: Linear low density polyethylene (LLDPE).

Propylene is rather insensitive to free radical initiators. With some cationic initiators low molecular weight atactic polymer can be produced, which has no practical interest. When it is polymerized stereospecifically, usually to highly isotactic poly(propylene), it yields highly crystalline material, exhibiting excellent mechanical properties and a high melting point, which has found a large number of applications (water pipes, drains, textile fibers, ropes ...).

Non-isotactic random copolymers of ethylene and propylene are also of great interest because of their low glass transition temperature and of their tendency to crystallize upon extension, whereby the modulus is enhanced. Such polymers, usually referred to as ethylene-propylene rubbers (EPR), are manufactured by non-stereoregulating Ziegler-Natta catalyst systems. In some instances a small amount of an unconjugated diene (hexadiene) is copolymerized with the two monomers quoted above, to fit the linear polymer backbone with some dangling unsaturations (EPDR). These double bonds contribute to the vulcanization of the elastomers. The chief advangate of EPR's and EPDR's with respect to polydiene rubbers is their enhanced chemical stability, especially with respect to oxidation (ozone).

In the case of dienes, systems were sought that could yield high 1,4-cis unit contents, to induce the rubberlike behaviour (after vulcanization). Anionic polymerizations carried out in non-polar solvents with lithium as the counter-ion can attain this goal, as noted in chapter 5. However, polymerizations involving solvents, inert atmosphere and metalorganic initiators are not easily handled in large reactors. That is why much interest was devoted to Ziegler-Natta systems able to achieve adequate steric control of the growth reaction, especially to yield polybutadiene and polyisoprene with high 1,4-cis unit contents.

Acetylene is also a monomer that can be polymerized under mild conditions by means of various Ziegler-Natta systems. The reaction results in a fibrous polyacetylene exhibiting long sequences of conjugated double bonds. This polymer can be doped by reaction with a variety of electron donors (n or π), and it thus becomes electrically conducting.

8.8 Other Stereoregulating Catalyst Systems

Several polymerization methods involving steric control of the growth reaction have already been mentioned in previous chapters, especially those relevant to ionic polymerizations:

• *Conjugated Dienes* can be polymerized to polymers exhibiting high 1,4-cis unit contents by using lithium alkyls as initiators in strictly apolar solvents. The incoming monomer gets complexed with the lithium counterion which is simultaneously attached to the chain end by some kind of a π-allyl linkage:

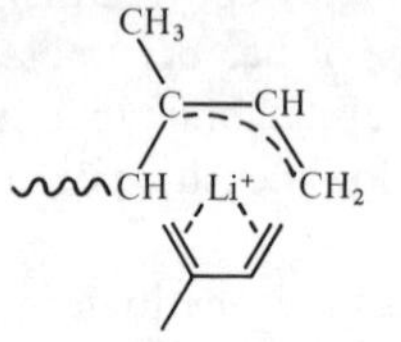

If any polar agent is present, the Li ion is solvated and the polymerization does not involve stereoregulation. The same is true if Li^+ is replaced by larger counterions (Na^+, K^+, Cs^+), since the coordination of the monomer to the metal is much weaker.

• *Methyl methacrylate* can be polymerized in non-polar solvents (toluene) at low temperature with either lithium alkyls or diethyl diphenylamino aluminum to iso- or syndiotactic polymers; however the concentration of the polar monomer has to be kept low to avoid solvation of the ion-pairs by monomer molecules.

• When *vinyl ethers* are polymerized cationically at low temperature, isotactic growth is favored, provided the solvent is of low polarity, and consequently of low solvating power with respect to the active sites.

Apart from these quite classical examples, some more specific cases have led to industrial applications and should be mentioned here:

• Supported metal catalysis (also referred to as *Phillips Catalyst)* are made of an inert support (silica or alumina) of very high specific surface (several hundred square meters per gram) on which a metal oxide (CrO_3, MoO_3, V_2O_5) is deposited. This supported metal catalyst has first to be "activated" by lowering the oxidation degree of the metal, upon high temperature hydrogenation, or treatment with $LiBH_4$. Once activated, these catalysts proved extremely efficient for the polymerization of ethylene (and of other alkenes), as 1 *g* of catalyst can yield more than a ton of polymer. With ethylene no high pressure is required, and the polymers formed are essentially linear, with few branches and a high tendency to crystallize. In some cases a comonomer (for example 1-pentene or 1-hexene) is added on purpose to decrease the crystallinity.

● Other stereoregulating systems have been sought to get polydienes with high 1,4-cis-unit contents. Bis π-crotyl nickel iodide was found to be efficient, as well as π-cyclooctadienyl nickel/CF_3COOH systems; various rhodium salts proved satisfactory, as well.

8.9 Metathesis Polymerizations

Several transition metal complexes, especially those of tungsten, titanium, ruthenium, molybdenum, are known from classical organic chemistry to catalyze trans-alkylidenation reactions, which can be schematized as:

$$\begin{array}{c} R_1\text{—CH=CH—}R_1 \\ + \\ R_2\text{—CH=CH—}R_2 \end{array} \rightleftarrows \begin{array}{c} R_1\text{—CH} \\ \| \\ R_2\text{—CH} \end{array} + \begin{array}{c} \text{CH—}R_1 \\ \| \\ \text{CH—}R_2 \end{array}$$

This reaction does not affect the total number of double bonds present, but it obviously changes their distribution, i.e. the substituents attached to them. Known for more than 25 years, there reactions are governed by entropic factors. In cycloalcenes, the ring strain of the cyclic monomer may be the driving force for pushing the equilibrium towards the formation of polymers, the polyalcenamers. The polymer backbone contains many double bonds and reshuffling reactions are likely to occur. One could also expect macrocyclic oligomers to be formed. However, by a proper choice of the ligands of the transition metal, the catalyst activity can be adjusted in such a way that the unsaturations of the chain do not take part in the reaction. The ring opening polymerization of the monomer can proceed to completion, yielding linear chains with hardly any cyclic oligomers.

Among the catalyst systems that were found capable of initiating such metathesis polymerizations, some are insoluble, usually supported on silica or alumina. The most common of them are based upon WO_3, $W(CO)_6$, MoO_3 or Re_2O_7, with or without activating cocatalyst.

Other catalyst systems are soluble in aromatic hydrocarbons. They are constituted of a transition metal salt such as WCl_6, $RuCl_3$, $MoCl_5$, with a metal organic cocatalyst such as $AlEt_3$ or $AlEtCl_2$. An electron donating compound such as ethanol, butanol or even water is often added in minute amounts. More recently catalyst systems involving a metal-carbon double bond (metallocene or metallocarbene) have been developed. The most typical example is diphenylmethylene tungsten tetra-carbonyl, but a large variety of metal carbenes have been shown to be efficient as polymerization initiators.

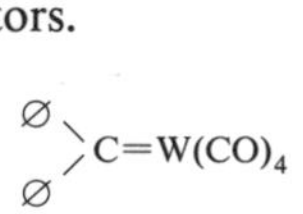

Additional interest for metathesis polymerization is provided by the fact that the propagation reaction sometimes involves stereocontrol, under well chosen conditions.

An example is given by the ring opening polymerization of cyclopentene. With the system $WCl_6/AlEt_3$ the process yields an all-trans polypentenamer, whereas an all-cis polypentenamer is obtained with $MoCl_5/AlEt_3$, under similar experimental conditions.

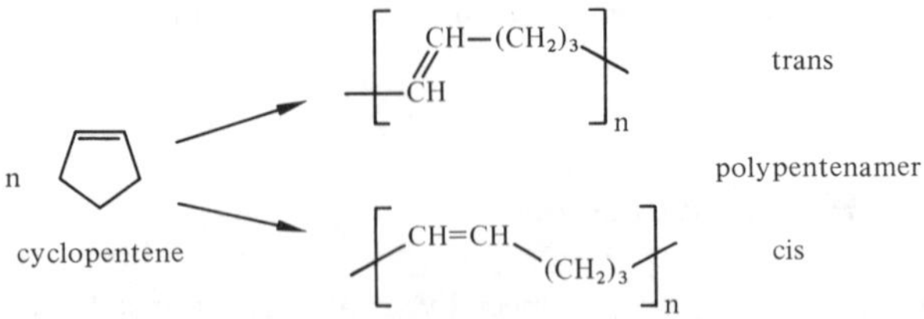

The polymerization of norbornene initiated with $RuCl_3$ in the presence of some butanol yields a polymer that contains about 75% trans units, the cycles of the chain being cis.

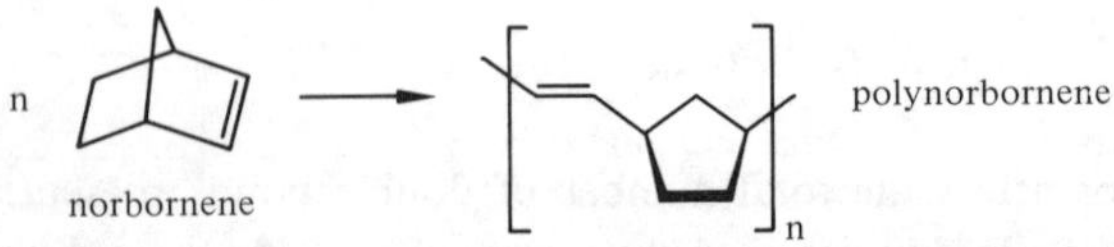

The mechanism of methathesis polymerization was for a long time quite controversial. It is now elucidated, at least in the cases in which the catalyst involves a metallocarbene site, as the result of the work of several authors, especially Schrock and Grubbs. Some of these polymerizations can even be quoted as living: No spontaneous transfer or termination would occur; The molecular weight of the polymer formed is proportional to the conversion attained; Further addition of monomer results in an increase of the length of the existing chains. The polydispersity of the samples obtained is low. As a consequence, functionalization at chain end can be obtained by induced deactivation. Block copolymer syntheses were also attempted by sequential addition of cycloalcenes, and they were successful, at least in some cases.

The mechanism can be schematized as follows:

Initiation

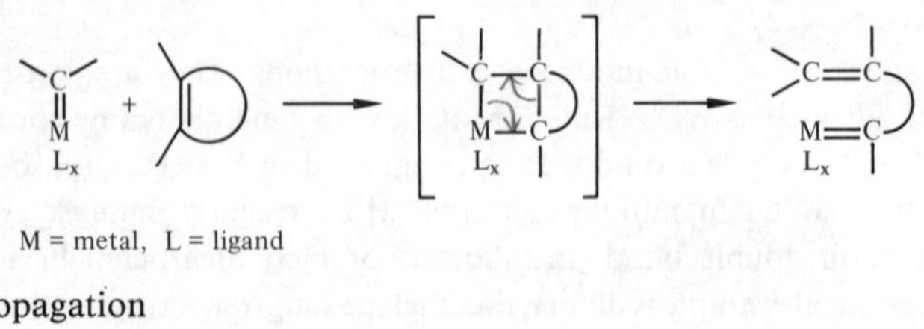

M = metal, L = ligand

Propagation

Induced deactivation can be carried out with carbonyl compounds by a Wittig type reaction, yielding an end-standing unsaturation.

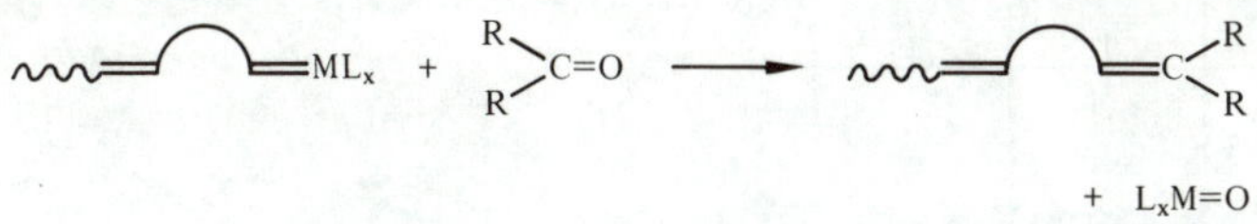

This mechanism is considered to apply to most ring opening metathesis polymerizations, although it is not clear yet how the initial metallocarbene is formed, when the catalyst system involves a transition metal salt.

Polyalcenamers are of great industrial interest, as the cyclic monomers, such as cyclopentadiene, dicyclopentadiene, norbornene, etc. are easily available. The polyalcenamers have found a number of applications, especially in the rubber industry. The existence of double bonds on the polymer backbone is essential, as it is a prerequisite for vulcanizations to be carried out.

Problem 8.1

The definitions of "triad tacticity" in respect to vinyl homopolymers that are potentially stereoregular, are as follows:

i = isotactic triad tacticity, fraction of triads having the configuration

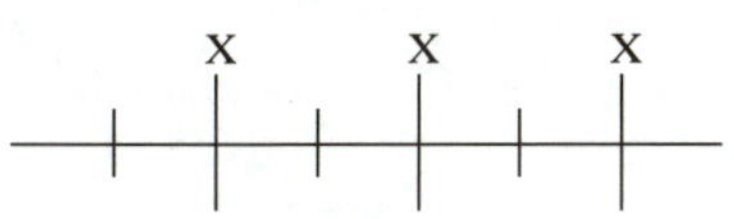

(lll)

or 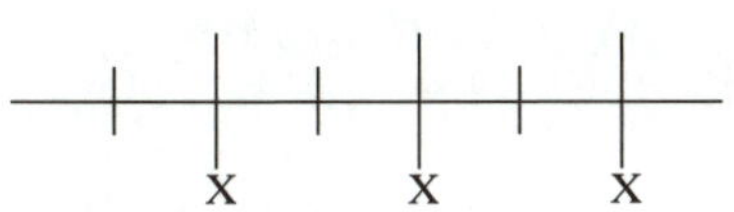

(ddd)

where  = CH_2 and  = CHX of a vinyl unit CH_2CHX.

s = syndiotactic triad tacticity, fraction of triads having the configuration

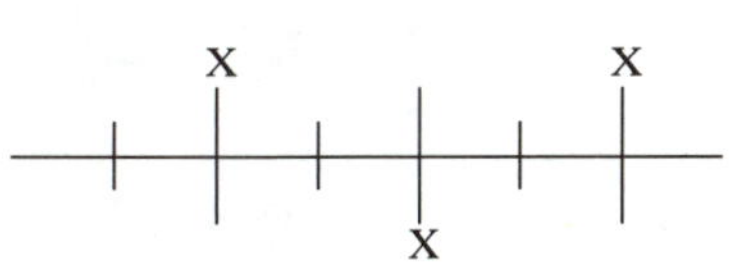

(ldl)

or 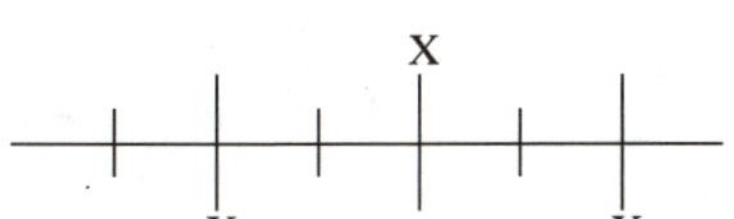

(dld)

h = heterotactic triad tacticity, fraction of triads having the configuration

(ddl)

or

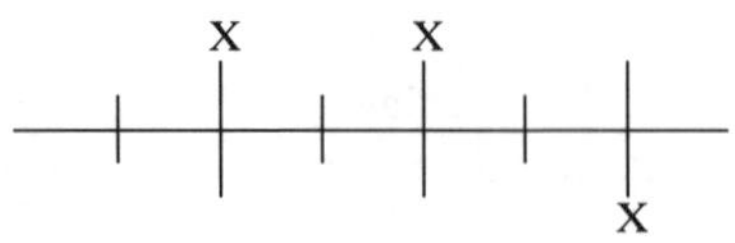

(lld)

Let the probability of an incoming monomer being added with its X on the *same side* of the skeletal chain as *the X on the preceding unit* be $k_m/(k_m + k_r)$. Let the probability of an incoming monomer being added with its X on the opposite side of the skeletal chain be $k_r/(k_m + k_r)$; m and r refer to meso and racemic addition, respectively.

Show the analogies between the copolymer triad probabilities P_{aaa}, P_{aab}, etc., and the above triad probabilities in P_{ddd}, P_{lll}, P_{ddl}. Show that, out of these eight corresponding probabilities, two comprise what is meant by i, two comprise what is meant by s, and four comprise what is meant by h.

Finally, write the equations for i, s, and h in terms of $k_m/(k_m + k_r)$ and $k_r/(k_m + k_r)$.

Chapter 9
Chemical Reactions on Polymers

9.1 Introduction

Historically, the chemical modification of natural polymers has been the determining factor in the development of their applications as materials, as textile fibers, as coatings, etc. Wood cellulose, for instance, cannot be used as such. Ways had to be found to chemically modify it, first in order to render it soluble, and also to have it fitted with appropriate functions. Cellulose derivatives originating from the various chemical transformations of the alcohol functions carried by the carbohydrate chain have found a large number of uses: cellulose acetate as textile fibers, cellulose nitrates as films, coatings, and even explosives, methylcellulose and carboxymethylcellulose as hydrophilic additives, etc. ... However, cellulose chemistry is a very wide field, and it cannot be treated extensively within the frame of this textbook.

Natural rubber (1,4-cis polyisoprene) is another example of a polymer material which gains interest by chemical modification. It is a liquid polymer which has no interest as such unless linkages have been introduced chemically between the individual chains. The resulting elastomeric networks are materials of utmost interest.

This chapter is primarily devoted to chemical reactions on synthetic polymers. The two chief aspects of the chemical transformation of a polymer will be discussed successively:

The first section is devoted to reactions involving the backbone chain: Scissions and bridge forming reactions. The existence of degradation reactions affect the durability of polymer materials, when they are submitted to heat, moisture, oxidation, and/or irradiation. These so-called ageing processes*) deserve good understanding, as chain scissions generally weaken the mechanical strength of the materials. Conversely, extensive inter-chain bridge formation will make them hard and brittle.

The second part of this chapter focusses on the chemical transformations involving the substituents carried by the monomer units of the polymer chain. Such reactions are also of major interest, for a variety of reasons:

- Some polymers are not accessible directly, since either the corresponding monomer does not exist, or it does not polymerize.
- A partial modification of a polymer is sometimes easier to carry out than a copolymerization of the corresponding monomers.
- Cyclization reactions involving two neighboring units along the chain are often free of side reactions, contrary to cyclopolymerization of a non conjugated bis-unsaturated monomer (see section 4.10.2).

---

*) "Chemical ageing" processes are to be distinguished from "physical ageing", which affects plastics, fibers and elastomers in a different way: Formation of cracks and crazes, loss of mechanical properties, etc. originating from mechanical constraints to which the polymeric material was submitted.

There are a number of differences between a classical reaction involving small molecules and the same reaction affecting functions hanging on a polymer chain:

- First, side reactions should be avoided, to prevent chain scissions and/or the occurrence of unwanted functions on the chain.
- Secondly, the reactivity of a function carried by the monomer units of a polymer chain is strongly influenced by neighboring units. Reactivity enhancement or depression can be observed, depending on the system considered, on the type of reaction and on the environment of the chain (solvent).
- Thirdly, cyclizations may occur preferentially upon reaction of a function with either of its closest neighbors along the chain.

It is important, therefore, upon studying chemical modifications of polymers, to provide adequate characterization of the polymers obtained. It is necessary to check for possible changes in the degree of polymerization, for possible unwanted functions on the chain, for cyclizations that may have occurred during the process. Moreover, the degree of conversion attained has to be determined and, whenever possible, the resulting distribution of modified and unmodified units along the chain should also be established.

9.2 Chemical Reactions Involving the Backbone Chain

The problem of *"ageing"* of polymeric materials is of major economic importance because it conditions their durability. It is also quite complex, because it involves the resistance of polymers to all factors a material can be exposed to during its lifetime: Oxygen, irradiation, heat, moisture, and bacterial attacks. However, the most important of them are atmospheric oxygen and irradiation (by UV especially), since they are most likely to induce chain scissions, causing the mechanical properties to be profoundly affected.

9.2.1 Oxidative Degradation

Oxidative degradation is in most cases a free radical chain reaction, which is greatly accelerated upon irradiation by UV-, γ- or X-rays.

A radical site is generally created on the polymer chain by removal of a tertiary hydrogen atom. This site binds oxygen to give a peroxy radical. The latter can in turn take away another tertiary hydrogen to yield a *hydroperoxide function* or it can induce *chain scission*. Calling PH the polymer and $P^\bullet$ the radical originating upon removal of a tertiary hydrogen, this process can be schematized as:

$$\mathrm{PH} \xrightarrow{h\nu} \mathrm{P^\bullet}$$

$$\mathrm{P^\bullet + O_2} \longrightarrow \mathrm{P{-}OO^\bullet}$$

$$\mathrm{POO^\bullet + PH} \longrightarrow \mathrm{POOH + P^\bullet}$$

The chain scission mechanism has been studied in great detail; it usually proceeds via a six-centered transition state, and a monomer molecule is released at each chain break.

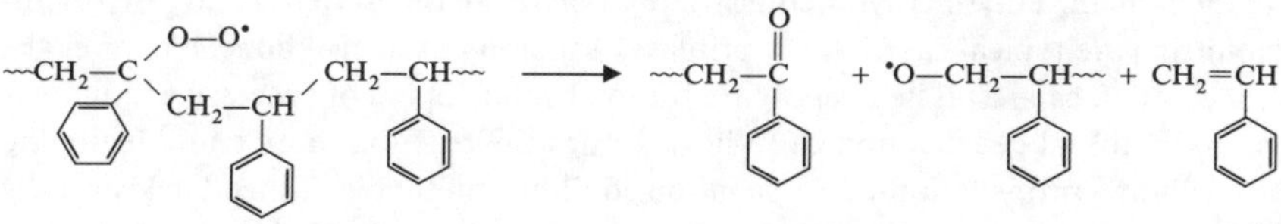

This general reaction scheme applies to most vinylic and acrylic polymers, to poly(ethylene oxide), and poly(butylene oxide) as well (in the latter cases ethylene is expelled at each scission). Polymers differ greatly in their sensitivity to such chain scissions.

The sensitivity to oxidative degradation upon irradiation sometimes originates from (or is enhanced by) oxidized functions *already present* in the polymer backbone. A typical case is that of poly(ethylene) chains containing some ketone functions randomly distributed along the chain. These functions are subject to the well known chain scissions generally referred to as *Norrish I* and *Norrish II* which can be schematized as:

Norrish I: $\sim\sim CH_2{-}C(=O){-}CH_2{-}CH_2 \sim\sim \longrightarrow \sim\sim CH_2{-}C^{\bullet}(=O) + {}^{\bullet}CH_2{-}CH_2 \sim\sim$

Norrish II: $\sim\sim CH_2{-}C(=O){-}CH_2{-}CH_2{-}CH_2 \sim\sim \longrightarrow$

$\sim\sim CH_2{-}C(=O){-}CH_3 + CH_2{=}CH{-}CH_2 \sim\sim$

Norrish type II scissions contribute to about 90% to chain degradation in poly(ethylene), when exposed to light or to any kind of irradiation (UV, γ-rays, X-rays, electron beams).

Carbonyl functions being *weak points* in a polymer chain, the above degradation process was applied to get *photo-oxidizable polymers,* that could be broken into small fragments upon irradiation (by sunshine for instance). For that purpose, ethylene has been copolymerized with a small proportion of carbon monoxide, the target being to get carbonyl functions distributed randomly at invervals of 20 to 40 carbons along the chain. The resulting fragments can subsequently undergo biological degradation. This *"disposable"* plastics could be of great help in environmental protection. This is a case in which degradation of the polymer chain has been intentionally sought.

Usually, the reverse is true, and one tries to prevent radical formation on the chain and/or the oxidative chain reaction that results in an increasing number of chain scissions, upon prolonged exposure to light and atmospheric oxygen. This is why polymers are often compounded with some additives such as *radical scavengers, UV-protecting agents,* or *antioxidants.* Sterically hindered amines, such as 2,6-di-t.butylpiperidine are of special interest to prevent oxidative degradation in some commonly used technical polymers.

9.2.2 Chain Unzipping

Chain unzipping occurs chiefly in polymers involving quaternary carbon atoms along their chain. Poly(methyl methacrylate), poly(α-methyl styrene) and poly(methacrylonitrile) are typical cases. Any "primary" scission in the backbone, whatever the way it occurred, can result in a depolymerization by unizipping of successive monomer units, until mutual deactivation of radicals brings the reaction to an end. Unzipping is the result of a propagation-depropagation equilibrium, that is pushed backwards by the fact that the monomer is volatile. Other polymers susceptible of giving yield to depropagation (or unzipping) reactions include poly(oxolane) and polyoxymethylene, originating from cationic ring opening polymerization, and the alternating copolymers of SO_2 and alkenes, generally called poly(alkylene sulfones) (see chapter 7).

9.2.3 Cleavage by Ozone

Double bond cleavage by ozone is known to occur in many organic compounds. This phenomenon is of special importance in polydiene rubbers that contain double bonds in their main chain.

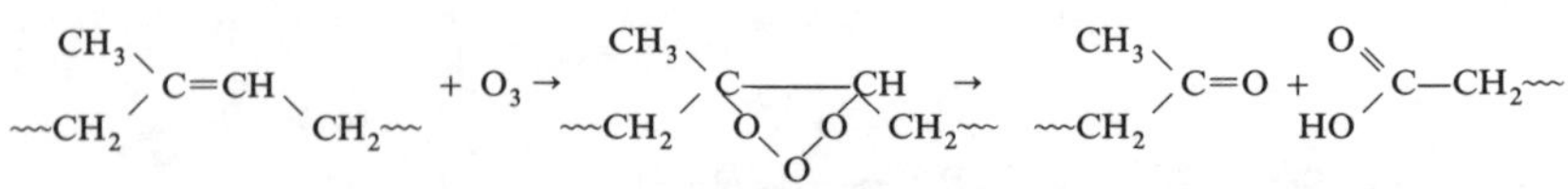

(primary ozonide)

Ozone greatly harms such rubbers by inducing degradation, even more rapidly when the material is under slight stress, because of enhanced diffusion of ozone into the rubber sample. As an increasing number of scissions occur, cracks are formed, generally perpendicular to the stress axis, thus weakening markedly the mechanical resistance. Carbon black, that is widely used as a filler, seems to act to some extent as a stabilizing agent, though the mechanism of that protection has not been elucidated yet. A marked advantage of ethylene-propylene rubbers with respect to polydienes is that they are far less sensitive to ozone because their chain does not contain double bonds.

9.2.4 Periodic Acid Cleavage

Periodic acid or sodium periodate is known to cleave quantitatively and rapidly α-glycols. This method can be used for instance to measure the amount of *head-to-head placement* of units in a poly(vinyl alcohol) sample, and consequently in the poly(vinyl acetate) it originates from

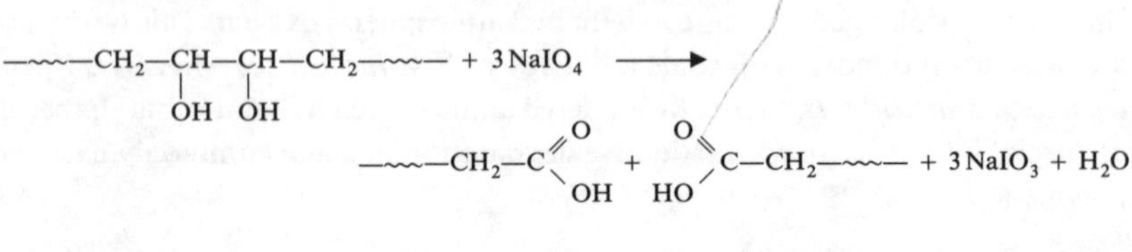

9.2.5 Free Radical Crosslinking

If a polymer is irradiated in the absence of oxygen, the reactions mentioned in section 9.2.1 cannot take place, and the free radicals created can only react between themselves (provided unzipping is excluded). Upon recombination, bonds are formed between chains, and the molecular weight increases until eventually crosslinking occurs. This network formation process fundamentally modifies the general behaviour of the polymeric materials.

Poly(ethylene oxide) when submitted to γ-irradiation in water solution yields a swollen network, which strongly retains the water that solvates the elastic chains. Crosslinking is far less effective if PEO is irradiated in the bulk state, as it is a highly crystalline polymer, and bond formation is expected to take place chiefly in the amorphous parts of the material.

An application of crosslinking upon irradiation, in the absence of oxygen, was developed in the technology of *integrated circuits.* A polymer film is submitted to irradiation by means of an electron beam through a mask that hides part of it. The irradiated parts crosslink and become insoluble; the non-irradiated fractions remain soluble and can be removed with an adequate solvent. This so-called *"polymer resist"* is referred to as negative, as the irradiated parts remain*).

9.2.6 Vulcanization of Elastomers

Linear 1,4-cis-poly(diene) chains are flexible and their glass transition temperature is far below room temperature. When submitted to a stress the chains have a tendency to crystallize, but "long-term elasticity" is not possible and the materials yield by viscous flow, even when their molecular weights are very high.

Permanent elastic properties are conveyed to such materials by building bridges between individual molecules, to tie them together at random intervals; the material will then recover its shape after release of the stress, as the yield is eliminated. The number of such bridges beyond the gel-point (the point at which viscosity becomes inifinite) determines the response of the material to a mechanical stress. The higher the degree of crosslinking, the higher the elastic modulus (ratio of tensile stress to tensile strain).

To induce crosslinking the *vulcanization process* first discovered by *Goodyear* in 1840 is still the most commonly used. It is an early example of a reaction carried out upon stirring a highly viscous mixture of a poly 1,4-cis-diene (or natural hevea rubber) and sulfur. The reaction can be accelerated by various activators such as zinc salts of benzothiazol or dithiocarbamates.

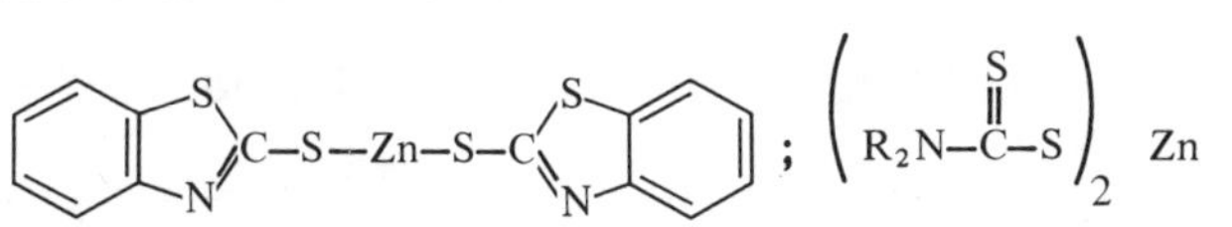

*) To get positive "polymer resists" the same technology can be applied with a polymer film that undergoes unzipping upon irradiation (such as PMMA), instead of crosslinking. In that case, the irradiated areas vanish as a result of depolymerization, whereas the others stay.

The mechanism of this bridge forming reaction is still not well understood. It is known however that sulfur can add to some kinds of double bonds, and can also react in allylic position. The bridges between two chains can contain in most cases 1 or 2 sulfur atoms, sometimes up to 5. The mechanism may exhibit ionic character.

There are alternate ways of introducing crosslinks into a polymer that contains double bonds. Irradiation can be used to create radical sites on the backbone and, upon recombination, bridges (or transverse bonds) are created between chains, as was described above. The use of free radical initiators such as dicumyl peroxide or di-tert.-butyl peroxide can also provide primary radicals, that will in turn attack double bonds, establishing transverse links to induce crosslinking.

With ethylene-propylene rubbers (EPR) free radical producing compounds can also be used: But in this case the radical fragments attack the polymer chain, abstracting hydrogen, and thereby producing radicals on the chain, which recombine. However, to facilitate vulcanization, the EPR chain is often fitted with a small proportion of pendent double bonds, arising from some 1,5-hexadiene (or another unconjugated diene) incorporated upon terpolymerization with ethylene and propylene. The efficiency of the vulcanization process is noticeably enhanced. The same technique can also be used for poly(dimethyl-siloxane) rubber vulcanization, if the PDMS chains contain a low amount of vinyl silane linkages (replacing randomly methyl silane bonds).

Many other crosslinking processes should be quoted, but as they originate from reactions of side groups (and not of the backbone), they are not considered in this section.

9.3 Chemical Reactions Involving the Functions

As indicated in the introduction, the reactivity of a function A with respect to an incoming reagent R to give a function B depends upon the *environment* of A along the polymer chain. It is generally different from that of a micromolecular model molecule, in which the A function can be considered isolated. In a vinylic or acrylic polymer chain, in which every other carbon atom carries a substituent, neighboring effects play a role, either enhancing or decreasing the reactivity of function A. In the case of ionic processes, the polarity of the solvent also influences the reaction; furthermore its solvation power towards the chain has to be taken into account, because the polymer chains expands when it is well solvated, thus allowing a better access of the incoming reagent R to the functions A carried by the chain.

9.3.1 Kinetic Aspects of the Chemical Modification of Polymers

Let us consider a homopolymer ...AAAA... and let us assume we want to transform the A function of each monomer unit into a B function. At the beginning, each A function is surrounded by two A functions in AAA triads, and the rate of transformation is governed by the rate constant k_0. But as the transformation proceeds, more and more A functions are surrounded by B (modified) functions, and two other rate constants have to be taken into consideration: k_1 (for A functions surrounded by one A and one B function in AAB (or BAA) triads) and k_2 (for the A function left in a BAB triad).

(a) There are systems in which $k_0 = k_1 = k_2$. In such cases the replacement of A by B groups is a regular *random process* to which *Bernouilli's statistics* are applicable, and complete conversion can be attained. The final product is a ...BBB... homopolymer. There are few cases of this type, like:
— Esterification of poly(methacrylic acid) by diazomethane (CH_2N_2).
— Hydrolysis of syndiotactic poly(diphenylmethyl methacrylate).

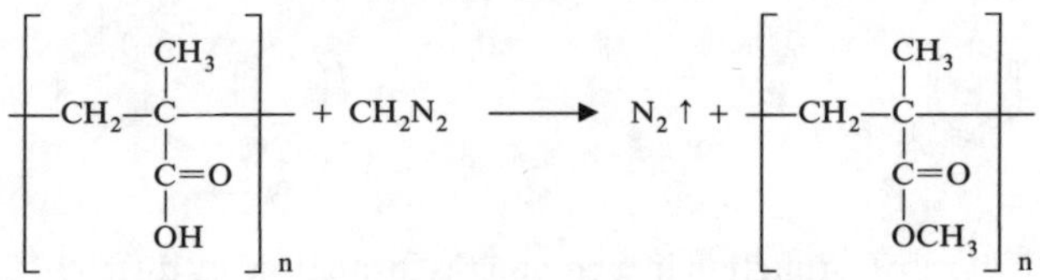

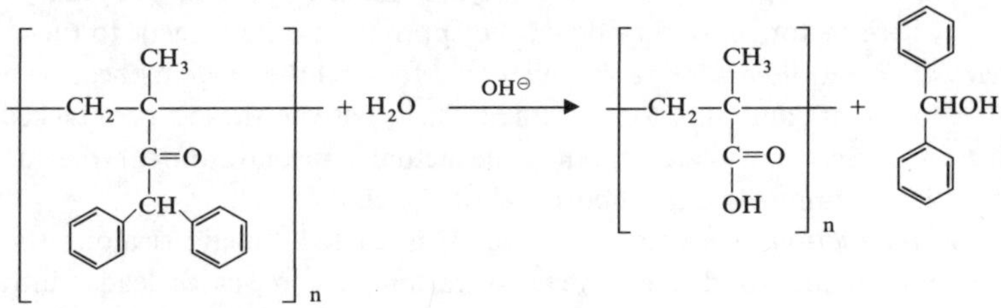

(b) There are also cases where $k_0 < k_1 \leqslant k_2$. The presence of neighboring B groups here *enhances* the reactivity of A functions. A marked tendency to form sequences of transformed (B) units is observed, whereas in other parts of the chain, sequences of A units remain untouched. This is the case of autocatalytic reactions of various kinds, such as the hydrolysis of isotactic PMMA, which involves the intermediate formation of cyclic anhydride.

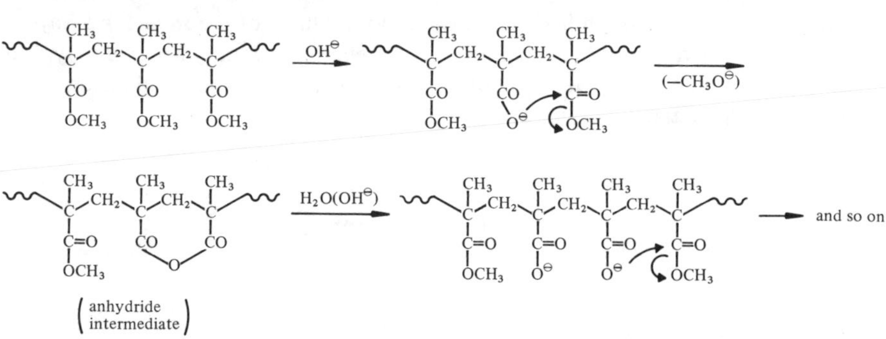

Obviously *Bernouilli's statistics* cannot account for the distribution of A and B units along the chain. It should be pointed out that the reaction can go to completion, as before.

(c) There are also cases in which $k_0 > k_1 \geqslant k_2$. In such cases, the effect of groups already reacted is to *decrease* the reactivity of the remaining A functions. A quantitative transformation is difficult to obtain, because the reaction slows down (or even stops) when most of the remaining A units are in BAB triads. Here again *Markov's statistics* have to be used to evaluate the proportions of the triads of various kinds formed at given degrees of conversion. Only for very low degrees of conversion can one use, as a first approximation, Bernouillian statistics.
Example: Quaternization of poly(vinylpyridine).

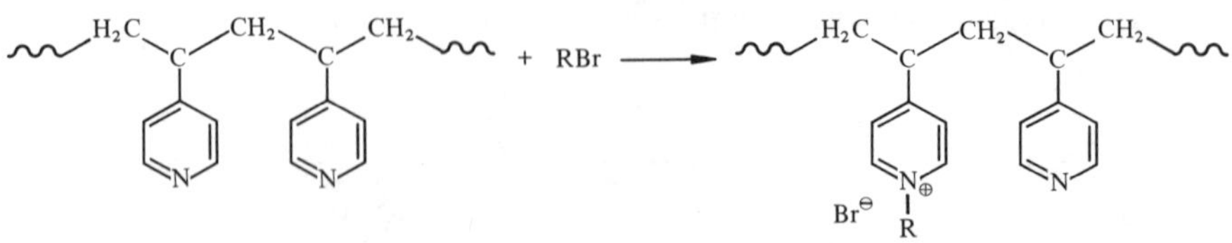

The chemical reactivity of a function in a polymer is influenced by different factors:

1) *Steric hindrance*, especially when the polymer chain comprises quaternary carbon atoms. These factors may slow down the approach of the reagent to the chain.
2) *Electrostatic* and *polarity effects* which tend to repel the reagents bearing electrostatic charges of same sign (or of same polarity) as the sites to be attacked.
3) *Solvation effects* which can modify the actual concentration of the incoming reagent in the immediate neighborhood of the chain.
4) *Specific interactions* between reacted and unreacted functions along the chain which may enhance (or decrease) reactivities, and, in some cases, lead to intramolecular cyclizations.

9.3.2 Cyclization Reactions

Cyclization reactions have to be taken into consideration, since there are a number of cases in which two neighboring units can react together to yield cycles in the presence of an appropriate reagent. A distinction is generally introduced between:

- intra-sequence cyclization involving two monomer units of the *same* kind (say A) in a homopolymer poly A, or in a AA... sequence on a copolymer chain. Typical examples include the acetalization of poly(vinyl alcohol) or the dehalogenation of PVC by zinc metal:

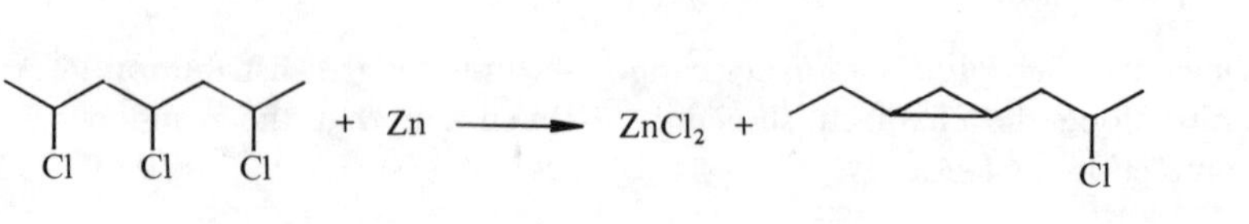

- inter-sequence cyclization between two neighboring units of *different* kinds (say A and B) in an AB random copolymer, or one modified and one non modified unit in a homopolymer A submitted to chemical transformation (A → B). Typical examples include the lactonization in vinyl chloride-methyl methacrylate diads, and the cyclizations that may occur upon nucleophilic substitutions on the ester functions of poly(methyl methacrylate):

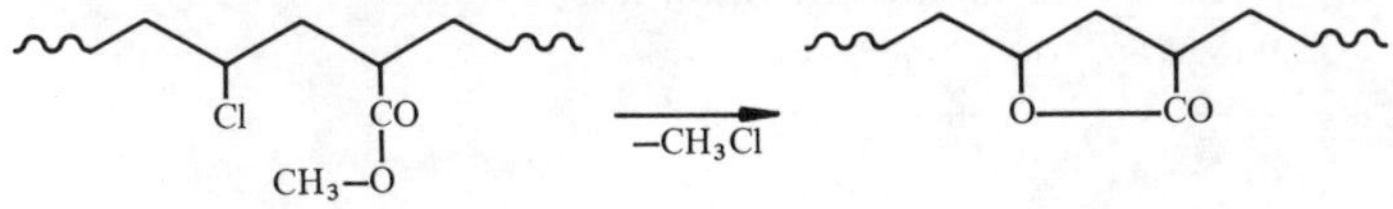

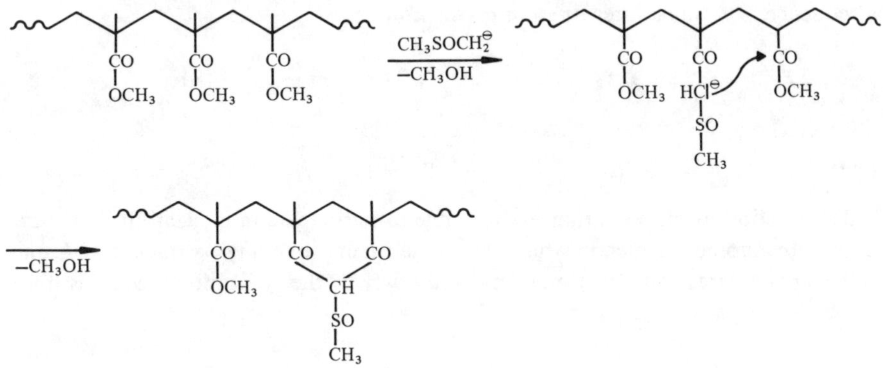

Irreversible cyclization reactions — whether intra- or intersequence — never go to completion. A number of units are left over, as they are isolated (trapped) between two ring structures. To study this feature in greater detail, let us consider a homopolymer A submitted to a reagent able to yield ring formation between two neighboring A units along the chain. We shall calculate the proportion of A functions that will be left over by the cyclizations process, as they get isolated between two ring structures.

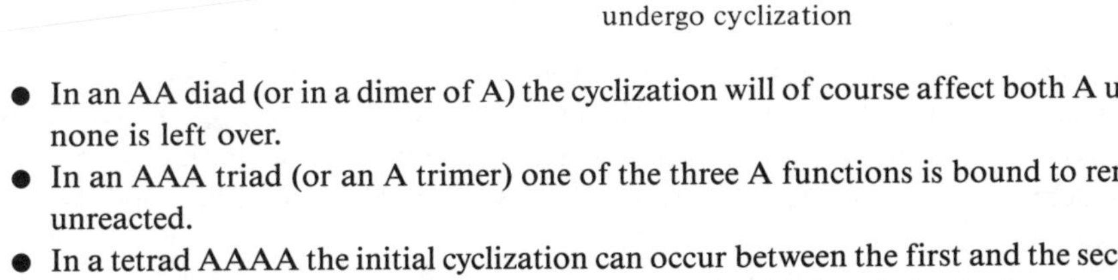

- In an AA diad (or in a dimer of A) the cyclization will of course affect both A units, none is left over.
- In an AAA triad (or an A trimer) one of the three A functions is bound to remain unreacted.
- In a tetrad AAAA the initial cyclization can occur between the first and the second, between the second and the third, or between the third and the forth units, with equal probabilities. When it occurs between A_2 and A_3, the functions A_1 and A_4

are unable to take part in further reaction, and they are left over. The number of A functions that will not participate in cyclization is therefore:

$$S_4 = \frac{1}{3} \cdot 2 = \frac{2}{3} .$$

- The same calculation carried out on an A pentad will show that one unit is bound to be left over ($S_5 = 1$) and for hexads the result is:

$$S_6 = \frac{1}{5} \cdot 2 \cdot \frac{2}{3} + \frac{1}{5} \cdot 2 \cdot 2 = \frac{16}{15} .$$

Flory established a recurrence formula to calculate the *number* of units left over in a sequence of *n* units, after *random cyclization:*

$$S_n = \frac{2}{n - 1} [S_1 + S_2 + S_3 + \ldots + S_{n-2}] = \frac{2}{n - 1} \sum_{i=1}^{n-2} (S_i) \tag{9.9}$$

(with $S_1 = 1$, $S_2 = 0$, $S_3 = 1$, $S_4 = \frac{2}{3}$, $S_5 = 1$)

The fraction of all units that are not able to participate in cyclization is the ratio S_n/n*). In a homopolymer, in which the whole chain is one single sequence of A units (n being very large), the fraction of units left over by the cyclization process is found to be:

$$f_n = \lim \left(\frac{S_n}{n} \right)_{n \to \infty} = e^{-2} = 0.135 \tag{9.10}$$

This means that upon carrying out a random intra-sequence cyclization on a homopolymer 13.5% of the units are unable to participate in the reaction, as they get isolated between cyclized units. The maximum possible conversion is 86.5% and this value has been checked experimentally in many instances.

In the case of a copolymer the fraction of units left over upon random intra-sequence cyclization can be expressed as:

$$f_n = \sum_n S_n P_n / \sum_n P_n \tag{9.11}$$

where P_n stands for the probability of finding a sequence of *n* A units along the chain and S_n is the number of units unable to take part in random cyclization in such a sequence.

The same relation applies to inter-sequence cyclization between neighboring A and B units, except that P_n is then the probability of finding alternating AB sequences with a total of *n* units. If the copolymer is strictly alternating the limiting conversion in random inter-sequence cyclization is 86.5% as it is in the case of homopolymers.

*) The values of $\bar{S}_n$ and of $\frac{\bar{S}_n}{n}$ are given in the appendix.

9.3.3 Examples of Chemical Reactions Affecting the Substituents of a Polymer Chain

A large variety of chemical modifications can be carried out on polymers by reacting their functional groups with adequate reagents, without changing their polymerization degree (neither chain scission, nor crosslinking). Some of these reactions have gained importance as they give access to new polymers of interest, others should be quoted only as typical examples of chemical modification of polymers.

9.3.3.1 Addition Reactions

Chlorination of polydienes is carried out in solution, to various degrees. The properties of the polymers obtained widely depend on the degree of chlorination. These chlorinated rubbers are used in coatings and in paints, especially in ship building industry. *Carbenes* are also able to add onto the double bonds of polydienes. Dichlorocarbene is formed in situ upon reaction of sodium hydroxide onto chloroform:

$$HCCl_3 + KOH \longrightarrow H_2O + :CCl_2$$

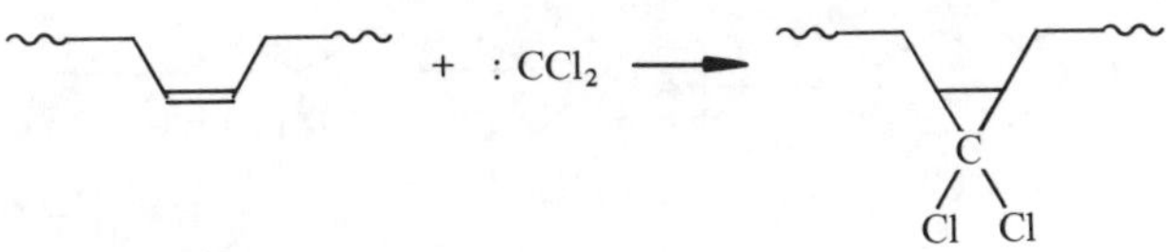

These polymers exihibit properties similar to those of chlorinated rubbers.

Maleic anhydride also reacts with polydiene chains. However, this reaction involves the allylic sites of the backbone. The reaction is used to introduce carboxylic sites into elastomeric molecules.

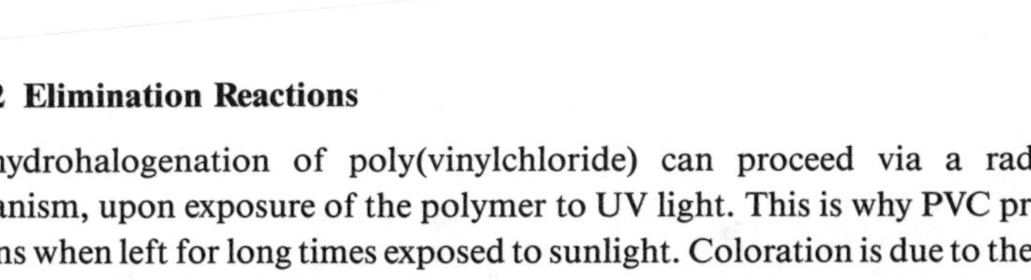

9.3.3.2 Elimination Reactions

Dehydrohalogenation of poly(vinylchloride) can proceed via a radical chain mechanism, upon exposure of the polymer to UV light. This is why PVC progressively darkens when left for long times exposed to sunlight. Coloration is due to the sequential dehydrohalogenation, brought about by *allylic activation* of the sites in α-position to a double bond already created.

Anionic dehydrohalogenation is induced by attack of bases upon PVC. However, two reactions compete: Elimination and substitution, as shown in the scheme below.

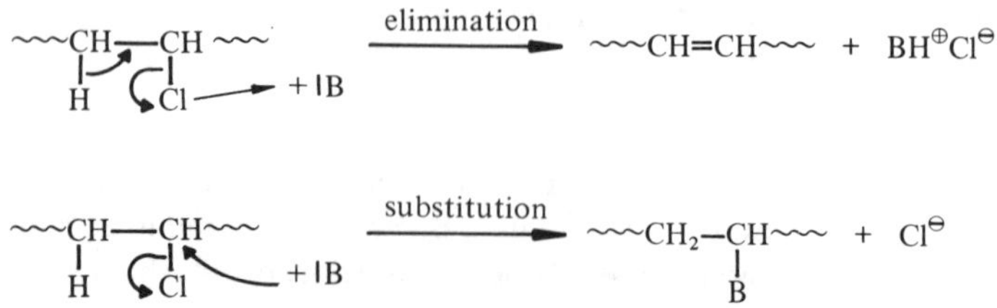

The stronger the nucleophilic character of the base | B, the higher the proportion of elimination. However even with the strongest nucleophiles some substitution occurs, whereby the sequences of conjugated double bonds are interrupted. A better way to obtain very long sequences of conjugated double bonds along the chain is to polymerize acetylene (by Ziegler-Natta catalysis), instead of dehydrohalogenating PVC.

Another kind of elimination reaction involving PVC, by means of zinc metal, involves cyclizations and its conversion is limited to 86.5%, as shown previously, owing to isolated C—Cl bonds, that cannot undergo reaction as they are located between two cyclopropane units already formed:

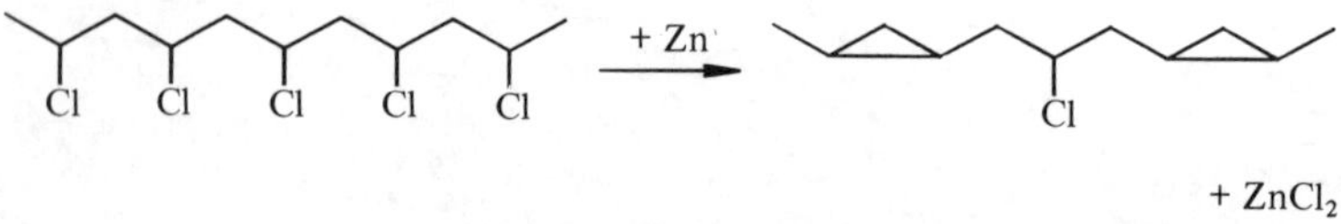

Cyclization followed by elimination via controlled oxidation has also been performed on two polymers bearing unsaturated side-chains, namely poly(acrylonitrile) and poly(1,2-butadiene) whereby thermostable *ladder polymers* are obtained. They have been used in space vehicles and also as reinforcing fibers in composites meant to resist high temperatures. Their carbon content is very high, and they are the precursors of the now wide-spread carbon fibers.

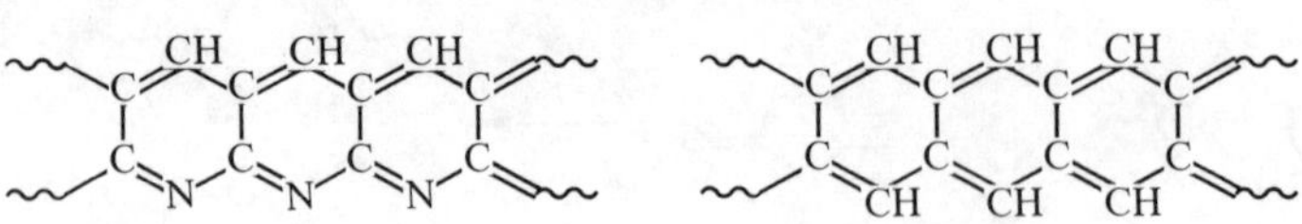

9.3.3.3 Substitution Reactions

Substitution reactions are by far the most commonly used reactions to get chemical modification of a polymer. Here are a few examples:

Polyvinylalcohol and derivatives

Hydrolysis reactions performed on poly(vinyl acetate) are the usual access to poly(vinyl alcohol), as the corresponding monomer (vinyl alcohol) does not exist. In fact the reaction is a transesterification with methanol, catalyzed by bases such as $OH^{\ominus}$ or $CH_3O^{\ominus}$, and it can be written as:

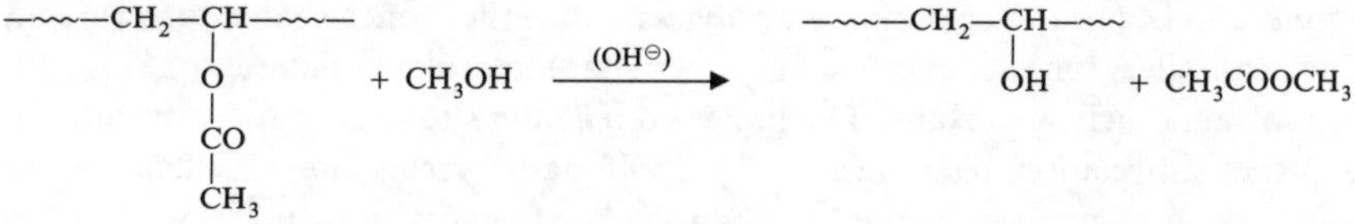

It can be carried to completion and without noticable degradation of the backbone. Poly(vinyl alcohol) has found many applications as a hydrophilic non-ionic polymer.

In turn, poly(vinyl alcohol) can react with aldehydes, such as butyraldehyde, whereby neighboring alcohol functions react to give acetal cycles, yielding poly(vinyl butyral).

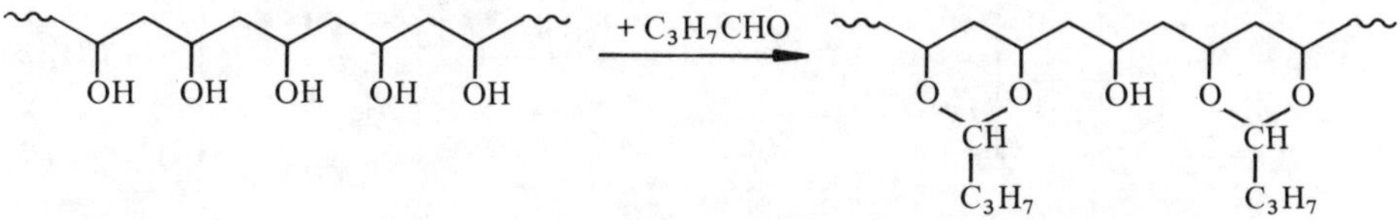

The reaction occurs at random, and its conversion is limited to 86.5%, as isolated OH functions cannot take part in the reaction. This process requires weak acid catalysts. It can be reversed in strong acid media where the acetal functions are hydrolyzed. Because of that, it has sometimes been possible to go beyond the theoretical limit of 86.5% by choosing adequate conditions, and allowing sufficient time for the reaction to equilibrate.

Poly(vinyl alcohol) can also undergo metalation by exchange with alcoholates or with dimsylsodium (the metalated derivative of dimethylsulfoxide DMSO), whereupon it can be reacted with alkyl halides to yield vinyl ether units, or with propane sultone to yield sulfonated units. None of these reactions is quantitative although they are of interest: The latter reaction, for instance, enables one to get low charge density polyelectrolytes:

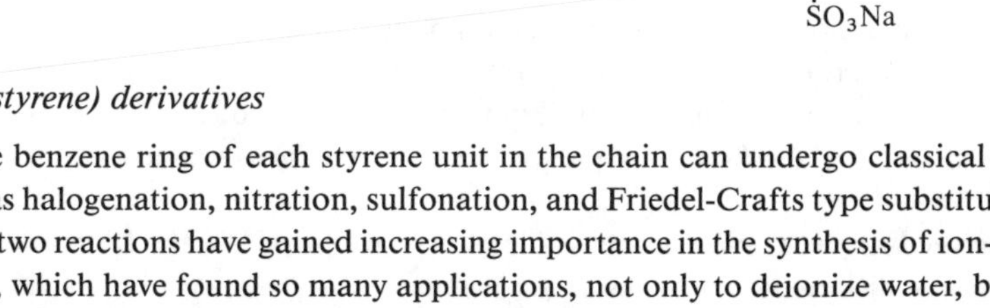

Poly(styrene) derivatives

The benzene ring of each styrene unit in the chain can undergo classical reactions such as halogenation, nitration, sulfonation, and Friedel-Crafts type substitution. The latter two reactions have gained increasing importance in the synthesis of ion-exchange resins, which have found so many applications, not only to deionize water, but also to carry out reactions. Generally the starting polymer is a tightly crosslinked, yet macroporous, random copolymer of styrene and divinyl benzene, but the reactions can be carried out as well on linear polystyrene.

Sulfonations occur upon reacting the polymer with either concentrated sulfuric acid (in which case silver ions act as efficient catalysts) or molecular complexes of SO_3 with organic solvents such as dioxane. The preferred reaction site is the para position, but some ortho sulfonation also occurs. The sulfonated resins are excellent cation exchangers, and can be regenerated easily upon treatment with hydrochloric acid to remove the ions that have been retained.

Cationic *chloromethylation* is the first step of the chemical modification of polystyrene to make anion-exchange resins. This reaction is best carried out with monochlorodimethylether in the presence of a Lewis acid ($SnCl_4$):

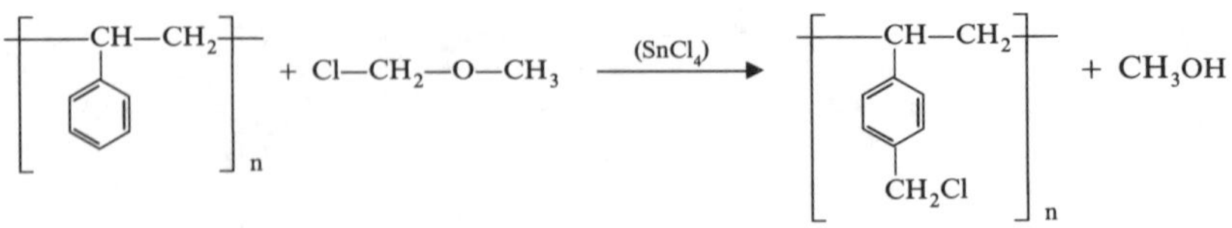

Some of the benzyl chloride sites formed can in turn react with phenyl rings to produce *bridges* between chains, which does not harm as the starting material is crosslinked anyway.

The second step involves reaction of the benzyl halide sites with a tertiary amine, whereby quaternary ammonium salts are formed on the polymer chains. The material is then treated with sodium hydroxide to remove the halide counter ions, and it is then ready to serve as a regenerable anion exchange resin:

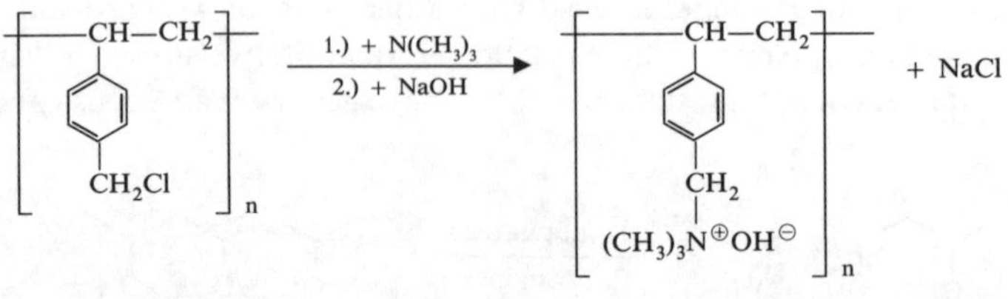

Reactions on poly(methyl methacrylates)

Hydrolysis of PMMA has been studied thoroughly, though it is of no great economical interest, because methacrylic acid can be polymerized readily. Poly(methyl methacrylate) does not hydrolyze easily, but its ability to undergo this reaction depends upon its microstructure. Under given experimental conditions isotactic PMMA reacts far more rapidly than its syndiotactic homologue. This is due to an intermediate anhydride formation, and as a consequence hydrolysis proceeds sequentially as already discussed.

The intermediate cyclization is impossible with syndiotactic PMMA, thus making its hydrolysis very difficult to carry out.

There are some easily hydrolysable poly(alkyl methacrylates), namely those with rather bulky alkyl substituents such as tert.-butyl, diphenylmethyl, or p-nitrophenyl. Acid catalysts are sometimes more efficient than the classical basic hydrolysis.

Nucleophilic substitution reactions on PMMA have also been studied. Various nucleophiles such as phenyl lithium, picolyl lithium*), and dimsyl sodium**), can react onto the carbonyl functions to yield substitutions. In the latter case a β-ketosulfoxide is formed, which can in turn get metalated and give yield to cyclization. As mentioned already these reactions are affected by the microstructure of the starting material; they are also strongly influenced by the polarity of the solvent used, as it can be expected for nucleophilic reactions involving ionic species.

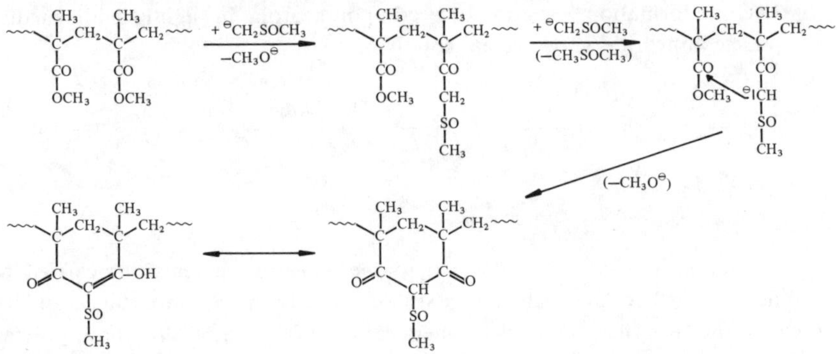

These nucleophilic substitutions (even in those cases in which no cyclizations can take place) do not go to completion, since the reaction gets more difficult as the degree of substitution increases, and the k_2 rate constant referring to BAB triads is equal (or close) to zero.

Lactonization reactions

The tendency to form 5 or 6 membered cyclic esters on a polymer chain is quite high, as the following example will show.

- Upon heating, *random copolymers* of *vinyl chloride* and *methyl methacrylate* loose readily CH_3Cl and lactone formation occurs at the expenses of VC/MMA diads:

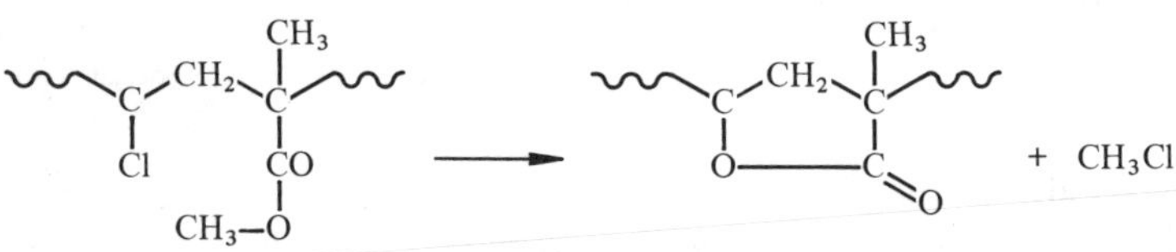

This rather unexpected reaction quantitatively uses up these diads, as it was shown by comparison with the probability of AB diad formation calculated from the reactivity ratios characterizing the system vinyl chloride-methyl methacrylate***).

- *Poly(methacroleine)* can undergo disproportionation upon treatment with dilute alkali hydroxide, whereupon cyclization takes place to form 6-membered lactones:

---

*) α-picolyllithium 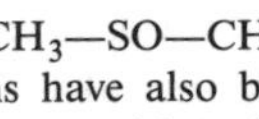

**) Dimsylsodium CH_3—SO—CH_2Na.

***) Similar reactions have also been observed upon heating poly(methyl α-chloroacrylate). Lactonization occurs with evolution of CH_3Cl.

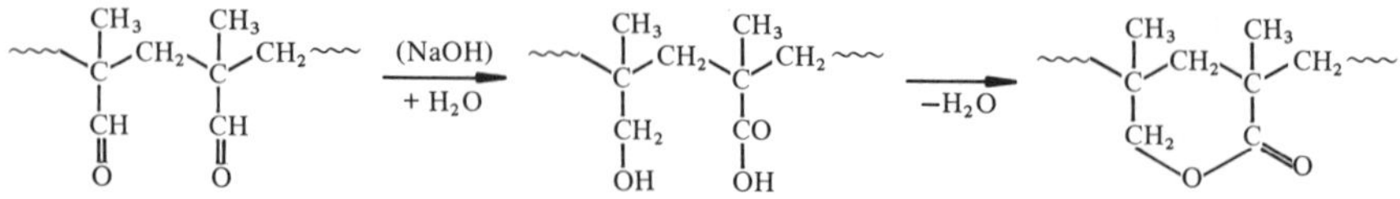

Similarly, poly(methacrylamide), when treated with NaOCl, undergoes lactamization.

- Lactone formation was also observed upon heating, in slightly acidic medium, of maleic anhydride/vinyl acetate copolymers:

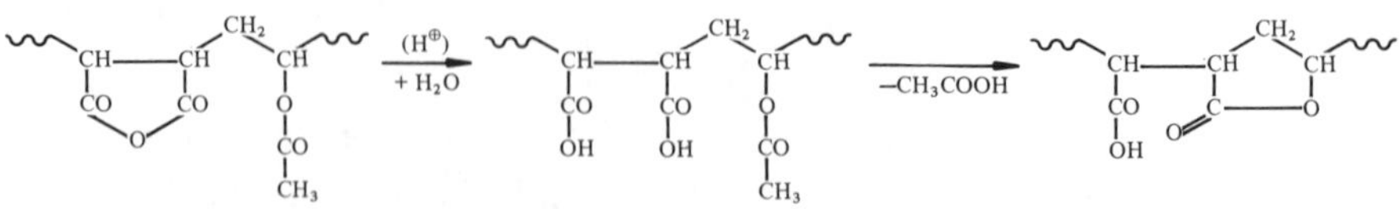

Some attention has to be devoted to the substitution reactions carried out on cellulose, since they have been and are still of major economic importance, and because they are the very first examples of chemical modification of a (natural) polymer.

Native cellulose is a highly crystalline hydrogen bonded solid that does not melt (until it decomposes) and does not swell in solvents. To be used, it has first to be modified chemically by *xanthogenation* (i. e. attack by rather concentrated NaOH and by CS_2) whereby it becomes soluble. Upon treatment with an acid the xanthogenate functions are removed, but the so-called *"regenerated" cellulose* can be cast as films or fibers. This is the basis of *"viscose"*.

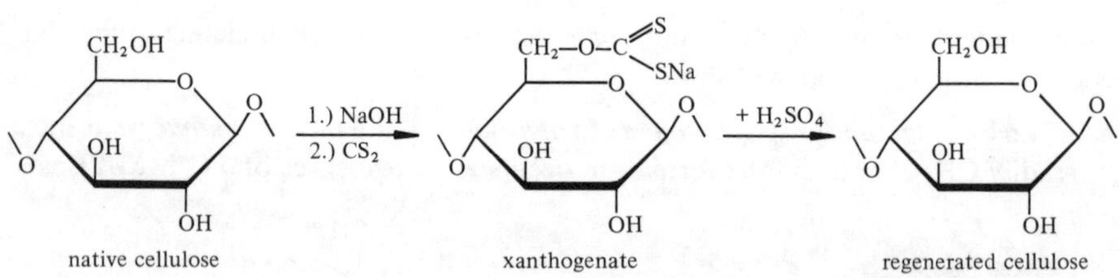

The primary alcohol function is the most reactive one. However, some reactions can also occur on the secondary hydroxyls. The probability of chain scissions appears to be rather small.

Other methods of solubilization of cellulose have been developed, and can also yield regenerated cellulose that is chemically identical with native cellulose.

Esters of cellulose have gained great importance. They are obtained by reaction of regenerated cellulose with a carboxylic acid. Sometimes the corresponding anhydride is added to help remove the water that is formed, as esterification is an equilibrium reaction. The reaction is heterogeneous, and only when 2 out of 3 alcohol functions of each unit have been esterified does the cellulose ester dissolve. Cellulose diacetate and cellulose triacetate are also fiber and film forming species and are still used as such.

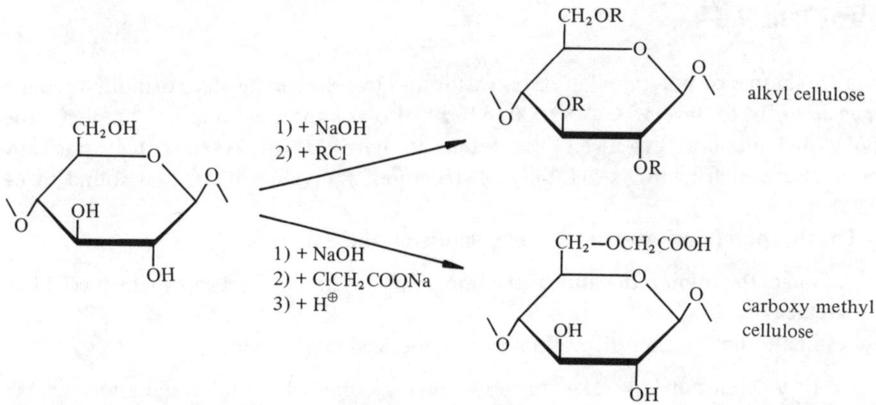

Cellulose nitrates are made by reaction of cellulose with concentrated nitric acid. Here again the esterification can affect one, two or all three sites of each repeat unit, and they have quite different applications, as films, coatings, or even as explosives.

Other important cellulose derivatives result from substitutions carried out on previously metalated cellulose. Reaction of methyl chloride will yield *methyl celluloses*, and reaction of mono chloracetic salts will result in *carboxy methyl cellulose*, a polyelectrolytic, water soluble derivative of cellulose, that can be used as suspension stabilizer for instance.

Appendix 9

Table of the Values of S_n and S_n/n

n	S_n	S_n/n
1	1	—
2	0	—
3	1	—
4	0.667	—
5	1	0.2
6	1.067	0.178
7	1.222	0.175
8	1.353	0.169
9	1.489	0.165
10	1.624	0.162
11	1.760	0.160
12	1.895	0.158
13	2.030	0.156
14	2.166	0.155
15	2.301	0.153
16	2.436	0.152
17	2.572	0.151
18	2.707	0.150
19	2.842	0.149
20	2.979	0.148
∞	—	$0.135 = e^{-2}$

Problem 9.1

a) A sample of poly(vinyl acetate) was found by reliable absolute osmotic pressure measurement to have $\overline{M}_n = 192000$. After hydrolysis (presumed to be complete), the poly(vinyl alcohol) produced was found to have $\overline{M}_n = 22000$. After attack by periodate, which hydrolyses 1,2-glycol structures, the value of $\overline{M}_n$ was found to be 7300.

For this poly(vinyl acetate) sample, and from these data

— estimate the minimum value of the branching coefficient and show why it could be greater,

— estimate the fraction of monomers polymerized head to head.

b) It is common to refer to poly(vinyl alcohol) by the designation: "95% deacetylated", or "99.8% deacetylated". Let f be the fraction of remaining acetate groups. Thus 90% deacetylated means $f = 0.1$.

Derive expressions for the eight triad probabilities AAA, AAO, AOO, . . . where, for example, AAO is the probability that a side group selected at random is an acetate group (A) which is followed by another acetate group (A) and then by a hydroxyl group (O). What assumption does this calculation imply? State qualitatively which triads would be favored if the deacetylation reaction were either autoaccelerated or autoretarded.

Problem 9.2

In the following transformations of pre-existing polymers, it is assumed that the main chain is not disrupted. Determine the fractional change in molecular weight of the polymer, occuring upon its transformation:

a) Cellulose is converted completely to cellulose triacetate by acetic acid/anhydride treatment.
b) Poly(vinyl chloride) experiences dehydrochlorination leading to perfect conjugate unsaturation.
c) Polystyrene is converted to 50% by treatment with chloromethylether and subsequently it is reacted quantitatively with trimethylamine.
d) Poly(vinyl acetate) is completely hydrolyzed by NaOH (branching of the initial polymer is neglected).
e) Poly(vinyl alcohol) is reacted with butyraldehyde to the limiting conversion, without any crosslinking.
f) Poly(tert.-butyl acrylate) is completely hydrolyzed by NaOH.

Problem 9.3

Butyl rubber containing 3 mole percent isoprene is crosslinked via its isoprene units. What is the minimum weight average length of the chains that will permit a network to be formed, provided all isoprene units become branch units.

Problem 9.4

A copolymer of methyl acrylate (A) and vinyl acetate (B) is generated from a mixture of monomers A and B, the mole ratio being maintained to 3:1. The radical reactivity ratios are: $r_a = 9.0$, $r_b = 0.1$.

This copolymer is then hydrolyzed by hot NaOH solution and washed with 0.1 N H_2SO_4. The polymer is then heated to 150 °C under vacuum, whereby esterification of neighboring carboxyl and hydroxyl groups produces lactones. Estimate the maximum number of acrylic acid residues that are to be involved in lactone ring formation.

Chapter 10
Functional Polymers, Block and Graft Copolymers, Model Networks

10.1 Introduction

This chapter is devoted to the methods of synthesis of well defined polymer chains; macromolecules fitted *quantitatively* at chain end with a known function (capable of further reactions); block copolymers, the polymer chain of which is made of a limited number of homopolymer sequences of different chemical nature (generally 2 or 3) and graft copolymers constituted of a homopolymeric backbone carrying a number of grafts of different chemical nature, randomly distributed. These species were defined in chapter 1. Also, some indications will be given on the pathways leading to "model networks", in which the elastically effective, linear network chains (connecting two branch points) are of known length, and of narrow molecular weight distribution.

The high interest devoted in recent years to block and to graft copolymers arises from the *incompatibility* between most chemically unlike polymer chains. If such homopolymeric segments are linked together in a block copolymer the chain conformations have to acknowledge for the repulsions originating from this incompatibility, which in most cases leads to intramolecular phase separation. As a result the material becomes biphasic and behaves like a *self-reinforced* material. In a triblock copolymer in which the center block is elastomeric and constitutes the continuous phase, and the two others, either amorphous vitreous or crystalline, become the dispersed phase, the material behaves like a vulcanized elastomer over a temperature range reaching to the glass transition temperature or to the melting point of the dispersed phase.

At temperatures higher than T_G (or T_M) the material flows or yields irreversibly, and it can be processed, extruded or molded. It recovers its elastic behavior upon cooling down to temperatures below T_G (or T_M). These materials are referred to as *thermoplastic elastomers**).

Two main pathways have been used for the synthesis of block and graft copolymers:

- Either a chain is fitted with active sites that are capable of initiating the polymerization of another monomer (scheme 1),
- or a polymer chain, fitted with end standing functions X, is allowed to react with another polymer chain fitted with antagonist functions Y, to establish links between the two chains (scheme 2).

---

*) It should be remembered that polymers other than block copolymers can exhibit the behavior of thermoplastic elastomers. It is the case, in particular, of the so-called ionomers, copolymers of ethylene and methacrylic acid, after "solidification" of the latter units by means of divalent ions, which thus build bridges between chains, yielding reversible "vulcanization".

Scheme 1

Scheme 2

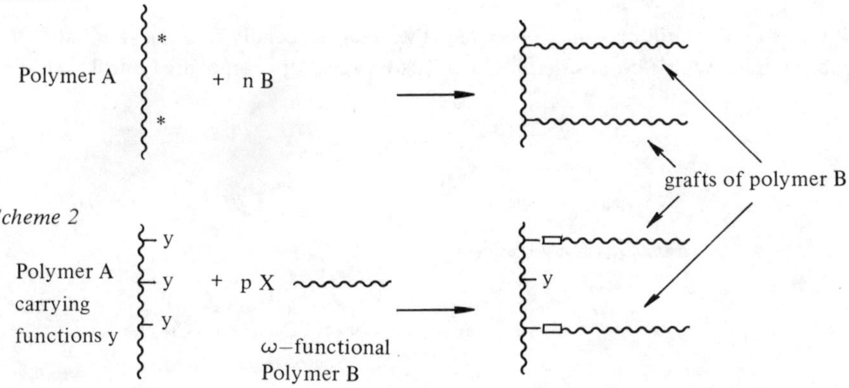

Block copolymers are obtained when the sites (or the functions) on polymer A are located at chain end, and graft copolymers result when the sites (or the functions) are randomly distributed along the poly-A backbone.

10.2 Functional Polymers

We shall first consider the various methods able to yield a polymer carrying one well defined function at the chain end. In the subsequent sections of this chapter it will be shown that several methods of synthesis of block copolymers depart from ω-functional polymers.

10.2.1 Functionalization by Step Growth Polymerization

Polymers originating from polycondensation reactions always carry functional groups at both chain ends (except for the case in which the limitation of the molecular weight is achieved by addition of some mono-functional compound).

Upon self-condensation of $\underline{a}\underline{b}$ type base molecules the polymers always carry one $\underline{a}$ and one $\underline{b}$ function.

The polymers arising from stoichiometric polycondensation of $\underline{a}\underline{a}$ and $\underline{b}\underline{b}$ base molecules carry $\underline{a}$ and $\underline{b}$ functions at chain ends, distributed at random (50% $\underline{a}$--$\underline{b}$ type, 25% $\underline{a}$--$\underline{a}$ type and 25% $\underline{b}$--$\underline{b}$ type molecules). Upon using an excess of $\underline{b}\underline{b}$ base molecules, one can have all macromolecules fitted at both chain ends with $\underline{b}$ functions, provided quantitative conversion of the $\underline{a}$ functions has been attained. An advantage of this procedure is that the number average degree of polymerization is efficiently controlled by the stoichiometric imbalance between $\underline{a}$ and $\underline{b}$ functions, i.e. the ratio r of $\underline{a}$ to $\underline{b}$ base units (chapter 2).

10.2.2 Functionalization by Free Radical Polymerization

Functional polymers can be obtained upon use of a functional initiator or by means of controlled transfer reactions.

10.2.2.1 Functional Initiators

Functional initiators have been synthesized especially among aliphatic azo compounds. Azobis-cyanovaleric acid and azobis-cyanopentanol are typical examples:

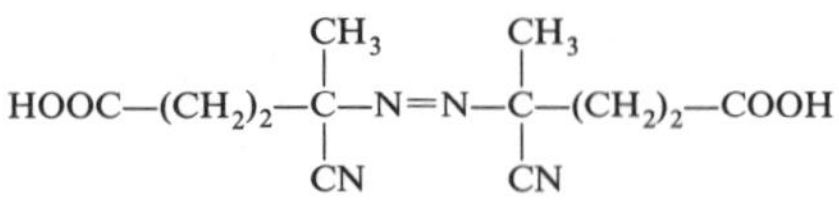

azobis-cyanovaleric acid

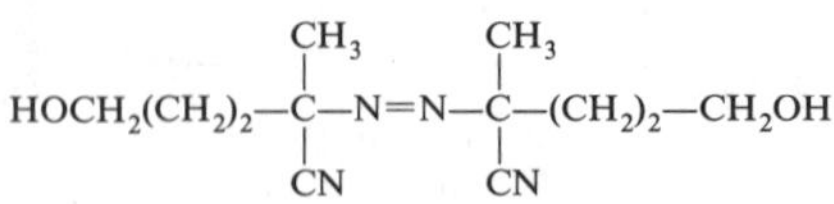

azobis-cyanopentanol

The homolytic cleavage of these initiators yields functional primary radicals, which in turn give rise to ω-functional polymer chains. However, this method has two drawbacks:

- First, termination may occur by disproportionation or by radical recombination. In the former case the polymer molecules will carry one endstanding function; in the latter case both ends of the macromolecules may carry functions. In most free radical processes, both modes of termination coexist.
- Secondly, any transfer reaction occurring during the polymerization process will yield unfunctionalized chains, unless the initiator itself acts as a transfer agent. But azo-compounds generally do not induce transfers, i. e., the transfer constant C_I is close to zero.

10.2.2.2 Functionalization by Transfer Processes

The prime role of a *transfer agent* in a free radical polymerization is to allow a control of the molecular weight of the species formed by a proper choice of the molar ratio [monomer]/[transfer agent]. The effectiveness of a transfer agent T (for a given monomer) is characterized by its transfer constant C_T. The problem is to find efficient functional transfer agents that would allow simultaneously the desired functionalization at chain end *and* an adequate control of the molecular weight. A large value of C_T is essential because it minimizes the proportion of polymer arising from primary radicals (which are not carrying any endstanding function)*).

Here are a few examples of functional transfer agents that meet the requirements stated above:

Mercaptoethanol, $HSCH_2CH_2OH$, and thioglycolic acid, $HSCH_2$—COOH, have been used with monomers such as styrene and methyl methacrylate to get end standing

*) Chain termination by recombination can be disregarded in such cases, as transfer is the dominant phenomenon, each initiating radical giving rise to many macromolecules before it terminates.

alcohol or acid functions. Iodoacetic acid, ICH_2COOH, has also been found to be efficient, especially with styrene. The transfer process can be schematized as:

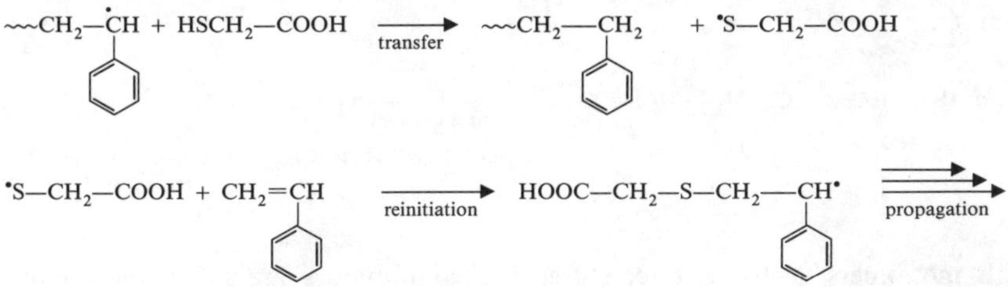

To further increase the proportion of functionalized macromolecules, use can be made of a functional initiator (carrying the same function as the transfer agent). Thus the macromolecules arising from primary radicals are also functionalized.

Photochemical reactions can also induce ω-functional polymer formation. A number of systems involving various metal carbonyls and halogenated compounds, such as CCl_4 have been found to give yield to radicals upon ultraviolet irradiation, as shown below:

$$Mo^{(0)}(CO)_6 + CCl_4 \xrightarrow[(UV)]{h\nu} CO\uparrow + Mo^{(I)}(CO)_5Cl + CCl_3^{\bullet}$$

The $CCl_3^{\bullet}$ radicals are able to initiate polymerizations of many vinylic and acrylic monomers, whereby the chains formed exhibit a CCl_3 group at chain end and most probably a CCl bond at the other chain end, as CCl_4 is known to be an effective transfer agent, too.

10.2.2.3 Telomerizations

Telomerizations have received increasing interest, as a special case of free radical transfer processes; they are especially useful for monomers such as vinyl chloride, vinylidene chloride, fluorine containing monomers and also styrene and vinyl acetate. *Telomers* are in fact functional oligomers. The degrees of polymerization range from 1 (adducts) to, say, 300, depending upon the systems. The end-standing function always originates from the transfer agent (often referred to as the telogen).

Telomerizations are initiated either by free radical initiators (such as benzoyl peroxide or azobisisobutyronitrile) or by various types of redox processes involving a metal salt (in many cases Fe^{+++} or Cu^{++} salts) and an organic reducing agent such as benzoine. Telogens include all kinds of trichloromethyl compounds CCl_3—R. Some of them originate themselves from telomerization reactions involving mere adduct formation, such as in the example quoted below:

$$CCl_4 + CH_2{=}CH{-}CH_2OH \xrightarrow[(benzoine)]{Fe^{+++}} CCl_3{-}CH_2{-}CHCl{-}CH_2OH$$

Vinyl chloride can be telomerized with this adduct (2,4,4,4-tetrachloro butanol) or with hydroxyethyl trichloro acetate ($CCl_3{-}COOCH_2CH_2OH$) in the presence of a redox initiating system to yield polymer chains fitted with terminal OH functions according to:

$$nCH_2{=}CHCl + CCl_3CH_2{-}CHCl{-}CH_2OH \xrightarrow[\text{redox initiation}]{} HOCH_2{-}CHCl{-}CH_2{-}CCl_2{-}\left[CH_2{-}\underset{\displaystyle Cl}{\underset{|}{CH}}\right]_n{-}Cl$$

With monomers such as styrene, free radical initiators are more appropriate. Similarly telomerization of vinyl chloride in the presence of trichloroacetic acid yields polymers bearing an endstanding carboxylic acid function, the molecular weight of which depends upon the monomer to telogen mole ratio (see section 3.10).

10.2.3 Functionalization by Anionic Polymerization

The *"living"* character of the sites in anionic polymerizations, carried out in aprotic solvents with organometallic initiators, offers a wide range of possibilities of functionalizations at chain end. All of them are based upon known reactions of metal organic derivatives towards electrophiles of various kinds, e. g. carbon dioxide, oxiranes, esters, nitriles, anhydrides, halides, etc. These reactions have been reviewed in chapter 5.

It should be added that the polymer chains can be fitted either at one end, or at both ends with functions, depending upon whether the initiator chosen was mono- or dicarbanionic. In any case the molecular weights can be chosen at will, and the molecular weight distributions remain narrow after functionalization.

10.2.4 Functionalization by Cationic Polymerization

The polymerization of vinylic monomers with cationic initiators is strongly affected by transfer reactions, implicating especially the monomers. These methods are therefore not well suited for the synthesis of functional polymers. However rather low molecular weight functional polymers have been made by the so-called *"inifer method"* which has already been discussed in chapter 6.

Quite different is the case of the cationic ring opening polymerization of monomers such as oxolane, dioxolane, and tert.-butylaziridine which can be considered as "living", provided the counterions are stable. In these cases functionalization reactions can be carried out with various nucleophiles such as alkali hydroxides (or even excess water), phenolates, alcoholates, and amines, as it has already been shown in chapter 6. Here again the chains can be fitted at one end or at both ends with adequate functions. The main disadvantage of this method is that it applies to a rather limited number of cases.

10.3 Block Copolymers

10.3.1 Block Copolymers by Free Radical Copolymerization Techniques

Free radical copolymerization techniques are *not* adequate to synthesize block copolymers, as was shown in chapter 4. The condition for spontaneous block copolymer formation upon copolymerization of a mixture of A and B monomers would imply that both radical reactivity ratios r_a and r_b be far larger than unity: No such system has ever been found.

Feeding successively growing radicals with monomers A and B has also led to failures, because of the very short lifetime of radicals.

There are however a few methods that are worth mentioning:

- One can make use of a difunctional initiator such as m-diisopropylbenzene dihydroperoxide.

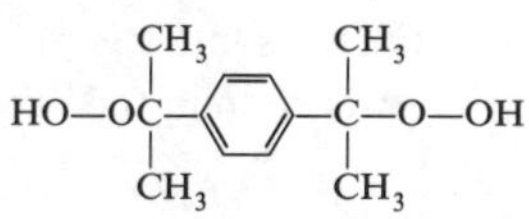

The first step involves polymerization of monomer A with this initiator, under mild conditions, in order to get polymer A chains fitted at chain end with a hydroperoxy function. The polymer is washed free of unreacted initiator and it is then mixed with monomer B. The polymerization of this latter species is initiated upon reduction of the hydroperoxide function with an iron salt.

$$(\text{polymer A}\!\!+\!R{-}O{-}O{-}H + Fe^{++} \longrightarrow Fe^{+++} + OH^{\ominus} + (\text{polymer A}\!\!+\!R{-}O^{\bullet}$$

- One can also make use of adequate *transfer reactions*. Free radical polymerization of monomer A is carried out in the presence of a transfer agent such as triethylamine, under conditions chosen such as to have almost all chains fitted with a diethyl amine end group:

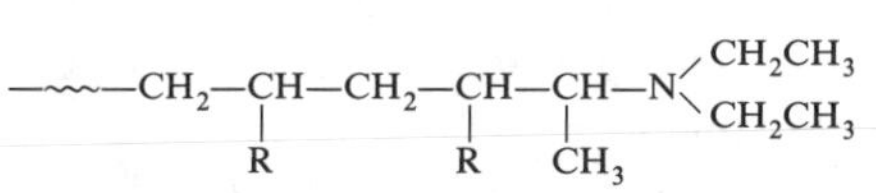

The second step involves the polymerization of monomer B in the presence of the polymer A made in the first step. If the transfer constant of the diethylamino group with respect to the radicals B is sufficiently high, transfer results, whereupon block copolymers poly-A — poly-B result. But further purification is necessary to make the copolymer free of the corresponding homopolymers.

- Attempts have been made to use the rather high transfer constant of trichloro methyl groups, as already mentioned. If the free radical polymerization of monomer A is carried out in the presence of CCl_4 the polymer molecules are likely to bear a CCl_3 end group. The amount of CCl_4 used also enables one to control the average molecular weight of the polymer obtained. To get the second block attached to the first one, the best way is to use the *photo reduction* method of *Bamford,* already mentioned, involving a metal carbonyl, whereby radicals are formed at chain end, which are able to initiate the polymerization of monomer B:

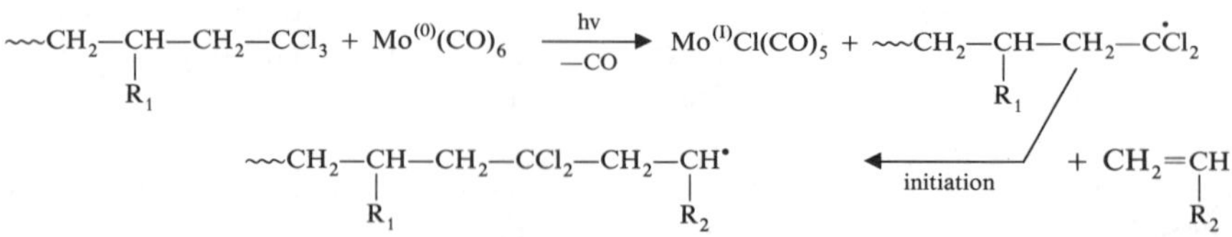

It should be remembered, however, that upon using transfer reactions the better the functionalization yields, the lower the molecular weight. These methods are fitted for the synthesis of *low molecular* weight block copolymers only.

Telomerization processes discussed above can also serve for the synthesis of *"block cotelomers"*, but in this case the molecular weights are still lower than in the two other cases described here.

10.3.2 Block Copolymers by Anionic Living Polymerization Techniques

As was briefly mentioned in chapter 5, the most obvious method to make block copolymers is to use the *"living" carbanionic sites* of a polymer A to initiate the polymerization of monomer B. If this second initiation reaction is fast and quantitative, each of the blocks will exhibit a *Poisson type* distribution of molecular weights, which implies that the fluctuations in composition within the copolymer sample should be of very small amplitude. To obtain such well defined "model" block copolymers, two conditions have to be fulfilled:

- Monomers A and B must be polymerizable anionically.
- The nucleophilicity of the anion $A^{\ominus}$ should be high enough to attack rapidly the incoming monomer B (but it should not give rise to side reactions).

Slow initiation would result in enhanced compositional heterogeneity. A scale of monomer electrophilicity has been set up in table 10.1.

The lower the position of a monomer A in the list, the higher the electrophilicity of the monomer, and the lower the nucleophilicity of the corresponding carbanion $A^{\ominus}$. A carbanion derived from a given monomer is in principle capable of initiating the polymerization of all monomers located below it in the list.

When the nucleophilicity of the carbanion $A^{\ominus}$ is very high, side reactions may arise if monomer B exhibits a marked electroaffinity. A proper way of avoiding this side reaction is to decrease the nucleophilicity of the carbanion by an intermediate addition of 1,1-diphenylethylene (DPE), or even sometimes of oxirane. DPE does not homopolymerize, so only one unit gets incorporated, and the diphenyl methyl carbanion that results is still efficient, but far less reactive than the original styrene, α-methyl styrene or diene carbanions. As an example let us quote the synthesis of styrene-methyl meth-

Table 10.1. List of anionically polymerizable monomers placed according to their increasing electrophilicity

— dimethylamino styrene, p-methoxy styrene
— p-methyl styrene, α-methyl styrene
— styrene
— butadiene
— isoprene, piperylene
— vinylnaphthalene (1- or 2-substituted)
— p-chloro styrene, p-bromo styrene
— vinyl pyridine (2- or 4-substituted)
— diphenyl ethylene (does not propagate)
— alkyl methacrylates
— acrylontrile, methacrylonitrile
— propiolactone
— ethylene oxide, propylene oxide
— ethylene sulfide, propylene sulfide
— vinylidene cyanide, α-cyanoacrylates

acrylate block copolymers, in which the intermediate DPE addition prevents side reactions of the ester function.

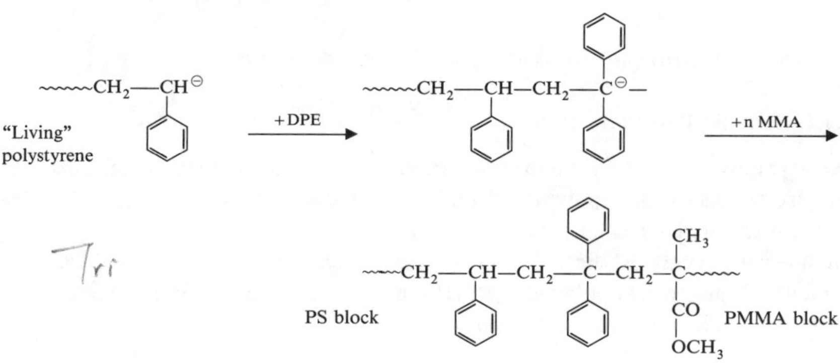

Triblock copolymers have been found of greater interest in a number of applications, especially as thermoplastic elastomers. A typical example is the triblock copolymer ABA in which a central elastomeric block (poly(butadiene) or poly(isoprene)) is surrounded by two blocks of an amorphous vitrous polymer. The most obvious way of synthesizing triblock copolymers is to use a bifunctional anionic initiator, and to polymerize monomer B to a dicarbanionic "living" polymer poly B, that will constitute the center block. Thereafter the polymerization of monomer A is initiated by both "living" sites of the poly-B precursor. Finally the sites have to be deactivated, for example by protonation. This method only applies if monomer B is less electrophilic than monomer A, as indicated earlier.

If this condition is not fulfilled another method has to be used. A monofunctional initiator is introduced to polymerize monomer A to a living polymer species. Thereafter monomer B is added, whereupon a "living" diblock copolymer results. Next a *coupling agent* is added in stoichiometric amount to yield the triblock copolymer wanted. Phosgene, α,α'-dibromo-p-xylene, and dichlorodimethyl silane have been shown to be efficient coupling agents.

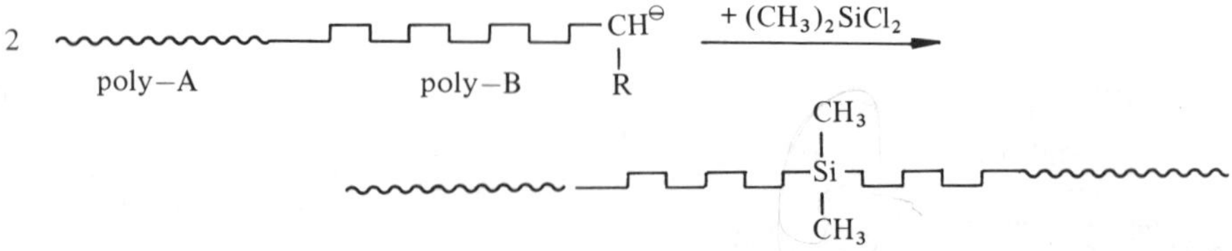

For industrial synthesis of triblock copolymers of styrene and diene to make thermoplastic elastomers, an alternate method was developed. Styrene monomer is polymerized first, in a non-polar solvent, using BuLi as initiator. Thereafter a mixture of styrene and diene is added to the "living" poly(styrene). By doing so, one takes advantage of the fact that the attack of styrene anions on a diene is much faster than that of a diene carbanion onto styrene: $k_{sd} \gg k_{ds}$. In spite of the higher rate of homopolymerization of styrene ($k_{ss} > k_{dd}$) the diene will polymerize first, and styrene will constitute the third block. This procedure, using $Li^{\oplus}$ as counterion and an apolar solvent, provides high 1,4-cis unit content in the central diene block, which therefore exhibits the expected elastomeric behavior*).

10.3.3 Other Techniques to Synthesize Block Copolymers

10.3.3.1 Cationic Polymerization

Cationic polymerization techniques are of interest in so far as living species exist, as in the case of ring opening polymerization of oxolane or tert.-butylaziridine. But even then, the attempts have been far from successful.

An interesting case is the *cross-termination* reaction between "living" cationic poly(oxolane), also called poly-THF (PTHF), and "living" anionic poly(styrene). The reaction yields block copolymers according to:

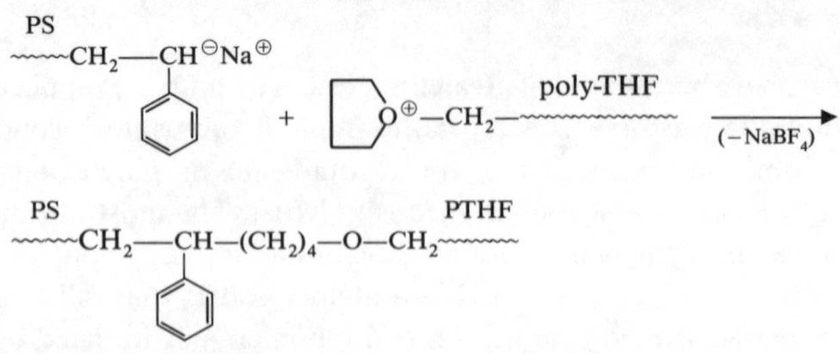

PS–PTHF block copolymer

*) These materials are thermoplastic elastomers because of the *intramolecular phase separation,* over a given range of compositions. The PS part is gathered in structural units (spheres or cylinders), whereas the elastomeric part fills the volume between these structural units, that act as crosslinks, below the T_G of the PS parts.

10.3.3.2 Site Transformation Techniques

Site transformation techniques have been studied carefully over the past years, with the aim of making block copolymers involving monomers that do not undergo polymerization with the same type of active sites.

An early example of *anionic to cationic* site transformation is shown below. A "living" carbanionic poly(styrene) is reacted with excess phosgene whereby an acid chloride function is formed at chain end. After removal of the excess phosgene, the polymer is treated with silver hexafluoroantimonate in THF, whereby a poly-THF block is formed, linked to the poly(styrene) block by an ester linkage:

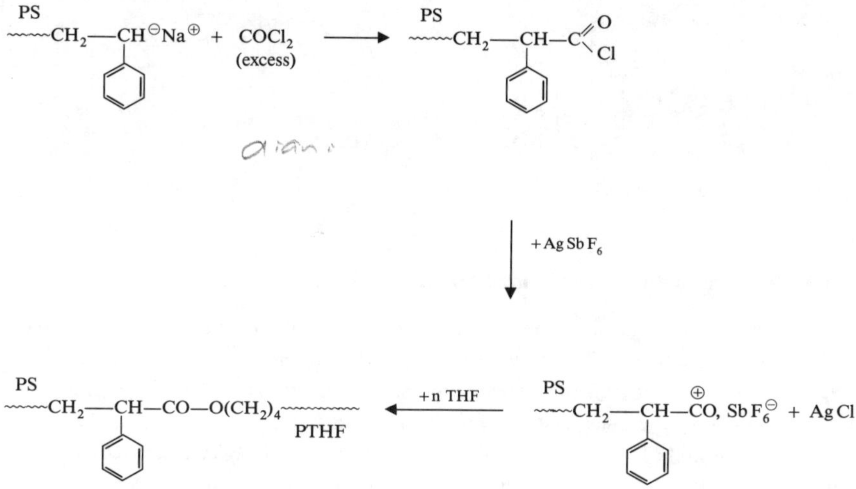

Anionic to free radical site transformations have been carried out successfully by means of intermediate carbon-lead linkages, which cleave homolytically, giving yield to radicals.

Of great interest is the *anionic to coordinated anionic* transformation, as it gives access to block copolymers comprising poly(ethylene), poly(propylene), and even poly(acetylene) blocks. These monomers can be polymerized only with Ziegler-Natta type systems. Typical ways include activation of the carbon-lithium bond by means of a transition metal halide, or replacement of the anionic site by a more active carbon-magnesium or carbon-aluminium bond, again activated by addition of a transition metal compound. Better results have been obtained upon choosing homogeneous Ziegler-Natta type systems, rather than heterogeneous ones.

A special case is that of block copolymers containing poly(amide) blocks, obtained upon ring opening polymerization of lactams. Here the most efficient way is to fit a polymer chain with an acyllactam end group, which may play the role of a "promoter" in the "activated monomer" lactam polymerization.

The process is straightforward and yields rather well defined block copolymers.

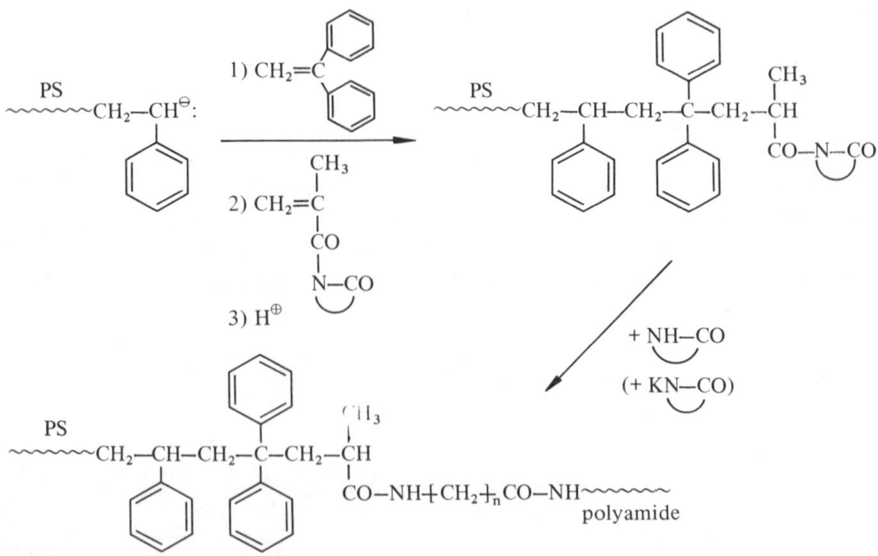

10.3.3.3 Coupling between Preformed Functional Blocks

ω-functional polymers obtained as shown in part 2 of this chapter can be reacted onto each other to give block copolymers, provided the yields are high enough, in spite of the low molar concentration of the reagents. Numerous examples of such reactions have been given in the past decade. However it should be recalled that the major difficulty encountered is precisely the *incompatibility* between the constituting homopolymers. Even in the presence of a common solvent the tendency to phase separation is high, and when this occurs the chances for the X and Y functions to react are decreased, because they are located in different phases. Once the reaction has started and some block copolymer is formed, the mixture becomes homogeneous, as the copolymer acts as a *compatibilizer* between the two homopolymers.

Here are a few examples of block copolymers formed upon reaction of ω-functional polymers:

- ω-carboxylic poly(styrene) can be reacted with polyesters, fitted at both chain ends with alcohol functions. This reaction can be speeded up by addition of dicyclohexyl-carbodiimide.
- ω-hydroxy poly(dienes) have also been reacted with various ω-functional polymers. In some cases it is advantageous to first end cap the polymer chain with a reactive function and to mix it in a second step with the other functional polymer. This can be done by reacting for instance the poly(butadiene) with a two-fold excess of a diisocyanate. The isocyanate functions are then reacted stoichiometrically with the OH-functions carried by the other polymer (polyester for instance).
- Block copolymers of poly(dimethylsiloxane), fitted at both chain ends with hydrogenosilane functions, can be reacted with poly(styrene) carrying vinylsilane functions at chain end. The latter species results from deactivation of "living"

poly(styrene) with chloro dimethyl vinylsilane, at low temperature. The hydrosilylation reaction is carried out in toluene solution in the presence of some catalyst (H_2PtCl_6), and high yields of block copolymers are obtained.

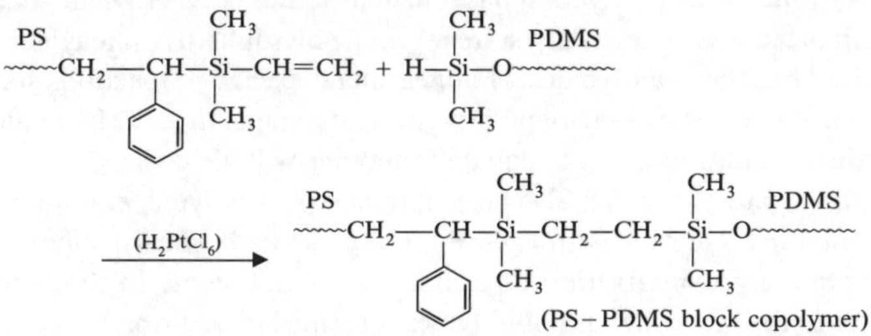

Triblock (or multiblock) PS/PDMS copolymers exhibit the characteristic properties of thermoplastic elastomers.

10.4 Graft Copolymer Synthesis

10.4.1 "Grafting from" Methods

As already indicated there is no major difference between block copolymer synthesis and graft copolymer synthesis, except for the location of the sites or functions to react. *"Grafting from"* methods are those which involve polymerization of monomer B starting from sites *created* on the polymer A.

Free radical sites can be formed on a polymer chain upon *irradiation* with γ-rays or electrons. If this irradiation is performed on a polymer swollen by a monomer, grafting generally results. This method of grafting has found a number of applications to modify polymers for special uses, to render them hydrophilic, to increase their swellability or their biocompatibility. Irradiation can be performed in the presence or in the absence of oxygen. The latter method is slower, but does not involve as many side reactions (especially chain scissions) as the former one.

Chemical initiation has also been attempted. Here a polymer backbone is made, that contains some cleavable bonds. For instance methyl methacrylate is copolymerized with some methacrylic chloride, and the acid chloride functions on the backbone are treated with tert.-butyl hydroperoxide. Upon heating this polymer in the presence of a second monomer, initiation takes place. But only half of the chains formed are grafted, the others arise from tert.-BuO• radicals.

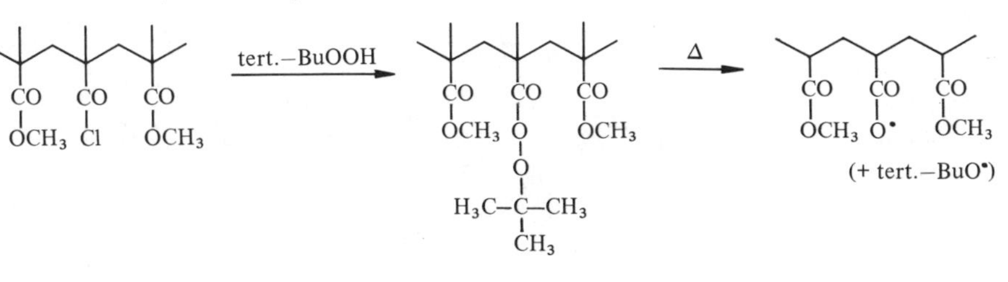

Anionic sites result from metalation reactions carried out on polymer backbones. The metalation can be performed in various fashions, but the reaction is generally difficult to control adequately. BuLi is often used as the metalating agent, and in most cases activation by means of a solvating base, such as tetramethyl ethylene diamine is required for enhanced efficiency. Poly(styrene) and poly(dimethyl phenylene oxide) have been used as backbone polymers. Halogen metal exchange reactions have also been performed in some instances with poly(p-chlorostyrene). Once the metalated sites have been obtained, addition of an adequate monomer will yield the grafts.

Cationic grafting by initiation has also been obtained. A copolymer containing some acid chloride functions is reacted with $AgSbF_6$. The generated oxocarbenium sites can be used to initiate the polymerization of oxolane or of isobutene. In the latter case, owing to transfer to the monomer, homopolymer is formed in addition to the expected graft copolymer. Partially chloromethylated poly(styrene) was also used successfully as a polymer backbone.

10.4.2 "Grafting onto" Methods

"Grafting onto" methods involve reaction of functions X sitting at chain end of a polymer species with antagonist groups Y distributed randomly on the backbone of the other polymer.

The anionic "grafting onto" method is based upon the reaction of "living" polymers like polystyrene, polybutadiene, or poly(vinyl pyridine) onto electrophilic functions carried by a polymer backbone (such as ester, anhydride, nitrile, benzyl halide).

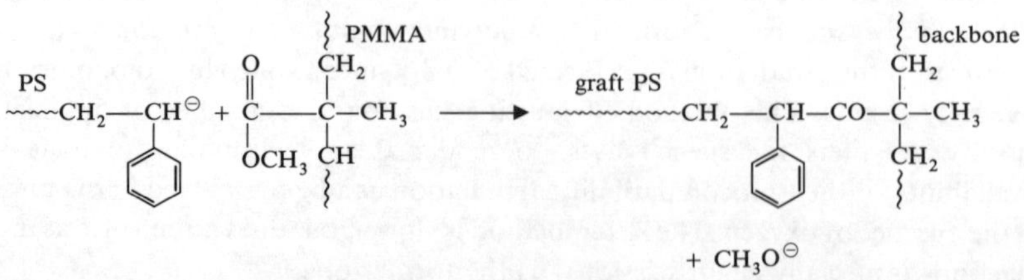

This method yields well defined polymers, as backbone polymer and grafted chains can be characterized independently.

Special interest has been devoted to *amphiphilic* graft copolymers that comprise hydrophilic grafts attached to a hydrophobic backbone. A typical case is that of polystyrene fitted with poly(ethylene oxide) grafts, made upon reaction of living PEO onto a partially chloromethylated polystyrene backbone.

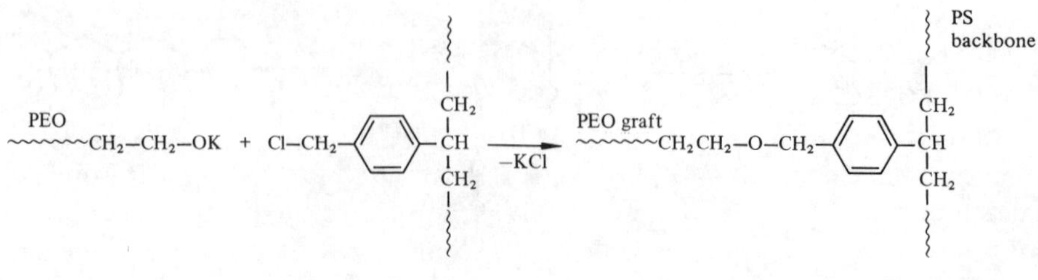

If these graft copolymers are placed in a solvent, they have a strong tendency to form micelles, except when the medium is capable of solvating both the grafted polymer and the backbone species. The term *"amphiphilic"* refers to the very different solubility properties of the two constituents of the graft copolymer.

10.4.3 Graft Copolymer Synthesis via Macromonomers

Macromonomers, also referred to as *macromers®* , are short polymer chains fitted at one end with a polymerizable unsaturation. A great variety of methods have been developed lately to synthesize such species, many of them deriving from functionalization methods. In particular, macromolecules fitted with one end standing OH-function can be reacted with methacrylic chloride, and ω-carboxylic polymers can be reacted with glycidyl methacrylate. In both cases, macromonomers bearing end-standing methacrylic ester functions result:

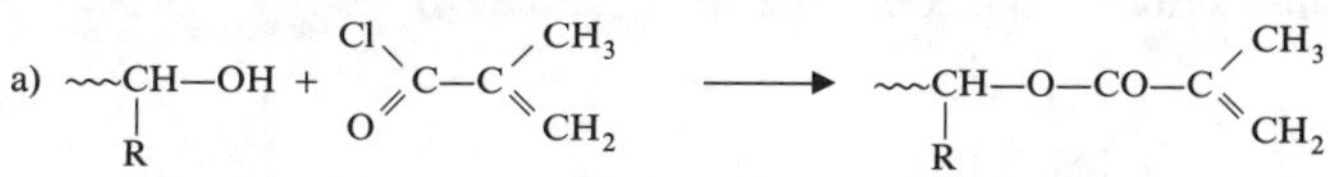

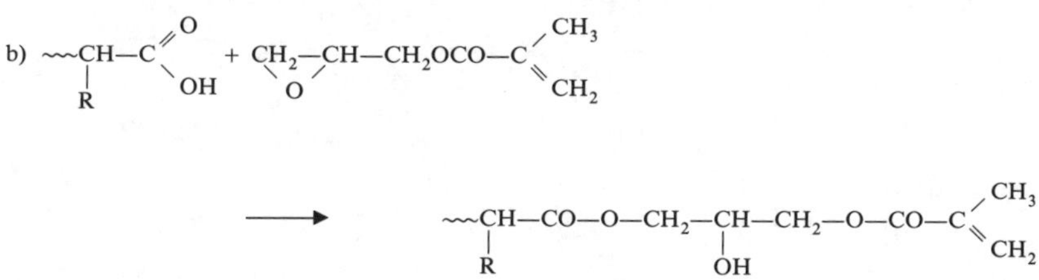

Ionic "living" polymerization techniques have also been used successfully to synthesize macromonomers. They provide adequate control of the molecular weight and low polydispersity of the samples. Two pathways can be used:

(1) Either an unsaturated initiator is used for the polymerization of a monomer, provided this initiator is efficient, and does not yield side reactions involving the unsaturation. A typical example is the preparation of poly(oxolane) macromonomers, using a methacryloyl oxocarbenium salt (MMA unsaturations are insensitive to attack by cations of any kind):

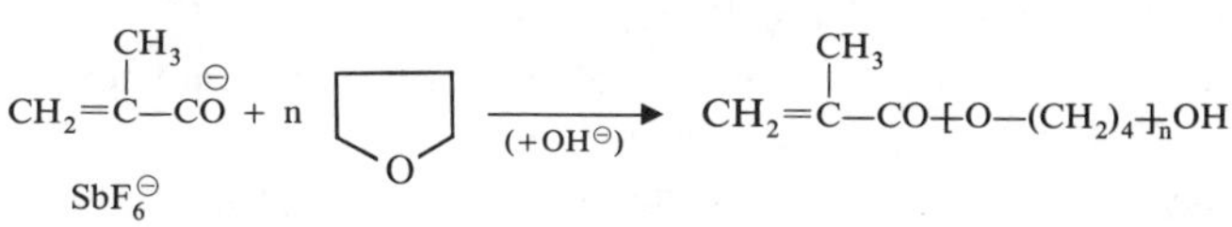

(2) Or, the "living" polymer, obtained upon initiation with a known amount of a standard anionic or cationic initiator, is reacted with an appropriate unsaturated deactivator (a nucleophile if the polymerization proceeded cationically, an electrophile if it was anionic). In a typical example, given below, a poly(styrene) macromonomer

is obtained upon reaction of the living sites (the nucleophilicity of which has to be reduced first) with methacryloyl chloride:

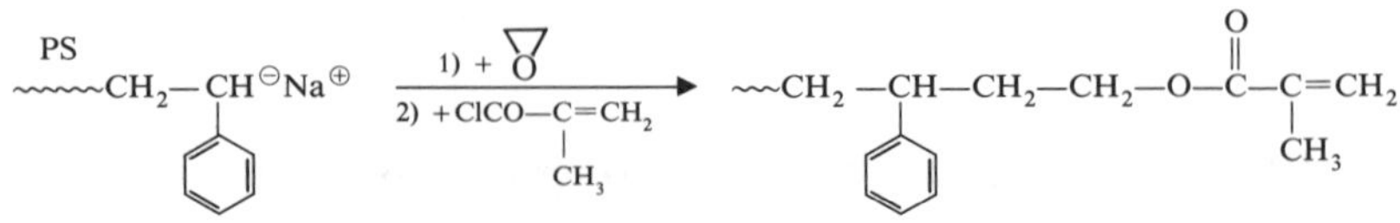

Copolymerization of macromonomers with a vinylic or acrylic monomer proceeds as a standard free radical copolymerization, except for the fact that the mole amount of macromonomer added to the reaction mixture is generally small, owing to its molecular weight, and to the fact that the graft copolymer should bear a limited number of grafts to be of any interest. If we call [A] the concentration of the comonomer and [M] the macromonomer concentration, whereby [A] > [M], the equation giving the instantaneous composition of the copolymer (see chapter 4)

$$\frac{d[A]}{d[M]} = \frac{[A]}{[M]} \cdot \frac{r_a[A] + [M]}{r_m[M] + [A]}$$

reduces to

$$\frac{d[A]}{d[M]} = r_a \cdot \frac{[A]}{[M]}$$

as in an "ideal" copolymerization. This implies that the reactivity ratio of the macromonomer does not play a major role in the process, and that the distribution of macromonomer units (i. e., of the grafts that are formed upon incorporation of these units into the growing chain) obeys *Bernouillian statistics.* Furthermore, in many cases, the magnitude of the compositional heterogeneity in such graft copolymers remains within narrow limits, even for rather high conversions, because of the high [A]/[M] ratio chosen.

Numerous examples of the free radical copolymerization of a macromonomer and a vinylic or acrylic comonomer have been given in recent years. Macromonomers of polystyrene, poly(ethylene oxide), poly(butylene oxide) (poly-THF), poly(methyl methacrylate), and polydienes have been copolymerized with monomers such as styrene, alkylmethacrylates, acrylonitrile, vinyl chloride, etc.

The chief interest of graft copolymers originates from their properties as surface modification agents. When a polymeric material is coated with a graft copolymer the backbone of which is of the same chemical nature, the grafts (that are incompatible) will accumulate on the surface, thereby modifying it. Such graft copolymers can be used as antistatic agents, surface humidifiers, dye binding intermediates, adhesives, lubricants, and so on.

10.5 Network Formation

10.5.1 General Methods — Random Crosslinking

Quite generally, polymeric networks are formed in polymerization processes involving monomers or base molecules that exhibit more than two patented functionalities (a double bond being difunctional in that context). The only exception is the step growth polymerization of $a\!-\!\!<^{b}_{b}$ type base molecules, with or without $\underline{ab}$ base molecules. This off-stoichiometry polycondensation obviously leads to highly branched, tree-like, yet uncrosslinked polymer molecules (cf. chapter 2).

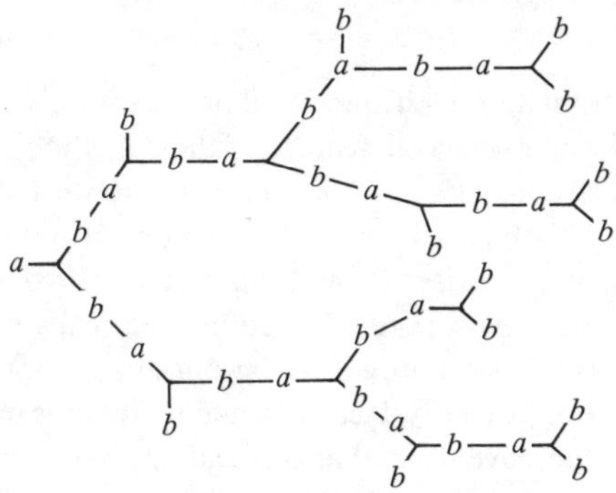

tree–like, uncrosslinked polymer

In all other cases recorded network formation eventually results from a polymerization involving units with functionalities higher than 2, provided sufficiently high conversion degrees are attained. Special cases are:

Step growth polymerization of $\underline{aa}$ and $\underline{bb}$ base molecules in the presence of some $b\!-\!\!<^{b}_{b}$ trifunctional molecules. A typical example is that of polyester formation between a diacid and a diol, with some triol (such as glycerol or better trimethylolpropane) to induce the branching, leading to network formation at a very precise conversion degree α_c that was calculated in chapter 2.

Free radical copolymerization between two vinylic or acrylic monomers, one of them exhibiting two double bonds. A typical case is that of styrene-divinylbenzene copolymerization — already mentioned in chapter 4 — that has led to numerous applications: The crosslinked poly(styrene) gels can be used as base materials for the synthesis of ion-exchange resins, and they can also serve as column fillings in size exclusion chromatography and in liquid chromatography. The amount of DVB the network contains determines its swellability, and the pore size commands the retention times in size exclusion chromatography.

"Vulcanization" processes, already discussed in chapter 9, are reactions aiming at bridge formation between already existing chains. Whatever reaction is used to form the links between individual chains, the process is more or less random. Once, *on the average*, each chain has got bridges connecting it to two other chains, the gel point has been reached. The more crosslinks are introduced thereafter, the tougher the material:

The modulus of elasticity (ratio of the stress applied to the resulting strain) increases as the number of elastically effective chains rises, i. e., as the number of branch points increases.

Yet all these methods involve random crosslinking which implies that no experimental access is provided to the length of the elastic chains, i. e., the linear chain elements connecting two successive elastically effective branch points. Furthermore, the distribution of the lengths of elastic chains is broad. Only when *Bernouillian statistics* applies to the crosslinking process — in step growth polymerizations and in some vulcanization reactions — this distribution is a "most probable" one. In many other cases it is far broader.

10.5.2 End-Linking Methods

Methods of crosslinking departing from an α,ω-difunctional precursor polymer are referred to as *end-linking methods.* The characteristic feature of these methods is that upon crosslinking the precursor chains become the elastically effective network chains.

If such a well characterized precursor polymer, bearing X functions at both chain ends, is reacted stoichiometrically with a plurifunctional compound fitted with f antagonist functions Y, crosslinking ensues. This type of reaction is basically a crosslinking polycondensation reaction. As conversion progresses beyond the gel point, the amount of extractable material (sol fraction) steadily decreases, and so does the number of loose chain ends (precursor chains that have reacted at one end and carry still one unreacted function). The sol fraction can be made negligibly small by proper attention to completeness of conversion. The number of structural defects such as pendent (dangling) chains, already mentioned, and loops (both ends of a precursor chains being attached to the same branch point) can thus be minimized.

If these conditions are fulfilled, the networks thus obtained are well defined. Not only is the number of elastic chains equal to the number of precursor polymer molecules n inserted, but also the number of branch points is given by $2\ n/f$, where f is the functionality of the crosslinking agent used.

Here are the few examples of endlinking processes, involving precursor molecules of various kinds.

- Poly(ethylene oxides) fitted with OH-functions at both ends can be reacted with plurifunctional isocyanates to yield water swellable, non-ionic networks. The same type of networks can also be obtained by first end capping the poly(ethylene oxide) with isocyanate functions (by reacting it with a two-fold excess of a diisocyanate) and getting thereafter nework formation by reaction with a stoichiometric amount of a triol, such as trimethylol propane.
- Similar end-linking processes can be carried out with other α,ω-dihydroxy polymers, such as poly(oxolane) (poly-THF), polyesters, and also dihydroxy poly(butadienes). The latter polymer can be obtained either anionically, using ethylene oxide as deactivator, or by free radical polymerization using H_2O_2 as the initiator (which also transfers to the chain).
- Polystyrene networks have been made by reacting α,ω-dicarbanionic living polymers with plurifunctional reagents, such as triallyloxytriazine, under strictly

stoichiometric conditions, whereby well defined networks of poly(styrene) were obtained. This method is also applicable to other bifunctional "living" polymers, provided the nucleophilicity of their carbanionic sites is sufficient.

● The same method should be also applicable to "living" bifunctional poly-THF chains, when reacted with plurifunctional nucleophiles such as hexaethylene tetramine; but no detailed studies have been carried out along this line (cf. chapter 5).

● Poly(dimethyl siloxane) can be synthesized under conditions such as to have all chains carry either two hydrogeno silane functions or two vinyl silane functions at chain ends (as indicated in chapter 6). On the other hand, modified cyclic tetramers (*D4*) with either HSi or $CH_2{=}CH{-}Si$ functions are readily available. It is well known that — in the presence of platinum containing catalysts (H_2PtCl_6) — hydrosilylation reactions proceed to high conversion and with no side reactions.

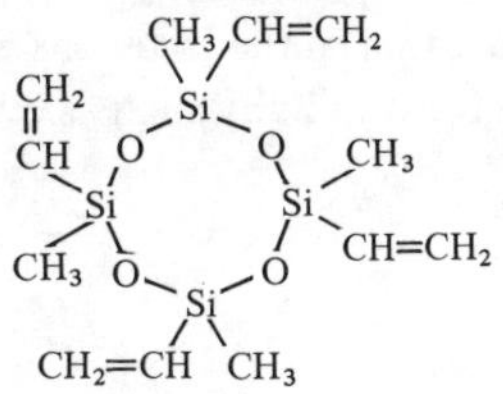

This reaction can be applied to the synthesis of PDMS networks, by reacting "tetravinyl *D4*" with α,ω-dihydrogenosilane PDMS, stoichiometrically and under appropriate conditions.

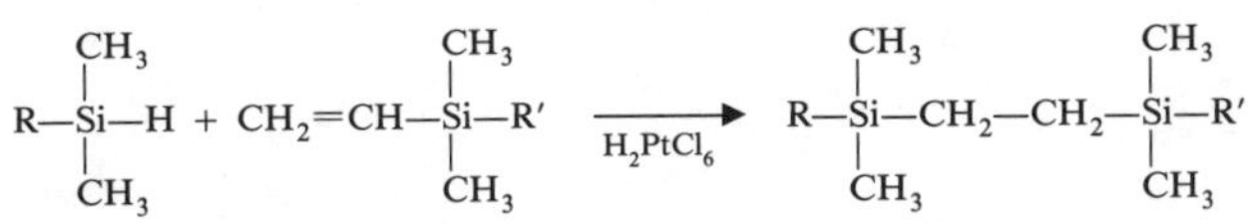

Alternately, "tetrahydrogeno *D4*" can be reacted with α,ω-divinylsilane PDMS. To obtain branch points of higher functionality, one can also react a poly(dimethyl siloxane) with end standing vinyl silane functions with PDMS oligomers carrying a number of HSi functions along their chain. The latter oligomer is obtained by ring opening polymerization of the "tetra-hydrogeno *D4*" (or tetramethyl cyclotetra-siloxane).

It should be pointed out that hydrosilylations can also involve allylic functions:

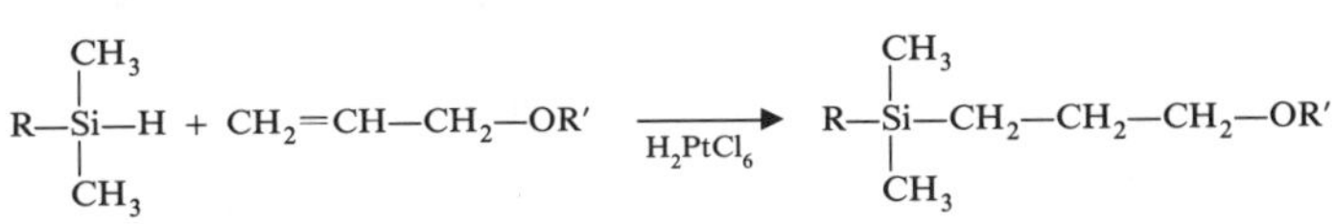

Thus PDMS "model networks" have been obtained upon reaction of α,ω-dihydrogeno-PDMS with tetra(allyloxy)ethane

$(CH_2{=}CH{-}CH_2{-}O)_2CH{-}CH(O{-}CH_2{-}CH{=}CH_2)_2$

in stoichiometric proportions, in the presence of some H_2PtCl_6 as the catalyst.

Another endlinking procedure derives from the anionic block copolymer synthesis. Styrene is polymerized using an efficient bifunctional anionic initiator (such as naphthalene sodium, or α-methylstyrene tetramer) to a living bifunctional precursor polymer. Subsequently the "living" sites of the precursor are used as initiators for the polymerization of a *small* amount of divinylbenzene, commonly between 3 and 10 molecules of DVB per living end. The polymerization of DVB yields tiny, tightly crosslinked nodules of poly-DVB that act as the branch points, as each of them connects the very number of chain ends of polystyrene precursors that have contributed to its formation. The polystyrene chains become the elastically effective network chains of the gel; the chief difference with respect to the preceding case is that the functionality of the branch points is higher (and depends upon the amount of DVB added), and that it may fluctuate within the sample.

Networks of polydienes, poly(vinyl pyridine), and poly(alkyl methacrylates) can be made according to the same procedure, except for the fact that in the latter cases DVB has to be replaced by a diunsaturated species of higher electrophilicity such as ethylene dimethacrylate.

Part II
Reactors and Processes for Manufacturing Polymers

Chapter 11
Interrelationships of Polymer Structure, Chemistry of Synthesis, and Process Engineering

11.1 Product Definition: Choice of Chemical Route: Engineering Decisions

The purpose of Part II is to provide a conceptual framework illustrating the engineering role in polymer synthesis. Tables 11.1 and 11.2 present one conceptual outline. The beginning is always product definition. Until this is fully established any further specification as to how to make it is fruitless. Product definition involves three separate concepts: **(1)** physical (molecular) properties, **(2)** scale of manufacture needed; **(3)** maximum allowable cost per unit of product.

Molecular properties to be specified include:

(a) Ranges of molecular weights M_w and M_n, and distribution.

(b) Range of copolymer composition in copolymers: $\langle F_A \rangle \pm \Delta F_A$, i.e. content of monomer averaged over the entire polymer produced with maximum and minimum values permissible.

(c) Specific configurational features of the chains, for example,

- ○ 1.4 versus 1.2 or 3.4 addition in dienes;
- ○ isotactic, syndiotactic, heterotactic sequence distribution in vinyls;
- ○ content of short branches, long branches (if any);
- ○ content of head to head coupled mers.

(d) Specific features of end groups

- ○ amino or carboxyl, or neither, in nylons;
- ○ terminal unsaturation in vinyls;
- ○ end groups intended for subsequent reaction.

These molecular properties will obviously determine the gross physical properties such as,

(e) Viscosity of the melt, and shear rate dependence;

(f) Mechanical properties in the solid state (ultimate tensile strength, toughness, loss tangent, etc.);

(g) Optical properties (refractive index, specific absorption in certain ranges of wave lengths);

Table 11.1.

Polymer
Product Specified
↓
Chemistry Decisions Needed
↓
Connection of Monomer Units

Only one kind — Two or more kinds

Want Alternation?
Random
Addition?
Block Sequences?
Specific "Coded"
Sequence?

Networks — Strings of Units

Some Monomer with
f > 2

Bifunctional
Monomer

If Step,
By-Product Removal
Needed

Step or Chain Reaction?
Kinetics

Homogeneous or
Heterogeneous

Appropriate
Temperature Level

Appropriate Pressure
if not Atmospheric

Where is Equilibrium?
Are We Near It?
Kinetic Control,
Diffusion Control, or
Both?
Undiluted or Solvent?
If Solvent, is it Crucial
to Chemistry, or only
Convenient?
Heterogeneous:
Obligatory or
Optional?
How Start Reaction?
Need to Stop Reaction?
Avoidable Competing
Reactions?
(Exclude O_2, H_2O?)
Unavoidable
Competing Reactions?
(How Cope With?)

(h) density, a function of percent crystallinity, or of time (long term relaxation) in non-crystalline glassy polymers.

In addition, the process of synthesis may introduce materials that will significantly affect optical and electrical properties, leading to turbidity or unwanted conductance. These include residual ions of redox initiation, soaps from emulsion polymerization, stabilizers from bead polymerization, silica, titania, alumina, chromia, etc., from stereoregulating (Ziegler-Natta, Phillips) polymerizations.

In polymers used in medicine and surgery, it is essential that no harmful residues of the polymer synthesis are present that could be transferred by diffusion from the polymer into fluids that will subsequently enter a patient, or into the patient by direct contact.

The second and third specifications for the product: scale of manufacture and cost, are arrived at with the assistance of experts in marketing. They can forecast, (or already know) the size of the market, the rate of growth, the competitors already in it, and the maximum cost of production per unit of polymer product which will permit the product to be sold at a profit.

After the product requirements are fully established, the list of Table 11.1 is followed. In most cases only a single chemical route is available. For example vinyl chloride is polymerized only by a chain reaction and only via a radical chain reaction; on the other hand, styrene can be polymerized via radical or anionic or cationic routes. Nylon 6.10 can be synthesized by the reversible slow, homogeneous step polymerization from diamine and diacid or the very fast irreversible heterogeneous step polymerization from diacid chloride and diamine. Styrene-butadiene co-polymers can be synthesized by free radical polymerization or by anionic polymerization, in which case the choice of a non-polar hydrocarbon as solvent is crucial, or by heterogeneous stereoregulating polymerization.

Thus the second task is to select the most appropriate chemical route to the polymer product, the specifications of which were the fist task.

After the chemical route is chosen (if there is a choice), and before the reactor type and process are selected, the changes in physical properties that occur as monomer is converted to polymer must be considered.

11.2 Physical Properties: Variation as Monomer is Converted to Polymer

Heat capacity, thermal conductivity, and density are properties which do not differ greatly between polymer and the monomer from which it is generated. In sharp contrast viscosity and diffusion coefficients vary by orders of magnitude, as monomer is converted to polymer in *homogeneous* systems.

Figures 11.1 to 11.4 illustrate schematically these trends. In Fig. 11.1, poly(methyl acrylate), a polymer with a glass transition T_G at 5 °C, is taken as an example. The viscosity *at 60 °C*, well above the glass transition, as a function of ϕ_2, volume fraction of polymer, is shown for three different polymer molecular weights, changing by orders of magnitude.

In Fig. 11.2, the viscosity at 60 *°C* of a solution of polymethyl methacrylate in its monomer is plotted vs. volume fraction of polymer ϕ_2; it tends towards infinity at the volume fraction ϕ_2^* which makes T_G equal to 60 *°C.* In Fig. 11.3 the diffusion coefficient of *monomer* through *polymer* is shown as a function of temperature, for the case of polymethyl methacrylate in its own monomer. At the value of ϕ_2 which corresponds to T_G, i. e. ϕ_2^*, D_{mp} drops towards zero. At temperatures well above T_G, the decrease of the diffusion coefficient is only of about two orders of magnitude. In contrast, the diffusion coefficient of polymer through polymer D_{pp} is much earlier decreased towards zero as ϕ_2 approaches ϕ_2^* at the glass transition. Furthermore, the

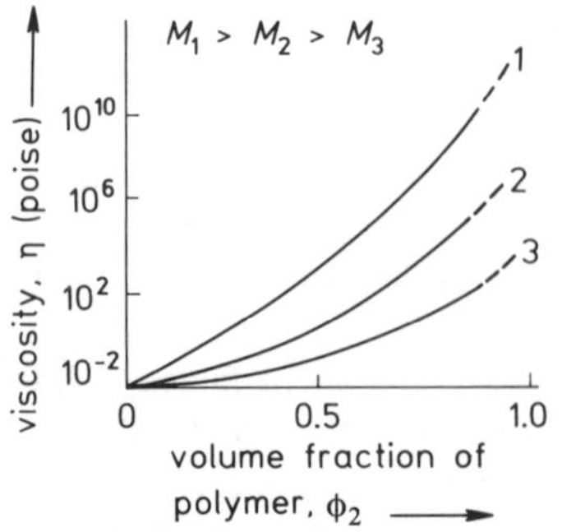

Fig. 11.1.

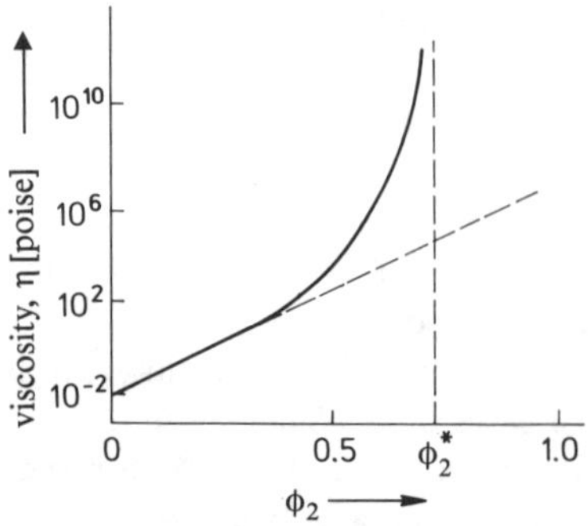

Fig. 11.2.

Fig. 11.1. Variation of the viscosity of poly(methyl acrylate)/methyl acrylate mixtures for 3 different polymer molecular weights, plotted versus volume fraction ϕ_2 of polymer

Fig. 11.2. Viscosity of PMMA-MMA mixtures plotted versus ϕ_2 at 60 °C. The viscosity tends towards infinity at the volume fraction ϕ^*, at which T_G of the system reaches 60 °C

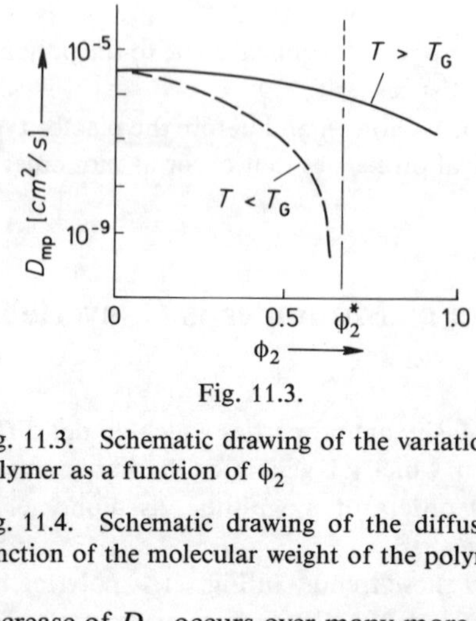

Fig. 11.3.

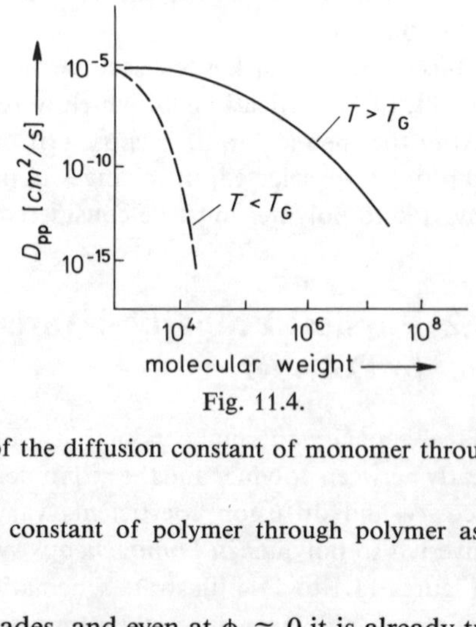

Fig. 11.4.

Fig. 11.3. Schematic drawing of the variation of the diffusion constant of monomer through polymer as a function of ϕ_2

Fig. 11.4. Schematic drawing of the diffusion constant of polymer through polymer as a function of the molecular weight of the polymer

decrease of D_{pp} occurs over many more decades, and even at $\phi_2 \cong 0$ it is already two or three orders of magnitude less than D_{mp}. Figs. 11.3 and 11.4 are entirely schematic; precise values of D_{mp} and D_{pp} can be quoted only if the molecular weight distribution of the polymer is fixed. However, the implications of increasing viscosity and decreasing diffusion coefficients in *single phase polymerizations* can be understood qualitatively.

Fig. 11.5 shows what would happen in the isothermal polymerization at 60 °C of pure monomer methyl methacrylate MMA to polymethyl methacrylate PMMA. Once the conversion had reached about 80% ($\phi_2 \approx 0.8$), PMMA dissolved in MMA would exhibit its glass transition at about 60 °C, and the chain addition of MMA would effectively cease. If the mixture is then heated to about 100 °C, which is slightly above the glass transition of pure PMMA without solvent, conversion of monomer to polymer can be carried to nearly 100%.

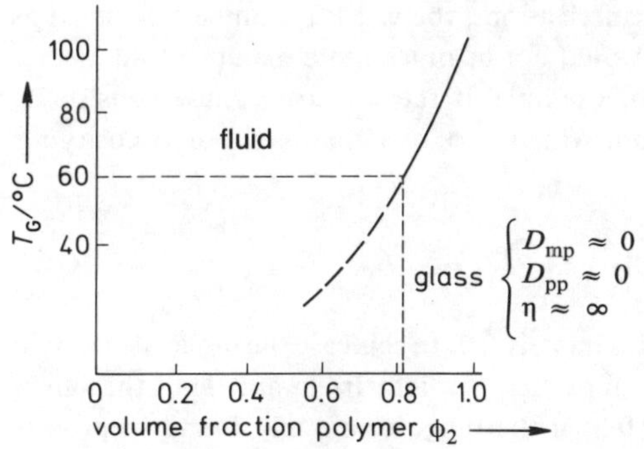

Fig. 11.5.

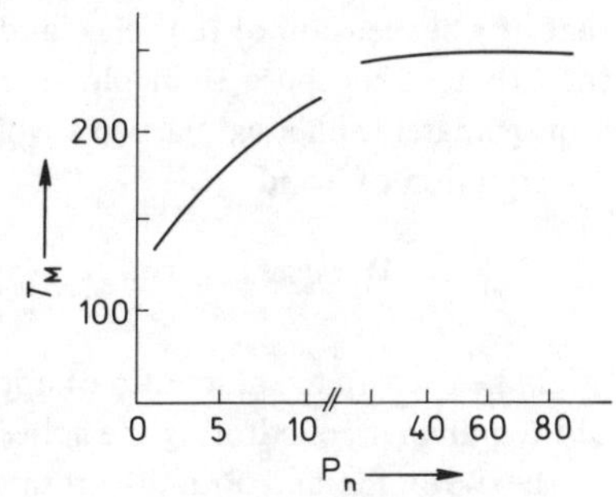

Fig. 11.6.

Fig. 11.5. Isothermal bulk polymerization of methylmethacrylate

Fig. 11.6. Variation of the melting point of nylon-6.6 versus its DP

Fig. 11.6 shows the variation of melting point T_M in the synthesis of nylon-6,6 from "nylon-salt", i. e. the stoichiometric mixture of hexamethylene diamine and adipic acid. If the nylon salt were maintained at a temperature of 150 °C, as a melt, the reaction could proceed to a degree of polymerization of about 6, whereupon the formation of a crystalline phase would ensue.

For solutions of polymers in monomer or in monomer and solvent, above the glass transition, the viscosity η varies with volume fraction of polymer ϕ and weight average molecular weight $\overline{M}_w$ according to:

$$\eta \propto \phi^{2.5} \, \overline{M}_w^{3.4} \tag{11.1}$$

In view of the relationships shown in Figures 11.1–11.4 and in eq. (11.1), the dimensionless groups of physical properties used in fluid mechanics, heat transfer, and mass transfer correlations take on values with polymer-monomer mixtures that completely eliminate the possibility of eddy (turbulent) transport in *homogeneous* polymerizations. For example,

$$\text{Reynolds number:} \quad \frac{\text{length} \times \text{velocity} \times \text{density}}{\text{viscosity}} \quad \text{approaches zero.}$$

$$\text{Prandtl number:} \quad \frac{\text{heat capacity} \times \text{viscosity}}{\text{conductivity}} \quad \text{approaches infinity.}$$

$$\text{Schmidt number:} \quad \frac{\text{viscosity}}{\text{density} \times \text{diffusion coefficient}} \quad \text{approaches infinity,}$$

more rapidly if the diffusion coefficient is D_{pp} rather than D_{mp}.

In contrast, there are *heterogeneous* polymerizations in which conversion occurs by the diffusion of monomer freely in a medium of *low* viscosity, *to* solid polymer having virtually infinite viscosity, but at the same time a very high surface to volume ratio with reactive *chain ends* accessible for reaction with the monomer.

In most of these *heterogeneous* polymerizations, the viscosity can be calculated as that of a suspension of particles, and would not be much more than two fold greater than that of the pure fluid phase. For example, if the dispersed phase consists of approximately spherical particles, volume fraction ϕ, in a continuum of viscosity η_0, the equation of Vand:

$$\eta = \eta_0 \{1 + 2.5\phi + 11.6\phi^2 + \ldots\} \tag{11.2}$$

night be used up to values of ϕ of approximately 0.2. In this case the molecular weight of the polymer constituting the spherical particle is totally irrelevant. Thus the values of the Reynolds and Prandtl groups fall in the range commonly found in process engineering of industrial chemicals.

The mass transfer issues implicit in the Schmidt group will be considered on a case by case basis.

11.3 Engineering Decisions Regarding Process and Reactor

The third task following selection of the chemical route is to address the list suggested in Table 11.2.

As to monomers, consider the range of properties from vinyl chloride: suffocant, carcinogenic, volatile (b.p. 14 °*C*), flammable, low surface tension, low viscosity, subject to runaway polymerization, to ε-caprolactam: non-volatile, melting only above 150 °*C*, non-hazardous.

Monomer reservoirs must be designed taking into account storage conditions relative to monomer properties and process conditions: vinyl chloride will be held as a liquid in a pressure vessel under a few bar; ethylene as a gas at several hundred bar; ε-caprolactam will have to be melted prior to reaction ($t > 150\,°C$); isobutene will be refrigerated to $t \approx -100\,°C$ prior to polymerization.

The scale of manufacture (tons per year to tons per hour) strongly affects the selection of reactor.

If the scale of manufacture is sufficiently large, the *continuous* process is usually less costly than a batch process for the synthesis of *non*-polymeric chemicals. While in principle any batch process in which reagents and products are fluid can be redesigned as a continuous process, there is a less obvious reason why in polymer manufacture batch processes may be preferred, even when the tonnage of product is large. This results from the following fact: Unlike "micromolecules" such as ethanol, which is a single chemical entity regardless of the process of its synthesis, most polymer products under a given *categorical* name — for example, poly(vinyl chloride) — differ substantially as a result of particular conditions of their manufacture. Variations in molecular weight distribution, degree of branching, isotactic-syndiotactic-heterotactic triad contents, and possibly small amounts of comonomers will change the material properties of the product considerably, making a particular batch useful for a particular service (e.g., rapidity of injection molding). In the example of poly(vinyl chloride), two percent of propylene as comonomer with vinyl chloride leads to effective

Table 11.2.

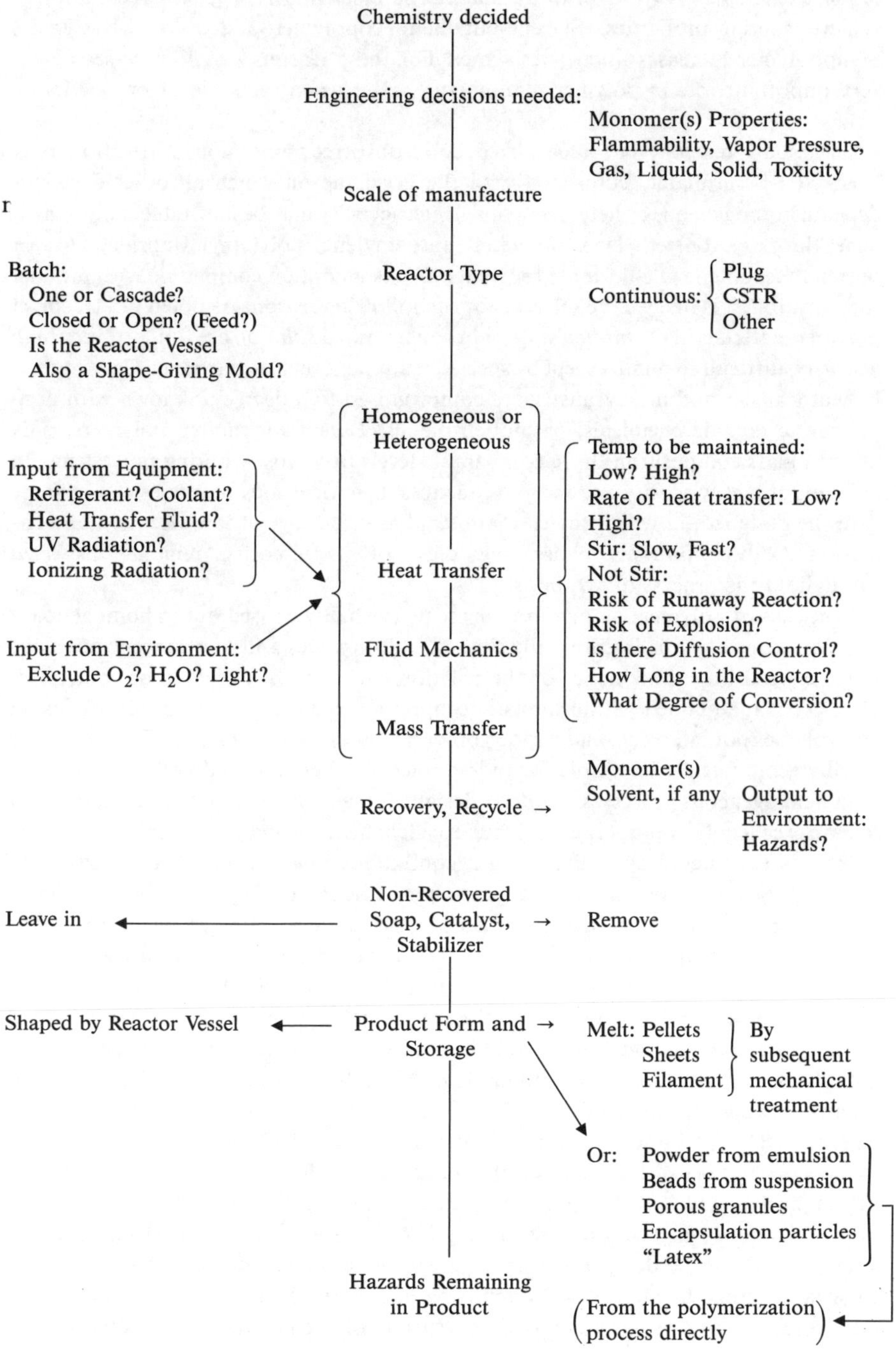

Chemistry decided
Engineering decisions needed:
Monomer(s) Properties:
Flammability, Vapor Pressure,
Gas, Liquid, Solid, Toxicity
Scale of manufacture
Batch:
One or Cascade?
Closed or Open? (Feed?)
Is the Reactor Vessel
Also a Shape-Giving Mold?
Reactor Type
Continuous:
Plug
CSTR
Other
Homogeneous or
Heterogeneous
Input from Equipment:
Refrigerant? Coolant?
Heat Transfer Fluid?
UV Radiation?
Ionizing Radiation?
Input from Environment:
Exclude O_2? H_2O? Light?
Heat Transfer
Fluid Mechanics
Mass Transfer
Temp to be maintained:
Low? High?
Rate of heat transfer: Low?
High?
Stir: Slow, Fast?
Not Stir:
Risk of Runaway Reaction?
Risk of Explosion?
Is there Diffusion Control?
How Long in the Reactor?
What Degree of Conversion?
Recovery, Recycle →
Monomer(s)
Solvent, if any
Output to
Environment:
Hazards?
Non-Recovered
Soap, Catalyst,
Stabilizer
Leave in
Remove
Product Form and
Storage
Shaped by Reactor Vessel
Melt: Pellets
Sheets
Filament
By
subsequent
mechanical
treatment
Or:
Powder from emulsion
Beads from suspension
Porous granules
Encapsulation particles
"Latex"
Hazards Remaining
in Product
From the polymerization
process directly

termination of kinetic chains, resulting in a lower molecular weight. Blending of one or more batches of polymer prior to final thermomechanical conversion is common to achieve special properties. Styrene-butadiene copolymer added to poly(styrene) homopolymer increases impact resistance. For these reasons *batch* processes play a very important role in polymer manufacture, even when the scale of production is large.

The issue of reactor type (batch stirred, batch unstirred, continuous stirred flow, plug flow, etc.) is intimately connected with the key issue on which all other decisions depend: homogeneous or heterogeneous. Heterogeneity may be inevitable: poly(vinylidene fluoride), polyethylene, polytetrafluoroethylene, poly(vinyl fluoride) etc., or optional: isobutene in chlorinated solvents. If it is an option compatible with product specification, it is likely to be selected for the following reason. As noted in 11.2, most polymer materials in the *molten* state or in concentrated *solution* are fluids of very high viscosity, difficult to pump except by special equipment such as extruders. Consequently heat transfer and mass transfer are compromised to a degree unknown with non-polymeric organic chemicals, or molten metals. Except for molten polymers, only mineral glass compositions have comparable levels of viscosity during processing. To circumvent high viscosity and its consequences, heterogeneous processes can in some cases be designed, in which the *continuous* phase is always of low viscosity, and the *dispersed* (discontinuous) phase, regardless of its viscosity, remains dispersed throughout the process of synthesis.

Synthesis of polymers on a *laboratory* scale is usually carried out in homogeneous solution, conversion is intentionally limited to low values and low concentrations of polymer, and thus the kinetics of the reaction can be readily studied or controlled, viscosity is maintained within limits appropriate to laboratory apparatus, diffusion control does not intervene, and various other advantages are obtained. This strategy is usually economically infeasible for the production of commercial polymers.

In heterogeneous processes, *properly* designed, the entire material can be controlled to have excellent flow properties and thus excellent heat and mass transfer rates: in most cases this is achieved by maintaining a continuous phase which is a low viscosity monomer or water, containing a limited volume fraction of dispersed polymer, but in one case it is almost the contrary: the porous, non-adherent fine grains of poly(vinyl chloride) containing monomer in the "mass" polymerization of poly(vinyl chloride).

Having established whether the polymer synthesis will be homogeneous or heterogeneous, one can now consider the set of questions (Table 11.2) regarding fluid mixing, heat transfer rate, mass transfer rate in relation to chemical kinetics. The enthalpy of polymerization will range from virtually zero for cyclic siloxanes to 20 kilocalories per mol for ethylene, 17 kcal/mol for vinyl chloride, to 11 kcal/mol for isobutene. The rates of conversion will range from the order of mols per litre per hour to mols per litre per second and thus the spatial rates of heat release will vary by orders of magnitude, and it will require the greatest attention to design in very fast vinyl polymerizations (vinyl chloride, isobutene). For these very rapid polymerizations safety strategies have to be devised to stop or greatly retard (slow down) the reaction if a runaway reaction threatens. Consequently redundant temperature and pressure sensors are mandatory to give early and reliable warning of such an impending situation.

In most polymerizations it is easy to exclude light, and more difficult to reduce oxygen and water content to any desired level (low level being mandatory for ionic polymerizations), and once continuous reactors are operating on purified feed streams the production is much easier than from batch reactors that must be purged before each new charge. In some cases UV radiation will be intentionally introduced (UV polymerization of thin layers) and oxygen will be supplied to ion plasma generators for the purpose of surface oxidation.

The problems of disposal of solvent after precipitation of polymer, and of removal of residual monomer under vacuum or otherwise are relatively trivial in the laboratory, but become major problems in industrial production. One major reason for not using a solvent is the cost of its removal from the polymer. How can the solvent be removed? Direct evaporation requires heat followed by condensation. Alternatively precipitation by addition of a non-solvent necessitates subsequent separation of solvent and non-solvent by distillation.

Removal of residual monomer from the polymer product is essential for all vinyl and diene monomers, from the viewpoint of odor and health hazard. While this can be achieved by a combination of high degrees of conversion followed by heating the polymer in vacuum ovens, for trace residues, there will be much more to cope with in styrene-butadiene copolymers and vinyl chloride polymers, in which there may be, respectively, 40% and 15% residual monomer at the end of polymerization. Whether the monomer be recycled to the reactor or passed through absorption towers, it cannot be discharged into the atmosphere in most countries.

The residues resulting from the polymerization process, for example the byproducts of homolytically dissociating free radical initiators, Ziegler-Natta catalyst residues, the soaps and polymeric stabilizers of emulsion and bead polymers in principle can be removed to any desired degree by dissolving the polymer in a solvent from which it can be precipitated, leaving the contaminants in the solvent non-solvent mixture, or by other costly and time consuming methods. Generally the cost of such purification is so great that it is not done. Indeed, with many commodity polymers, other materials will be *added*, including ultraviolet absorbers, free radical "getters" (scavengers), antioxidants, acid neutralizers, etc. Polymers destined for medical applications command a large premium in price, in view of the necessity of continually monitoring and certifying the absence of toxic extractable materials above levels established by such agencies as the Food and Drug Administration in the U. S. A. or their counterparts in other countries.

We have so far implied that output from the reactor would be stored in some form for subsequent use in a later manufacturing process. In Table 11.2 we note that on the one hand, the output of the polymerization reactor as a melt could be introduced into an extruder and extruded in rod form to be chopped into pellets suitable for blending and compounding prior to subsequent final thermomechanical processing. On the other hand the polymerization process itself may produce exactly what is needed in final form: beads to be used as ion exchange packing; a concentrated "latex", i. e., polymer particles in the fractional micron range, dispersed in water to be applied as a protective coating ("latex" paint).

We have also to take into account the fact that many important polymer products owe much of their value to a shape or form, permanently imparted by a *chemical* reaction, which took place in a specialized "reactor". This implies, of course, that the "reactor" must be capable of being opened or disassembled, so that the product can be harvested as a solid object of infinite "viscosity".

To conclude this overview, the material used to construct the reactor needs comment.

Most laboratory syntheses of monomers to polymers are carried out in glass vessels. Problems of glass degradation arise only under high pH (pH $>$ 12).

In scaling up these syntheses to pilot plant operation it is still possible to employ *glass lined autoclaves,* but the use of glass lining in large plant-scale reactors is usually avoided because of possible breaking, especially upon significant changes in temperature. If the total size of commercial manufacture is of the order of a few gallons, there is no problem using glass lining. Various alloys of *"stainless" steel* (iron-nickel-chromium) are preferred on account of general chemical inertness and mechanical strength.

"Stainless" steel resists corrosion because of the dense, nearly impermeable layer of mixed Ni-Fe-Cr oxides formed at the surface. However, this layer can be stripped off in the presence of hot aqueous solutions of high pH and thereafter all three metals may go into solution leading to corrosion and pitting of the reactor surface. It is conceivable that the metal ions might have bad effects on the polymer contents. For example, poly(ethylene oxide) is rapidly degraded in the presence of ferric ions. In polymer reactions carried out in stainless steel vessels, in the absence of water, for example in vinyl chloride bulk polymerizations, the problem would not exist. Certain alloys of stainless steel are susceptible to acid attack. Thus if chloroform were to be used as a solvent (since it might contain traces of hydrochloric acid), or if aluminium chloride were to be used in a cationic polymerization (accompanied by its cocatalyst HCl), it would be essential to verify that the grade of stainless steel in a proposed reactor system would be appropriate.

Copper is impressively disastrous with respect to some polymers, and in general direct contact of polymerizing mixtures with it or its alloys as bronze or brass is to be totally avoided. Relatively little copper ion in emulsion polymerized isoprene or butadiene is sufficient to catalyse oxidative degradation even in the aqueous emulsion of the poly(diene).

Given that the choice of metal is usually limited to stainless steel (steel is permissible in some anhydrous systems) heat transfer coefficients are mediocre because of the poor thermal conductivity of these materials. Thus heat transfer areas in some reactors may have to be augmented beyond simply jacketing the reactor walls — for example with internal coils.

Reactor walls, impeller blades, impeller shafts, valves and connecting pipes should be of the same alloy if the polymerizing mixture is aqueous and electrically conducting, in order to avoid the corrosion likely to occur when two disimilar metals are in contact with each other and an electrolyte.

Gaskets and shaft seals must be carefully considered in relation to the service asked for. Generally these materials are polymers or fiber reinforced polymers. It has to be established that at the pressures and temperatures of the process, and in the presence of monomers, solvents, catalysts, etc., to be contained, these sealing materials will not creep, expand, shrink, give off contaminants by elution into the reaction mixture, etc.

Tetrafluoroethylene polymers are remarkably inert and where their high modulus can be tolerated, they serve well in gaskets and shaft seals.

When a "reactor" also serves as a shape-giving mold (Appendix A), it is obviously important to insure that during the reaction, the confined monomer-polymer does not react with the surface of the mold so as to bind to it. Thus in addition to selecting surfaces of low reactivity (titanium, stainless steels) it may be necessary to pre-coat the surface with mold-releasing agents that prevent direct adsorption.

Chapter 12
Emulsion and Suspension Polymerization

Introduction

In this chapter we consider two polymerization processes having in common the following features:

(1) The monomers are polymerizable by free radical routes;

(2) The monomer and polymer are nearly completely insoluble in water and are dispersed therein as very small to small spherical particles;

(3) The continuous phase throughout the polymerization is water, at a volume fraction of usually 0.70 or more, and thus the effective viscosity of the entire mixture, calculable as a suspension of spheres is not much greater than twice the viscosity of water.

These processes are emulsion polymerization and suspension ("bead", "pearl") polymerization which — except for the above shared features — are completely different in kinetics.

12.1 Emulsion Polymerization

12.1.1 Introduction

As noted above, the monomer(s) must be **(a)** insoluble or sparingly soluble in water (acrylamide is a counter example) and **(b)** polymerizable by free radicals. The initiator is soluble *only* in the water phase, and it decomposes in this phase forming radicals which to an overwhelming degree arrive at dispersed *micellar* or polymer particles.

Fig. 12.1 shows the species present in emulsion polymerization. Fig. 12.2 shows schematically the three stages which occur in succession.

I. Micellar Stage: Soap, for example sodium palmitate, is dissolved in water at a concentration well above *critical micelle concentration.* Monomer(s) is (are) added and dispersed as droplets (0.1 *mm* – 1 *mm*), in the presence of a water soluble initiator. Monomer diffuses from droplets through water into micelles. While some initiation occurs probably in the water phase, most polymerization ensues in the *micelles.* The micelles become polymer particles, i. e., polymer dissolved in monomer. They grow from micellar dimensions, a few tens of Ångström units, eventually to hundreds of Ångström units in diameter (up to ~0.4 μm).

II. After about 15 percent conversion of the monomer initially charged, the micelles are exhausted, no new particles are generated, and monomer diffuses into the growing polymer particles from the droplets, to maintain thermodynamic balance of the monomer and polymer.

III. After 40 to 60% conversion the monomer droplet phase disappears (centrifugation no longer yields a monomer layer) and the polymer particles now remain about constant both in number and diameter as their monomer, roughly 40 to 60% of the volume, is converted to polymer.

Throughout the three stages in the *Smith-Ewart model,* each micelle or ensuing polymer particle alternately receives from the aqueous phase a radical, which starts polymerization within, and another radical, which combines with the growing chain radical end, terminating the reaction. This is referred to as "red light-green light" polymerization. On the average, the interval between capture of radicals is between 1 and 100 seconds (set by initiator concentration and soap concentration). Thus the red light or green light period may be of the order of 1 to 100 seconds.

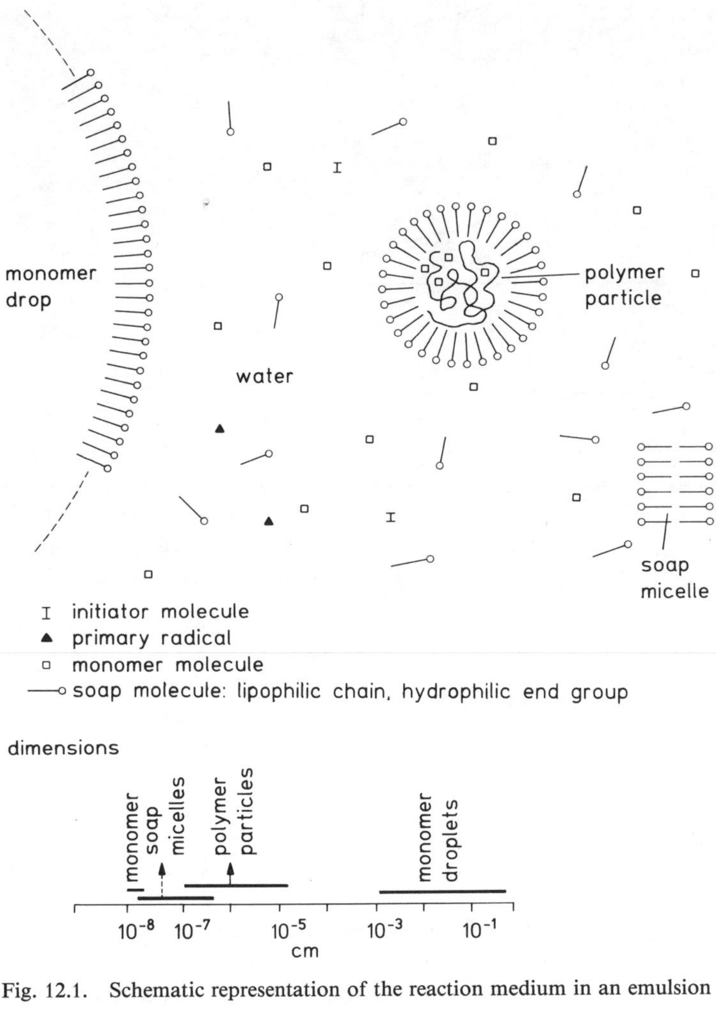

Fig. 12.1. Schematic representation of the reaction medium in an emulsion polymerization

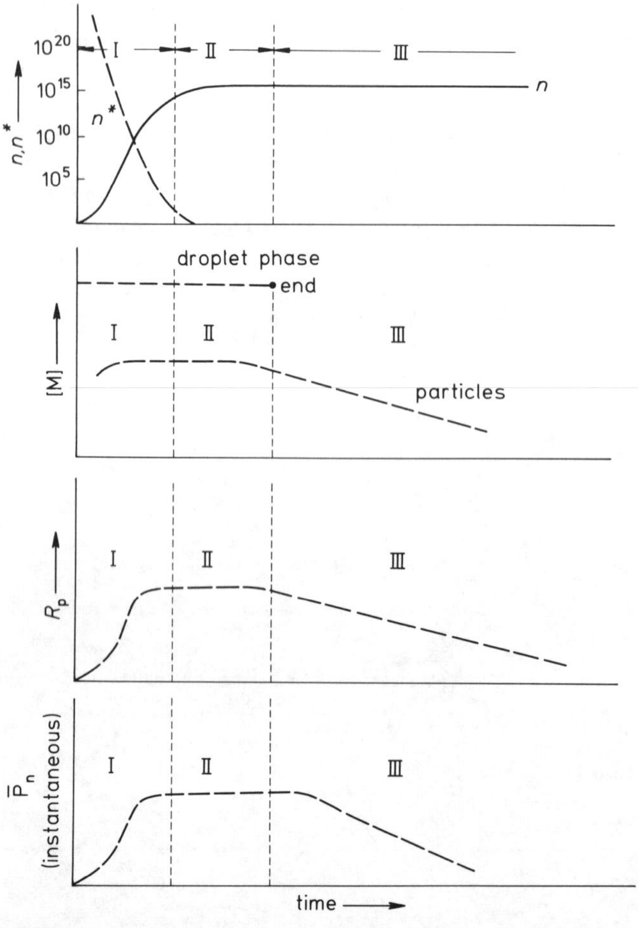

Fig. 12.2. Diagrams showing the three stages of emulsion polymerization: (I) Micelles increasing; (II) micelles exhausted, droplet phase remains; (III) droplet phase exhausted.
n number of particles per unit volume
n^* number of micelles per unit volume
[M] monomer concentration in the droplets and in the particles
R_p overall rate of polymerization
$\bar{P}_n$ instantaneous degree of polymerization

12.1.2 Advantages and Disadvantages

Advantages

(1) Termination is controlled externally by reason of the initiator's insolubility in the organic phase. The rate of arrival of radicals at the polymer particles is established *independently* of propagation, by regulation of soap content, which sets the number

of micelle/polymer particles at 10^{14} to 10^{17} per *ml* of emulsion, and by the redox initiator concentration in the aqueous phase, which establishes the rate of generation of radicals.

Thus for example butadiene, which cannot be economically polymerized *homogeneously* by radical initiation (k_t being very high compared to k_p), is readily polymerized by emulsion polymerization.

(2) Because at least 50% of the volume is water, *viscosity is always low,* and heat removal rates are easily managed. It would be difficult to provoke runaway, explosive polymerizations in emulsion polymerization.

(3) In many cases the emulsion finally produced, ~0.1 *µm* polymer particles at around 50% volume fraction in water, is a product *immediately applicable* as a coating, e.g., a paint or barrier coating layer: poly(vinylidene chloride) on polyethylene.

Disadvantage

The polymer will contain *residues* of the soap which are hard to eliminate, even if the polymer has been precipitated from the emulsion and washed. If the emulsion is used directly as a coating, all the soap will be present.

12.1.3 Initiators

Reduction-oxidation (redox) initiators are particularly useful in emulsion polymerization because they can produce high rates of radical generation at the moderate temperature levels common to emulsion polymerization. A primary requirement is that they be insoluble in the monomer droplet phase and insoluble in the polymer particles: Thus, they must be highly water soluble.

As examples of redox systems, cumyl hydroperoxide reacts with ferrous ion to produce one radical:

$$(C_6H_5){-}C(CH_3)_2OOH + Fe^{++} \longrightarrow Fe^{+++} + OH^{\ominus} + (C_6H_5)C(CH_3)_2O^{\bullet}$$

Persulfate reacts with ferrous ion to give:

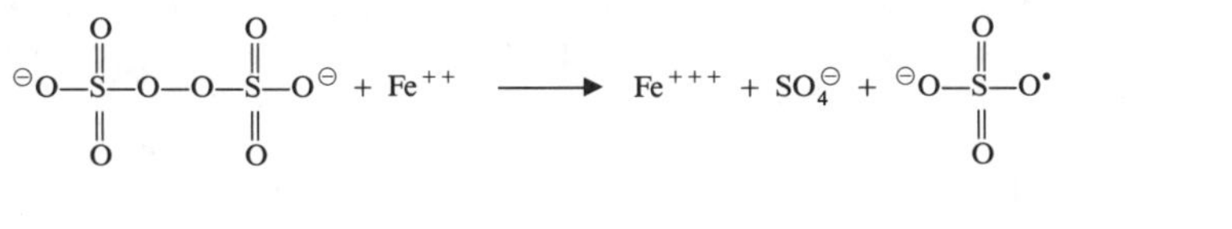

Another redox system is the ferrous-hydrogen peroxide mixture:

$$Fe^{++} + H_2O_2 \longrightarrow Fe^{+++} + OH^- + OH^{\bullet}$$

In general, the rate of production of radicals by redox initiators, $+d[R^{\bullet}]/dt$, given in $[mole \cdot l^{-1} \cdot s^{-1}]$, in a single phase medium, is given by the rate equation:

$$+\frac{d[R^{\bullet}]}{dt} = k_{d,r}[X][Y] \tag{12.1}$$

wherein [X] = molar concentration of peroxide
[Y] = molar concentration of reducing agent (e.g., Fe^{++})
$k_{d,r}$ = dissociation constant for redox pair

The factors 2 and f (efficiency factor), found in the equation for homolytic dissociation of a *single* initiator I,

$$+ d[R^{\bullet}]/dt = 2fk_d[I] \tag{12.2}$$

are absent in redox initiation.

Because in each initiation step only one radical is produced, recombination of this radical species is not favored.

As applied to emulsion polymerization, in which the volume fraction of organic phase is ϕ, and that of water $(1 - \phi)$, the rate ρ of production of initiator radicals (number of radicals per *cm*^{3} and per *s*) must therefore be, from eq. (12.1):

$$\rho = \frac{N_A}{10^3}(1 - \phi)k_{d,r}[X][Y] \tag{12.3}$$

wherein [X] and [Y] are calculated as moles per litre in the aqueous phase, and N_A is Avogadro's number.

12.1.4 Smith-Ewart Kinetic Scheme

There are three premises central to the Smith-Ewart theory:

(1) All radicals generated in the aqueous phase arrive at the surface of micelles, or of polymer particles derived therefrom.

(2) When a radical arrives at a micelle or a particle surface, it instantly initiates a single chain propagation, if there is not already radical activity within.

(3) If there is a chain propagation underway, the arriving radical stops it instantly by mutual termination.

These premises were reached by considering typical values of k_t for bulk polymerization (10^6 to 10^8 [$l \cdot mole^{-1} \cdot s^{-1}$]), and calculating the time that two radicals could coexist in a typical particle of 1 000 Å diameter (between 10^{-2} and 10^{-3} *s*). *In the aqueous phase*, the average time between radical formation from initiator and arrival at a micelle or polymer particle is very much shorter than the average time required to terminate with another radical in the aqueous phase.

These three central assumptions concerning radical generation and termination are combined with the additional assumption that monomer *homogeneously dissolved in water* is at *negligibly* small concentration, therefore not polymerizable in the water phase. With these four assumptions, the following results are obtained:

(1) The number of monomer molecules converted to polymer per second, $-dm/dt$, in a single particle in which monomer concentration is [M], is:

$$-\frac{dm}{dt} = \frac{1}{2}\, k_p [M] \tag{12.4}$$

The factor 1/2 arises from the "red light-green light" effect (on the average, *one* chain is growing half the time), and k_p [M] has the dimensions: monomers per second per radical.

Thus if there are n particles per *cm*^{3}, the rate per *cm*^{3} is $-n(dm/dt)$.

(2) *The degree of polymerization* $\overline{P_n}$ is

$$\overline{P_n} = k_p [M]\, \Delta t \tag{12.5}$$

where Δt, the "green light" period, is n/ρ, ρ being the rate of generation of radicals per *cm*^{3} per *second.* (The killing radical is assumed to terminate by *coupling* with the previously growing chain.)

(3) The number of particles per *cm*^{3}, n, finally generated depends on the rate ρ of generation of radicals per *cm*^{3} and on the number of soap molecules available per *cm*^{3} of total volume, thus, upon the initial concentration of soap, c_s, given in [*g/cm*^{3}], and the area of a soap molecule in a packed monolayer a_s (as deduced for example in a Langmuir trough). The result is:

$$n \approx 0.53\,(c_s a_s)^{3/5}\, \rho^{2/5} \left(\frac{dv}{dt}\right)^{-2/5} \tag{12.6}$$

The rate of generation of radicals ρ in the aqueous phase per *cm*^{3} per *s* is set by concentration and type of initiator, and by temperature (cf. eq. 12.3). The volumetric rate of growth dv/dt of the particles is:

$$\frac{dv}{dt} = -\, v_u q\, (dm/dt)\;, \text{ where} \tag{12.7}$$

v_u = volume of a monomer unit as polymerized.

$-v_u \cdot \dfrac{dm}{dt}$ = volume increase of polymer chain per *s*.

q = volume of (polymer + monomer)/volume of polymer alone, which during stage I and II is around 1.5 to 3.0, depending on thermodynamic affinity of monomer/polymer.

Solubility of Monomer in Water

When a monomer like vinyl acetate is soluble to the extent of a fraction of a weight percent, the initiator generated in the aqueous phase in effect sets off a *solution* polymerization of the dissolved monomer, but the growing polymer chains quickly become insoluble in the water and tend to cluster as colloidal aggregates. Since the soap molecules can diffuse from particle to particle throughout the medium, these can rapidly become stabilized and thereafter act as growing polymer particles, being fed with monomer from the monomer droplets, together with particles which were generated from true micelles.

Equilibrium between monomer drops and particles

One might speculate that the rate of growth of the polymer particles could be diffusion limited, that is, that the rate of diffusion of monomer from the droplets into the growing polymer particles through the aqueous phase would limit the rate at which polymerization could occur. One can readily show, assuming a diffusion coefficient (for the monomer through water) of around 10^{-5} [cm^2/sec], that, under any conceivable conditions and even with concentrations of monomer in the aqueous phase as low as 100 parts per million, the rate of flux between the monomer droplets and the polymer particles would be several orders of magnitude greater than the rate at which the monomer could be used up by polymerization in the polymer particles.

If this is true, it would seem that the monomer in the droplet phase (which, as noted above, is free of polymer) should diffuse indefinitely into the growing polymer particles.

This does *not* occur because the polymer particles are in fact in equilibrium with monomer droplets. The reason is the following: **(a)** the interfacial tension σ on these very small particles of radius *r* produces an effective excess pressure $P_{excess} = 2\sigma/r$ according to the Laplace equation. **(b)** This excess pressure P_{excess} increases the thermodynamic activity a_1 of the monomer of molar volume V in the particles, according to the Poynting relation:

$$\ln a_1 = P_{excess} \cdot V/RT$$

(R, T gas constant, temperature), but **(c)** this value of $\ln a_1 > 0$ is exactly balanced by the reduction of the thermodynamic activity of the monomer by reason of dilution from the pure state by the contained polymer of volume fraction ϕ_2, according to the Flory Huggins equation:

$$\ln a_1 = \ln(1 - \phi_2) + \phi_2 + \chi_1\phi_2^2$$

where χ_1 is the monomer-polymer interaction parameter.

Thus in combination equilibrium occurs when:

$$\frac{2\sigma}{r}\bar{V}/RT + \ln(1 - \phi_2) + \phi_2 + \chi_1\phi_2^2 = 0 \qquad (12.8)$$

Because the rate of polymerization is found to be constant in many emulsion/polymerizations during stage II (see next section), it is to be inferred that the concentration of monomer in the particles: $[M_0](1 - \phi_2)$ is constant, therefore that ϕ_2 is constant, and therefore that σ/r is approximately constant. Representative values are for $r = 5 \times 10^{-6}$ cm, $\sigma \simeq 30$ dyn/cm; $\bar{V} \sim 100$ cm^3/mol; χ_1 ranges from ~ 0 to ~ 0.4, and ϕ_2 is found to range from ~ 0.25 to ~ 0.5.

12.1.5 Rate of Conversion According to Stage

Fig. 12.3 illustrates the three stages of a typical batch emulsion polymerization. In Stage I, the micellar stage, the rate of polymerization increases as particles are generated from micelles, but the increase stops at about 15% conversion of the monomer, depending of course on the total amount of soap charged to the system initially.

In Stage II, "constant rate", the monomer droplets continue to exist and the monomer concentation [M] in the polymer particles remains almost constant as noted in the preceding section.

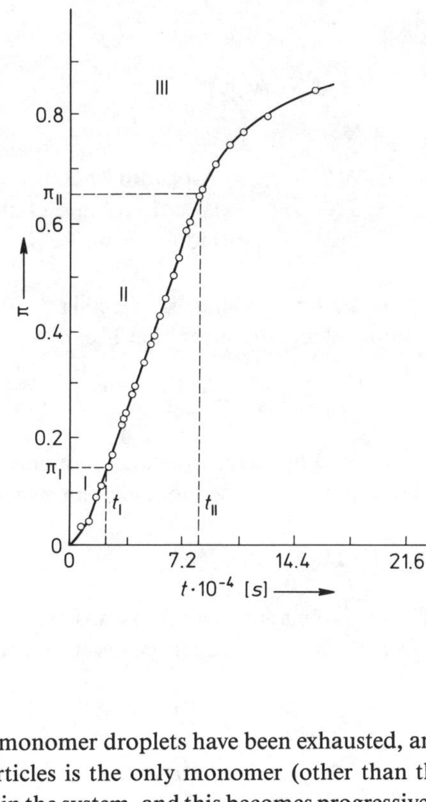

Fig. 12.3. Conversion versus time diagram for a typical batch emulsion polymerization

In Stage III, the falling rate period, the monomer droplets have been exhausted, and the monomer present in the polymer particles is the only monomer (other than the trivial amount dissolved in water) now left in the system, and this becomes progressively converted into polymer. Define π as the number of polymer units formed, divided by the total number of polymer units plus monomer units present. Thus, π is percent yield divided by 100.

Let t_{I} represent the time necessary to reach the end of the micellar period, at which time the fraction of monomer polymerized is π_{I}, and t_{II} the time at which the droplet phase is exhausted, at which time the fraction of monomer converted is π_{II}. Thus, in Stage II the difference $\Delta t \equiv t_{\mathrm{II}} - t_{\mathrm{I}}$ is the total period of operation at the constant rate condition. Consequently,

$$\int_{\pi_{\mathrm{I}}}^{\pi_{\mathrm{II}}} \mathrm{d}\pi = \frac{1}{\mathrm{C}} \int_{t_{\mathrm{I}}}^{t_{\mathrm{II}}} k_{\mathrm{p}} [M_0] (1 - \phi_2) \frac{n}{2} \, \mathrm{d}t \tag{12.9}$$

wherein k_p = propagation rate constant, $l \cdot mol^{-1} \cdot s^{-1}$
$[M_0]$ = molar concentration of undiluted pure monomer, $mol \cdot l^{-1}$
$(1 - \phi_2)$ = volume fraction monomer in particles (ϕ_2 vol. fraction dissolved polymer)
n = number of particles per cm^3 of total mixture
t = time, s
C = number of monomer molecules initially present per cm^3 of total mixture

In turn,

$$C = (f\, N_A)/(\bar{v}\, m_0) \tag{12.10}$$

where f = cm^3 of monomer charged per cm^3 of total mixture
N_A = Avogadro's number
$\bar{v}$ = specific volume of monomer, $cm^3 \cdot g^{-1}$
m_0 = monomer molecular weight, $g \cdot mol^{-1}$

If as is usually found, ϕ_2 remains constant in Stage II, (droplet phase present), equation 12.9 upon integration becomes:

$$\pi_{II} - \pi_I = \frac{k_p [M_0] (1 - \phi_2)}{2C} (t_{II} - t_I) \tag{12.11}$$

At the end of stage II the only monomer present is in the particles.
Thus *at this time* the monomer concentration in the particles is:

$$[M_0]\,(1 - \phi_2) = [M_0]\,(1 - \pi_{II}),$$

whereas throughout stage II, ϕ_2 was assumed constant while π increased.
Thus in stage III the conversion π as a function of time becomes:

$$\int_{\pi_{II}}^{\pi} d\pi = \frac{1}{C} \int_{t_{II}}^{t} k_p [M_0] (1 - \pi) \frac{n}{2}\, dt \tag{12.12}$$

which can be rearranged to give:

$$\int_{\pi_{II}}^{\pi} \frac{d\pi}{1 - \pi} = \frac{k_p [M_0]\, n}{2C} \int_{t_{II}}^{t} dt$$

from which we find

$$-\frac{\Delta \ln (1 - \pi)}{\Delta t} = \frac{k_p [M_0]\, n}{2C} \tag{12.13}$$

To illustrate the application of these calculations, we can determine from Fig. 12.4 — in which $\ln(1 - \pi)$ is plotted versus t (from the data of Fig. 12.3) — that the slope $\frac{\Delta \ln(1 - \pi)}{\Delta t}$ is -1.04×10^{-5} [s^{-1}]. Let us assume that in this experiment the number of particles n was $10^{14}/cm^3$ and that ϕ was 0.3 cm^3 of organic phase per cm^3 of total

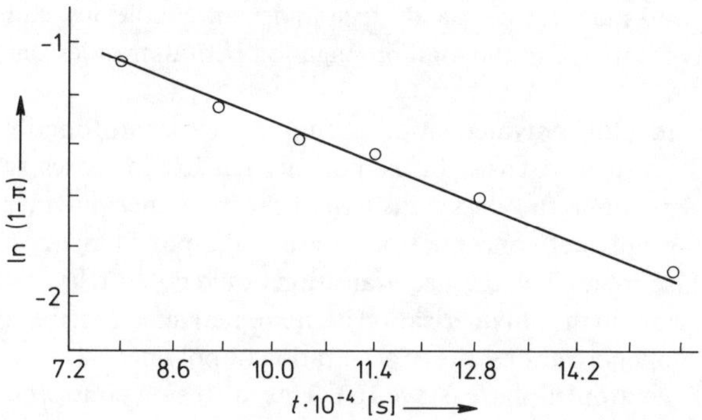

Fig. 12.4. Semi-logarithmic plot, ln(1 − π) versus time, of the data of Fig. 12.3

phase. The specific volume $\bar{v}$ for isoprene is about 1 [cm^3/g] and with a molecular weight of 62, the initial molarity [M_0] would have been about 1,000/62 = 16 [$mol \cdot l^{-1}$]. From eq. 12.13. and these numerical values the propagation rate constant k_p should be 36 [$l \cdot mol^{-1} \cdot s^{-1}$].

12.1.6 Multiple Radical Effects in Emulsion Polymerization

The *Smith-Ewart* theory leads to the conclusion that in every polymer particle there is a *"green light" period* in which only one radical is present and is propagating, forming an increasingly long polymer chain, and that, following the entry of a second radical into this particle, the second radical immediately finds the growing macroradical and terminates it, whereupon a *"red light" period* sets in, in which there is no propagation at all in the particle. Considering typical values of the termination constant k_t as deduced from *bulk* polymerizations and calculating how long two radicals could coexist in a growing particle, if the growing particle were part of a bulk substance, one can show that the period of coexistence would be of the order of milli-seconds, thus very short compared to the typical "green light" or "red light" periods, which are of the order of 1 to 100 seconds.

In reality, the polymer particle has an interface containing surfactant molecules, and the initiating radical which arrives from the aqueous phase is partially or strongly hydrophilic (a persulfate ion-radical is insoluble in an organic phase such as an isoprene-polyisoprene mixture, whereas a hydroxyl-radical might have limited solubility). In the case of a persulfate ion-radical, the arriving radical must start polymerization at the interface without invading the polymer particle, and thus the alleged termination step would have to be preceded by the growth of a short polymer chain which finally becomes long enough to drag the ionic end with its counterion into the interior of the polymer particle, and then find the end of the growing chain. This restriction might not be so severe with the hydroxyl radical, but since this radical would form an alcoholic hydroxyl, it is unlikely that it would immediately diffuse deeply into the polymer particle, until the chain which it initiated becomes sufficiently long, perhaps several ten's of monomer units. Thus, the assumption of instantaneous killing of the first radical by the second may not be exactly true under all conditions, but ordinarily the relative delay, as compared to the total propagation period or quiescent period, would be small.

However, as already noted in bulk polymerization, termination*) is profoundly affected as viscosity increases. In turn, viscosity in the polymer particle increases as polymer concentration increases, and as the glass transition of the monomer-polymer mixture rises (due to increasing polymer concentration) towards the polymerization temperature. In the example of isoprene cited, the glass transition would not be relevant, whereas it would become important in the polymerization of monomers such as styrene or methyl methacrylate. In all polymerizations, the concentration of polymer is bound to rise after the exhaustion of the droplet phase (stage III). Thus diffusion control of termination can lead to the existence of multiple growing radicals in a polymer particle. Monomer can still diffuse within the particle to arrive at (and be propagated by) macroradical chain ends, but the macroradicals cannot find each other because of the high viscosity of the continuum. This would correspond to the *autoacceleration* called the *Trommsdorf-Effect* in bulk polymerization. When the polymer concentration becomes such as to make the reaction temperature that of the glass transition, monomer diffusion to the radicals will diminish finally to zero.

Another way that multiple radicals could exist in a growing polymer particle, simultaneously propagating, would be if the particles have a diameter significantly greater than around $1/2\ \mu m$. In this case, because of the sheer volume of the polymer particle, a second radical arriving at the other side of the particle in which the first radical is growing may have some opportunity to propagate a chain before the two radicals find each other, even if the system is not encumbered by excessive viscosity. Ordinarily, polymer particles do not grow to diameters of this magnitude, but one can always expect to have complications if the polymer being produced has a glass transition above the temperature of polymerization.

In emulsion polymerization the temperature of polymerization effectively must exceed the freezing point of water and, at least in the micellar phase, it must be significantly less than the normal boiling point of water, since micelles are disrupted and stabilization is not obtainable if one operates close to the boiling point of water. In the third stage of an emulsion polymerization (post-micellar, and post-droplet), it is of course feasible to increase the temperature of the system above 100 °*C* by using a pressure vessel so as to overcome the problem of the glass transition, but this is not a problem of great industrial consequence, since most of the glassy polymers are made by methods other than emulsion polymerization and most of the polymers made by emulsion polymerization are homopolymers or copolymers whose glass transition lies well below 100 °*C*.

---

*) We have noted before that the true chemical reactivity is unchanged and thus k_t is not changed, but now diffusion becomes controlling.

12.1.7 Emulsion Copolymerization

Emulsion copolymerization is particularly appropriate for the following monomers:

1. styrene-butadiene
2. styrene-acrylonitrile
3. acrylonitrile-butadiene-styrene
4. vinylidene chloride-vinyl chloride
5. vinyl acetate-methyl acrylate
6. acrylates-methacrylates

In order to maintain constant polymer composition (constant mole fraction of each type of unit), the monomer composition must be kept constant at a value determined by the reactivity ratios r_A and r_B.

As with emulsion homopolymerizations, this could be done using a single batch reactor, with a *feed of monomer* adjusted over the time of the process. An alternate way, used in production of butadiene styrene rubber, is illustrated schematically in Fig. 12.5.

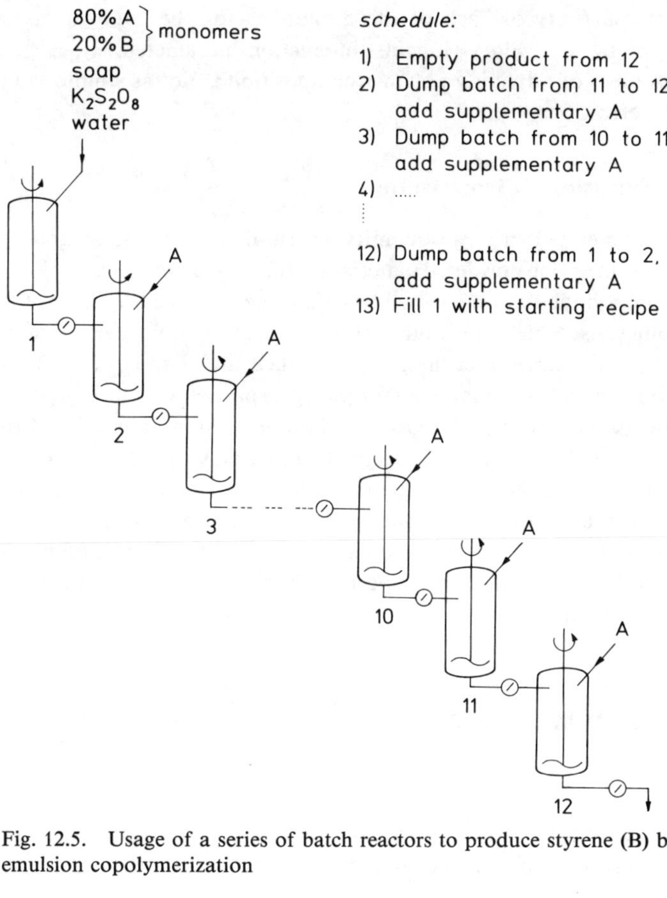

Fig. 12.5. Usage of a series of batch reactors to produce styrene (B) butadiene (A) rubber by emulsion copolymerization

A *series of batch reactors,* as many as twelve, are operated as shown. A batch of emulsion, starting with 80% butadiene and 20% styrene as mixed monomer, added to water, soap and initiator, is carried to a low degree of conversion, then transferred to a second reactor, to which some make-up butadiene is added, and the conversion is continued to a higher degree.

Small amounts of mercaptan may also be added to control molecular weight by promoting chain transfer. Sequential transfer of a batch of emulsion from reactor to reactor, accompanied by *correction* of the monomer composition through addition of butadiene, as the conversion is advanced, leads to a product copolymer in which all molecules have a ratio of A to B units (butadiene to styrene units) narrowly distributed around the intended average (e.g., 70:30).

It is not necessarily desirable to attain complete conversion. In diene copolymerization, branching and network formation will occur at high conversion. Thus polymerization is stopped usually at *60% to 70% conversion,* and the mixed residual monomer is recovered by steam stripping.

If two or more monomers are supplied in the *droplet* phase, their incorporation into polymer in the polymer *particles* will depend not only on reactivity ratios r_A and r_B but also on the relative *affinity* of each of the monomers for the copolymer, as measured for example by the Flory-Huggins interaction parameters $\chi_{A,\,polymer}$, $\chi_{B,\,polymer}$. Thus the problem of adjusting polymer composition is not as simple as in homogeneous copolymerizations.

12.1.8 Anomalous Emulsion Polymerization

Vinyl chloride is polymerized in large quantity by emulsion polymerization to produce *"paste grade"* resin, a fine powdered material useful in *"plastisol"* technology. Like vinylidene chloride, vinyl fluoride, vinylidene fluoride, and acrylonitrile, the polymer formed becomes insoluble in its monomer at low degrees of polymerization, and a *precipitation polymerization* ensues, as described in chapter 13. From consideration of equilibrium between monomer droplets and particles, if the monomer inside a polymer particle did not have its activity somehow reduced by interaction with polymer, diffusion from the drops could not occur. Thus we must postulate that the polymer in the particle is somewhat like a microcosmic network with some of the polymer segments being in an amorphous state, interacting with monomer.

Because of these profound differences, polymerization of the above listed monomers by emulsion polymerization differs in every respect from those in which the polymer particles are of a single phase.

12.2 Suspension Polymerization

12.2.1 Introduction

Suspension polymerization of water-insoluble monomers that can be initiated by radicals is preferred whenever it is desired to produce small spherical beads. This method is also called *pearl* or *bead polymerization*. Bead diameters range from 5 μ*m* (thus one or two orders of magnitude greater than in emulsion polymerization) up to 1 *cm* (in rare cases). The organic phase, containing a single monomer (or mixed monomers) and a radical initiator, which is soluble in the monomer but not in water, is progressively added to a reaction vessel containing water, fitted with an impeller, until the volume fraction of organic phase is of the order 0.2 to 0.35. The impeller breaks this phase into drops, each of which becomes a miniature "chemical reactor", in which monomer is converted to polymer by the radicals *generated within*. Thus in terms of kinetics, those applicable to bulk radical homopolymerization or copolymerization must apply. There is no feature of emulsion polymerization present, since radical generation occurs in the same phase as the termination.

12.2.2 Problems of Fluid Mechanics

The process is one whose success depends primarily on fluid mechanics and on "fortification" of the organic-water interface. As a counter-example, if benzene were added to a volume of water progressively with stirring, droplets would result, but upon cessation of stirring the droplets would rise through buoyancy and coalesce into a supernatant layer.

Furthermore, as has been demonstrated with dye tracer experiments, if red-dyed benzene is introduced into a stirred tank of water, containing already blue-dyed benzene droplets, collision of droplets and coalescence, followed by resplitting of large drops into small, will soon lead to a population of drops uniformly purple.

If nothing were done in the case of polymerizable monomer dispersed as drops in water, the drops should become increasingly viscous during polymerization of monomer to polymer, but *redivision* after coalescence would become increasingly difficult, and finally impossible. Thus, small drops would grow into larger drops or clusters, defeating the purpose of the process.

To circumvent this, two kinds of additives are commonly used, each of which greatly enhances the stability of the organic-water interface against coalescence upon contact of two drops. One type of additive includes platelet-like mineral particles, that preferentially concentrate at the organic-water interface. It is of course essential that the proposed stabilizing particle *not* become dispersed inside the organic phase droplets, since as such it will serve no protective purpose and only contaminate the polymer.

The second kind of additive is macromolecular. These are polymers which are soluble in water, insoluble in the monomer, and capable of concentrating at the interface between the two phases. The literature abounds with examples, but *poly(vinyl alcohol)* and starch products are mostly cited.

With either type of additive, the interface is converted from one characterized by high mobility (fluidity) and low interfacial tension, to one characterized by a quasi-elastic, quasi-rigid surface of high effective interfacial tension.

Thus, once formed at the proper size, the stabilized droplets with such interfacial coatings cannot readily coalesce upon mutual contact, or further subdivide.

Sometimes both mineral and polymeric additives are used together. It seems probable, but has not been conclusively demonstrated, that macromolecular additives function in yet a different way. Any solution of random coiling polymers (thus including poly(vinyl alcohol) in water) reacts so as to strongly oppose *extensional* flows. When a volume of fluid is pulled, rather than sheared, resistance is proportional to the product of extensional strain rate and of extensional viscosity (cf., resistance in shear flow is proportional to the product of *shear* strain rate and *shear* viscosity). The two viscosities are strain rate dependent but in *opposite* directions: Shear viscosity drops as strain rate increases, but extensional viscosity increases. Furthermore, even in the dilute aqueous polymer solution that constitutes the continuous phase in suspension polymerization, the *extensional viscosity* can increase by one or more orders of magnitude in regions of high strain rate. Owing to the extension of originally randomly coiled macromolecular chains, they expand to a length ultimately near that of their full contour value and thereby exert much greater drag on the solvent adjacent to it. This means that in the fluid volume near the moving tip of the impeller, where the extensional strain rate otherwise might be very high, the polymer dissolved in the water (in excess of that adsorbed at the drop-water interface) acts to reduce this strain rate, and thus the velocity (energy) with which one droplet can be impelled toward another is reduced.

12.2.3 Preferred Sequence of Operations

In view of these considerations, it is clear that for success in a suspension polymerization process, where regular spherical particles within a specified range of diameters are desired, the following sequence should be used:

(1) Premix initiator and monomer(s) to form initial organic phase.
(2) Add the organic phase to water (up to 35 *vol.-%*) under conditions of impeller agitation sufficient to produce the particle diameters desired — the smaller the particle, the higher the impeller speed needed.
(3) Add the stabilizers (mineral and/or polymeric), which should concentrate at the interface between organic drops and water.
(4) Reduce impeller speed so as to maintain gentle circulation in the vessel, adequate to prevent gravitational separation of particles from water.
(5) Raise temperature to that intended for polymerization (typically 50° to 80 °*C*) and maintain it until conversion has brought the glass transition of the bead mixture (polymer-monomer) near the polymerization temperature.
(6) Increase temperature towards 100 °*C*, sufficient to complete conversion.
(7) Lower temperature to ambient, and dump the batch onto a screen, retaining the polymer particles.

As one of the final processing steps, the collected beads may be cleared of stabilizer. For example, if carbonates were used, dilute hydrochloric acid rinse will destroy them. Polymeric stabilizers are *not* so easily removed, but hydrolysis can be employed in certain cases: For example, amylase digestion of starch.

In virtually every case, the product of suspension polymerization is glassy, rigid, and non-adhesive at room temperature.

12.2.4 Special Bead Polymerized Materials

Three products deserve special comments: "Foamable" poly(styrene), ion exchange beads, and chromatographic packings.

If styrene is suspension-polymerized in the presence of a limited quantity of cyclohexane, a poor solvent for the polymer, beads are produced which are glassy at room temperature, but contain residual *entrapped* cyclohexane, in significant percentage.

The beads can subsequently be charged into a mold cavity and rapidly heated, which causes their expansion into a closed cell foam that forms an insulating layer, after cooling to the glassy state.

In this example, it is particularly convenient to use suspension polymerization, because large cavities can be readily charged with the free flowing beads, and because the "packaging" with entrapped cyclohexane is so readily accomplished.

Uniform spherical beads of *controlled* and *narrow size* distributions are particularly desirable for packing *chromatographic* or *ion exchange columns,* since uniformity of filling space without entraining voids is essential for high efficiency. Particle diameters range from 5 μm in chromatographic colums to 1 to 2 *mm*, in ion exchange columns.

The polymers are invariably crosslinked for dimensional stability against pressure and changing solvents in chromatography, and to prevent dissolution in the case of ion exchange resins, following reactions which produce fixed ionic groups on the polymer.

Many of these materials are generated from divinyl benzene, or divinyl benzene and styrene. Others are generated from acrylates and divinyl benzene. Thus, to the extent that *copolymerization* is involved, these systems correspond to single charge reactions carried to completion without adjustment of monomer composition.

Inherent to many of the chromatographic and ion exchange polymer bead syntheses is the technique of introducing internal porosity by *syneresis.*

According to this principle, any polymeric network can hold an amount of solvent limited by its degree of crosslinking. The higher the degree of crosslinking, or the more adverse the solvent, the less the network can hold. Thus, for example, in a mixture of divinyl benzene and a poor solvent, e. g., cyclohexane, as polymerization and crosslinking proceeds, the polymeric material reaches a critical condition after which the cyclohexane is expelled, as the polymer becomes concentrated in strands that form the walls of pores. Further polymerization consolidates the polymer-rich phases, from which even more solvent is expelled, until in the end the polymeric material is distributed as shown in a typical scanning electron micrograph Fig. 12.6.

Control of pore size and pore volume is gained by choice of solvent/monomer ratio and the thermodynamic relation of solvent to the polymer segments. The greater the solvent content, the higher the final porosity. The poorer the solvent, the earlier syneresis occurs and the larger the pores become.

While the generation of porosity in a network by syneresis follows thermodynamic principles, unrelated to any particular form of copolymerization, the practical consequences are most commonly realized in connection with the production of chromatographic and ion exchange materials. It is also important to note that it is impractical to attempt a rigorous application of network swelling theory.

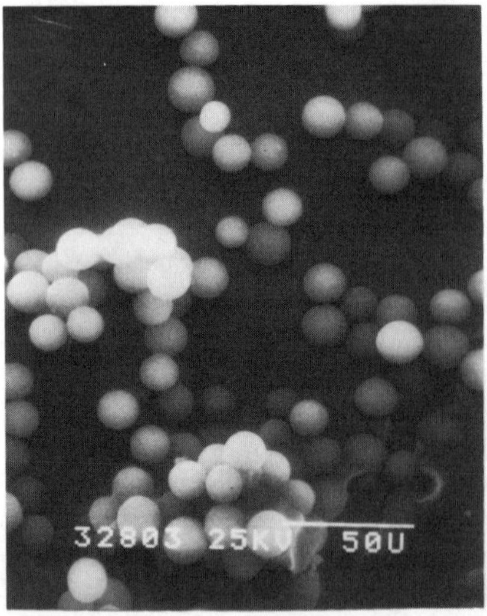

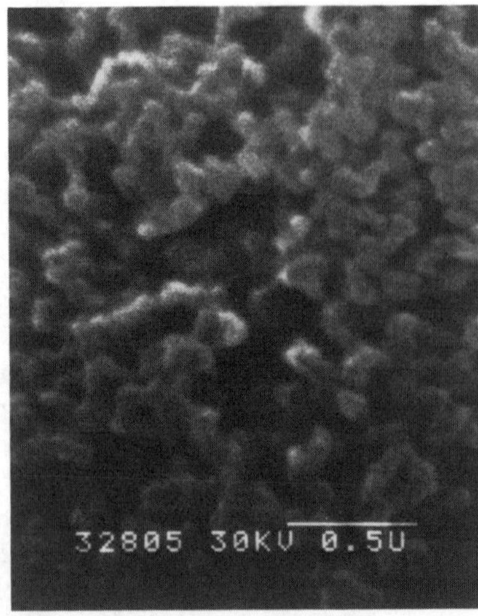

Fig. 12.6. Scanning Electron Microscope Photograph of Chromatography Packing made by suspension polymerization

Problem 12.1

Compare a homogeneous, single phase radical polymerization (*SP*) of butadiene with an emulsion polymerization (EP), and for convenience consider 1 *ml* of volume. The molar concentration of *undiluted* butadiene $[M_0] = 14.8$. Let us consider in each process a time, at which 25% of the initial butadiene has be converted to polymer. In each case the temperature will be 60 °C at which $k_p = 100\ [l \cdot mol^{-1} \cdot s^{-1}]$ and $k_t = 36 \times 10^7\ [l \cdot mol^{-1} \cdot s^{-1}]$.

To make a reasonable basis of comparison, we will say that in the *SP* case we start with 1 *ml* which contains 0.33 *ml* butadiene and 0.67 *ml* benzene as solvent, and we use sufficient azo compound to produce by decomposition 1×10^{12} *effective* radicals per second and per *ml*.

In the EP case, we will start with 0.33 *ml* butadiene and 0.67 *ml* water, enough sodium palmitate soap to create 1×10^{14} polymer particles per *ml*, and sufficient potassium persulfate to generate 1×10^{12} radicals per *ml* per second.

When 25% of the monomer has been converted, the micellar phase has disappeared, droplets still exist, and in the growing polymer particles the monomer concentration [M] remains steady at 7 [*mol/l*].

In both cases let us assume termination is exclusively by coupling. At this time in the respective processes,

— determine the rate of polymerization, i. e., the consumption of monomer per *ml* and per *s*.
— determine $\overline{P_n}$, assuming C_M and C_s are zero.

Problem 12.2

Consider two hypothetical cycles B and C relevant to emulsion polymerization, in the frame of reference of what can go on in a single particle. It is assumed that, at the instant under consideration, k_p [M] = 500 monomer units/second are being polymerized.

Time in [s]	*Cycle B* *Event*	*Time in [s]*	*Cycle C* *Event*
0	Initiating radical enters and propagation starts	0	Initiating radical enters and propagation starts
5	Chain transfer to monomer occurs	15	Second radical enters, cannot find first radical end
10	Chain transfer to monomer occurs	20	Second radical, having grown by addition, couples with end of first
15	Second radical enters, terminates chain by coupling		
30	Cycle starts again	30	Cycle starts again

Compare cycle B and C with an ideal cycle A in which *no* transfer occurs, and the red light period is exactly equal to the green light period, each being 15 seconds.

a) How does $\overline{P_n}$ differ?
b) How does the rate of polymerization differ, this being reported as monomers polymerized per particle per second of total elapsed time (red light and green light combined)?

Problem 12.3

In an emulsion polymerization of butadiene at 60 °C, let us assume the green light-red light period of Problem 12.1. Further assume that at the time in question a typical particle has a diameter of 1000 Å and has a concentration of butadiene monomer of 7 molar, i.e., half of the volume is monomer and the other half polymer.

Case I: The time is just *before* exhaustion of droplet phase. Droplets exist. As a result of one "green light" period, by what percentage will the volume of the particle increase?

Case II: The time is just *after* exhaustion of the droplets phase. What fraction of the monomer molecules existing in the particle will be formed into polymer during one green light period?

Problem 12.4

In the emulsion polymerization of 80% butadiene (A) and 20% styrene (B), at 60 °C, to produce SBR, the rate of generation of radicals is 10^{13} ml^{-1} s^{-1} and the steady number of particles is 10^{14} ml^{-1}. At a certain time, half the volume of a particle is occupied by polymer, the rest is monomer.

Derive an expression in terms of k_{aa}, k_{bb}, r_a, r_b, [A], and [B] for the number of monomer units incorporated during a "green light period".

Given that for butadiene (A) at 60 °C k_{aa} = 100 $l/mol \cdot s$, for styrene (B) at 60 °C k_{bb} = 176 $l/mol \cdot s$, and r_a = 4/3, r_b = 3/4, what is this number?

Problem 12.5

Reconsider problem 12.4, the following assumptions being added: the typical polymer particle at the time under consideration has reached a diameter of 1 000 Å, and a droplet phase mixture still exists, which will keep the mole ratio of monomers in the particles correct (3 butadiene/1 styrene).

a) What is the rate of flux of monomers into the particles, time averaged, in monomers per second?

b) Given a number average degree of polymerization of 1 200 for the polymer already formed, how many macromolecules are in a particle?

c) When the particle size was one third of its present volume, the micellar phase had just been exhausted and the soap molecules on the surface were close packed at 25 square Ångström units. What is the area occupied per soap molecule at this point?

Chapter 13
Heterogeneous Polymerizations and Interfacial Reactions

13.1 Introduction

In the first part of this chapter we survey certain industrial polymerizations in which, by choice or necessity, the polymer as it is generated from homogeneous monomer undergoes phase separation, and we seek to understand conceptually how propagation can continue under such conditions.

In the second part (13.5), we consider interfacial reactions either to create polymers or to transform previously-formed polymers.

13.2 Polymerization of Monomer without Solvent Resulting in a "Precipitated" Polymer

It seems contrary to intuition that a polymer should be insoluble in the undiluted monomer from which it is generated, but this is the case for all of the following "crystallizing" *polymers* at *temperatures appropriate* to *polymerization* (melting temperatures of pure polymers shown in parentheses):

vinyl chloride (140 – 200 °C)
vinyl fluoride (200 – 230 °C)
vinylidene chloride (200 °C)
vinylidene fluoride (185 – 200 °C)
acrylonitrile (~ 317 °C)
tetrafluorethylene (327 °C)

From a thermodynamic point of view the reason is fundamentally that the "solvent"-polymer interaction parameter (Flory-Huggins χ parameter) is near zero, not negative and thus insufficient to lower the melting point of a polymer-monomer mixture to the level of the polymerization reaction temperature.

Beyond a very low degree of polymerization or above a very low concentration, the polymer *"precipitates"* as a separate phase out of the monomer. It is thus to be expected that the kinetic features of the reaction would be very different from those found for homogeneous polymerizations. "Precipitation" in this case does *not* correspond to the solvent-polymer *incompatibility* that results in precipitation of poly(styrene) out of benzene by ethanol. In the case at hand, the monomer is unable to dissolve the crystalline portion of the polymer, but it does solvate the portions of the polymer chains in the "precipitate" which extend between the crystalline regions, i. e., the "amorphous" portion of the precipitate.

First, in most of these polymerizations, the rate tends to increase with conversion, and then becomes constant, resembling the features of an emulsion polymerization. This is illustated in Fig. 13.1, showing the course of poly(vinyl chloride) synthesis from bulk monomer.

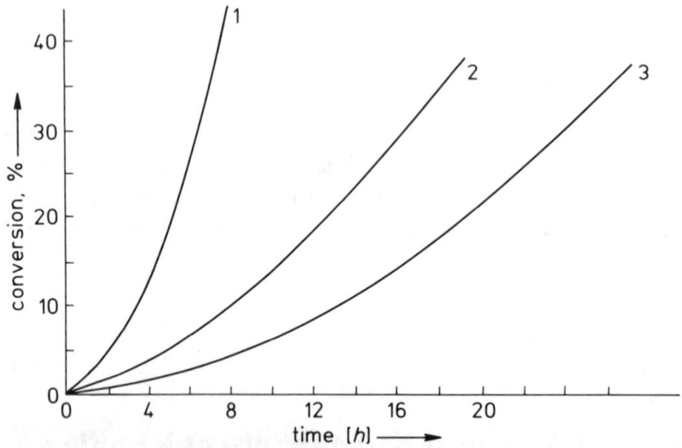

Fig. 13.1. Bulk polymerization of vinyl chloride; conversion versus time.
1) T = 54 °C; initiator = 0.06%
2) T = 40 °C; initiator = 0.12%
3) T = 40 °C; initiator = 0.06%

It is observed that after very low conversion (*around 5%*), the system becomes turbid and then a more dense opaque slurry of fine particles is formed. To explain this, the reaction model shown in Fig. 13.2 advanced. It has the following central features:

(a) Initiation starts (for example from azo-bis-isobutyronitrile) in the liquid monomer phase.

(b) Propagation proceeds, oligomer radicals remain soluble until they reach a chain length at which association by crystallization occurs.

(c) These crystalline nuclei remain active, and some of the polymeric radicals incorporated therein continue to add monomer, causing the nuclei to expand.

(d) Coalescence of nuclei occurs, leading to fine microscopic particles.

(e) During the growth of the nuclei it appears (from transmission electron micrographs) that they retain considerable porosity, which stays on in the particles; thus monomer and oligomeric radicals are still capable of diffusing into the nuclei and particles.

(f) The mode of termination appears to be analogous to that of emulsion polymerization, i. e., radicals arrive from the liquid phase at the surface of the nuclei (or particles). The bound radicals are terminated by coupling or by disproportionation.

For these reasons, it is virtually useless to attempt to construct detailed kinetic schemes, since the meaning of concentration becomes ambiguous — there being two phases, liquid and solid surface. For a given percentage conversion of monomer (and thus weight fraction of polymer), the surface area [cm^2/g] can be widely variable, depending on conditions of agitation, temperature, rate of polymerization etc.

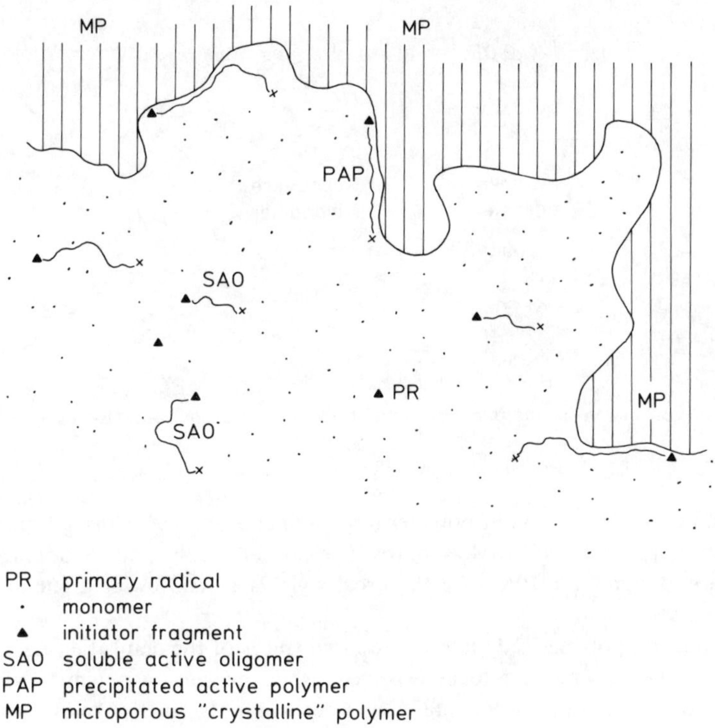

Fig. 13.2. Reaction model for the bulk polymerization of vinyl chloride

Any of the monomers listed above can be polymerized via emulsion or via suspension polymerization. Vinyl chloride is polymerized by both processes. The very fine particles (in the range of a few micrometers) produced as agglomerates from emulsion polymerization are particularly useful for *"plastisols"*, i. e., suspensions in "plasticizing" liquids such as dioctyl phthalate containing up to 60 *wt.*-% polymer. These fluids, when subsequently heated to >120 *°C*, evolve into viscoelastic media as the polymer particles, heated above the glass transition of PVC, absorb the suspending solvent.

Emulsion polymerized PVC is less porous than PVC produced by suspension polymerization, possibly as the result of the fact that in emulsion polymerization the monomer from which the particle is formed arrives from the external milieu — i. e., the particle grows progressively — whereas in suspension polymerization, the particle (bead) is of constant size, but its polymer content increases at the expense of monomer. A flimsy structure of precipitated PVC can be established throughout each bead early in the polymerization. If high porosity is desired, the polymerization can be stopped early and the monomer can be volatilized easily and recycled, since (1) insignificant polymer is dissolved in the monomer, so that the monomer has nearly the thermodynamic activity of pure monomer, and (2) the polymer is microporous, not continuous, so that escape occurs by capillary flow rather than by molecular diffusion. This is shown schematically in Fig. 13.3.

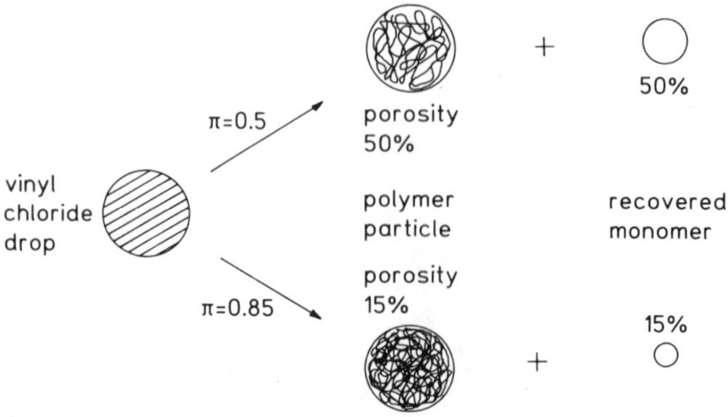

Fig. 13.3. Effect of conversion on the porosity of the poly(vinyl chloride) obtained by bulk polymerization

Thus, at the same weight fraction of polymer (e. g., 0.6) and of plasticizing solvent (0.4), emulsion polymerized PVC of low porosity yields an easily poured liquid, whereas suspension polymerized PVC of high porosity yields a free-flowing granular solid.

As with any emulsion polymerized material, residual traces of the emulsifier, e. g., soap, entrapped in the polymer, are to be expected, while suspension polymerized material may contain traces of suspending agents such as poly(vinyl alcohol), or minerals like carbonates.

Poly(vinyl chloride) having neither class of contaminant is produced on a large scale by a process referred to as *"mass" polymerization.* The only materials involved are bulk undiluted monomer, initiator, and polymer, the polymer in a separate phase from the monomer.

The existence of the two phases is essential to the viability of the process. In the *"prepolymerizers"* (relatively small autoclaves stirred by high speed turbines), the monomer and initiator are introduced at around 50 °*C*, and the polymerization is carried to about 15% conversion, producing a slurry of crystalline porous polymer particles in bulk monomer.

The slurry is transferred from the autoclave to a much larger vessel fitted with a jacketed cooling shell and an internal rotating helical blade. By slow rotation the blade moves the polymer against the walls. As polymerization proceeds, a substance analogous to high porosity suspension PVC results, i. e., microporous particles, free flowing as a solid, but containing internally initiator and monomer yet to be reacted.

Since there are only two phases at any time: liquid monomer of low viscosity and partly crystalline polymer well below its melting point (in excess of 140 °*C*) and its glass transition (~100 °*C*), stirring is at all times easily accomplished.

Because the normal boiling point of vinyl chloride is 14 °*C*, all polymerizations around 50 °*C* must be carried out under *modest pressure*, regardless of the process. Advantage is taken of this fact to control the rate of polymerization. For example, if for some reason the rate of polymerization increases beyond the planned value, which could lead to rapid increase in temperature, pressure on the reactor can be abruptly decreased. Immediately, evaporation of monomer out of the porous granules decreases the temperature by abstraction of latent heat of vaporization. Simultaneously the supply of monomer available for polymerization is decreased, and thus the rate of polymerization, at constant temperature, decreases.

Branching is of concern with PVC since it usually occurs on an α-carbon, which thus becomes a tertiary carbon, capable of subsequent reversion to a radical under elevated temperature or exposed to ultra violet radiation, as mentioned in chapter 9.

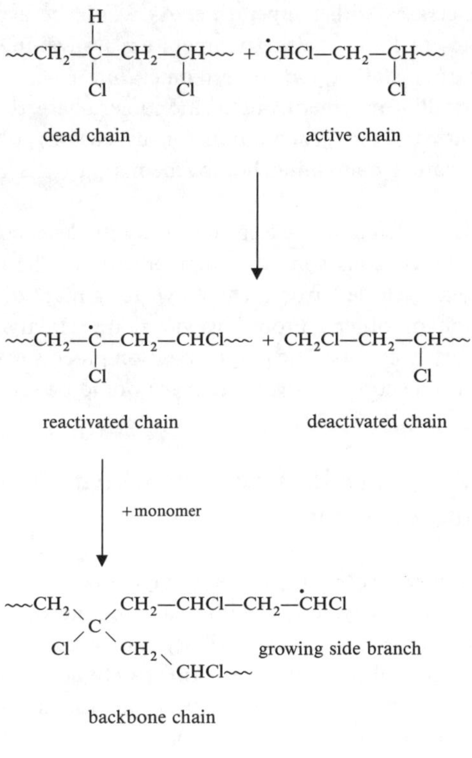

Thus usually conversion is carried to 80–86%, then stopped, and residual monomer is evaporated and recycled.

Earlier it was noted that as temperature of polymerization is decreased from $\sim +50\ °C$ to $-40\ °C$, the propagation step becomes more discriminating, favoring racemic (r) placement as compared to meso (m), and thus the polymer has an increasing syndiotactic triad content. Ultimately, a highly syndiotactic species of PVC is produced, so that no solvent will disrupt the crystallites, nor will crystalline melting occur before degradation.

Among the other "precipitating" polymers derived by free radical polymerization, poly(vinylidene chloride) has such a high melting point that it decomposes. Thus vinylidene chloride is invariably copolymerized with some (10–20 *mol-%*) vinyl chloride to *introduce disorder* and lower the equilibrium melting point. The latter is further depressed to practicable temperatures of blown film extrusion by the addition of *plasticizing solvents.*

Vinylidene fluoride, unlike vinylidene chloride, can be produced successfully as a homopolymer by conventional methods. It possesses the curious attribute of an unusually high content of head-head placement of units, ranging from ~6 to around 11%, as determined by nuclear magnetic resonance spectroscopy.

It is presumed that poly(vinylidene fluoride), which is produced mostly by emulsion or suspension polymerization, owes its variable head-head content to the polymerization temperature: The content increases with temperature. As would be expected, increasing head-head content leads to lower melting temperature (range 169 °*C* – 160 °*C*) and to lower percentages of crystallinity at room temperature.

Because of the separation of crystalline polymer from the monomer phase, the usual equations governing the kinetic course of homogeneous bulk (or suspension) polymerizations, or of emulsion polymerization containing homogeneous particles *do not apply.*

Finally, all of the vinyl monomers listed above, except tetrafluorethylene, could be polymerized by a single-phase homogeneous solution polymerization. The solvent would necessarily be strongly polar, selected from: Dimethyl acetamide, dimethyl formamide, hexafluoroisopropanol, or others. From the viewpoint of large scale production, this is impractical on two accounts: The polymerization process would be more complicated, not simpler, and the cost of *solvent recovery* would be very large.

13.3 Polymerization of Monomer in a Solvent which Causes "Precipitation" of the Polymer

As discussed in chapter 6, isobutylene (isobutene) is homopolymerized via cationic pathways to yield the homopolymer, or copolymerized with small amounts of isoprene (~5–10 *mol-%*) to yield "butyl rubber", i.e., poly(isobutylene) macromolecules containing a few double bonds by which subsequent crosslinking can be accomplished.

The polymerization of isobutylene *could* be carried out homogeneously in a solvent such as hexane which would dissolve the polymer. In most solution polymerizations, high levels of viscosity, several fold greater than that of the initial monomer-solvent mixture, would be produced far short of complete conversion of monomer, unless initial monomer concentration is low (<5%). Problems of *heat transfer* would become important, especially because of the high rate of reaction found generally in cationic polymerizations.

A second factor makes slurry polymerization particularly desirable: The necessity for low temperature levels, achieved by efficient mechanical refrigeration cycles, around −100 °*C*. As noted in chapter 6, transfer reactions are the predominant mode of limitation of chain length, and polymerizations at around 0 °*C* produce low molecular weight materials useless as rubber, but having some value as "tackifying" additives.

Chloromethane (b.p. −14 °*C*) is preferred as a liquid continuum for isobutylene polymerization, because (a), while it is a solvent for monomer, it is a precipitant for the polymer, so that a slurry of low viscosity results. (b) In case of "run-away" reactions it can be vented at $-14\ °C$ abstracting latent heat from the slurry, thus serving a role parallel to that of vinyl chloride monomer in the "mass" polymerization synthesis of poly(vinyl chloride).

Coalescence of the polymer in the slurry, if it occurred, would produce a large intractable mass in the reactor capable of seizing the agitator and defeating ordinary means of removal.

That coalescence does not ordinarily happen can be attributed to two factors:

(1) At $-100\ °C$, in the presence of methyl chloride, the polymer is not far above its glass transition, and thus long term contact between slurry particles would be needed to permit interdiffusion of macromolecules (the basis for coalescence).

(2) Though a very poor solvent thermodynamically, methyl chloride "wets" the slurry particles effectively and thus interferes further with the already slow process of coalescence.

The situation would be drastically different, of course, if the reaction temperature were not $-100\ °C$ but near zero, and if the polymer molecular weight were low.

The nature of the process of slurry polymerization of isobutylene (and for butyl rubber) is schematically indicated in Fig. 13.4. Soluble oligomer, carrying a catalyst

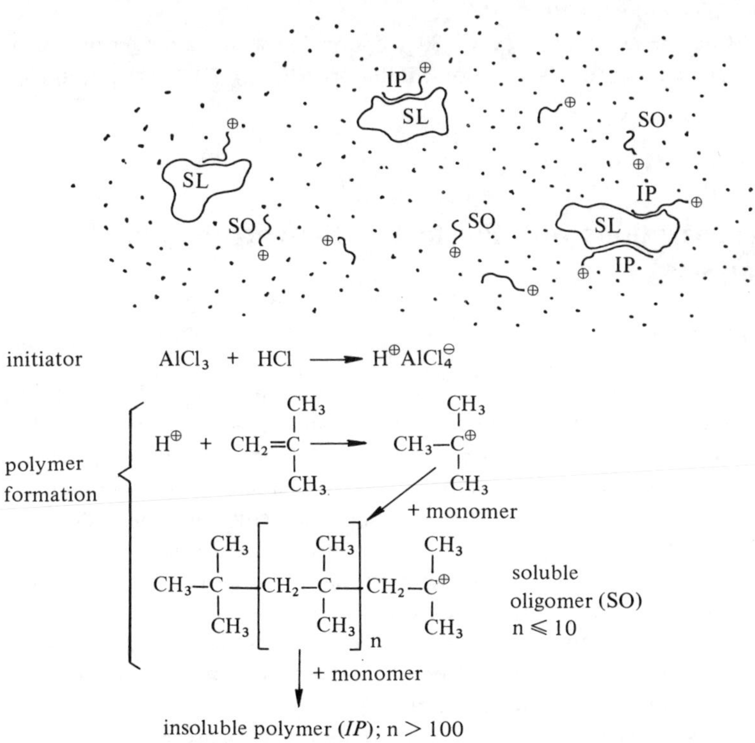

SL = slurry particle: aggregate of *IP*

Fig. 13.4. Slurry polymerization of isobutene

fragment at one end and a cation (+) at the other, grows until it exceeds the solubility limit; it then goes out of solution onto slurry particles *SL* where it continues to grow as insoluble polymer *IP* until some transfer reaction stops the chain growth.

As in the case of vinyl chloride polymerization, the formation of a separate phase of condensed polymer on the surface of which some macromolecules are propagating, defeats the application of kinetic equations appropriate to homogeneous systems. These equations are very complicated in the case of the cationic polymerization, as pointed out in chapter 6.

Because of the high reaction rate already mentioned, refrigerated slurry polymerization of isobutylene can be carried out as a continuous process, since a low residence time is required.

The possibility of allowing essentially adiabatic polymerization to occur, beginning at -100 °*C*, is illustrated by the following calculation:

Assume as a basis 1 *mol* of isobutylene (56 *g*), which is in a 10 *wt.-%* solution of chloromethane (total mass therefore 560 *g*). Assume 80% conversion of isobutylene to poly(isobutylene) in the continuous reactor. With an enthalpy of polymerization of 10000 [*cal/mol*], this would lead to heat release of 8000 *cal* into 560 *g* of heat capacity 0.5 [$cal \cdot K^{-1} \cdot g^{-1}$]. Thus the temperature rise ΔT will be: $(8000)/(0.5)(560) \approx 30$ °*C*. This would lead to a broad molecular weight distribution, with the peak between degrees of polymerization of 10^4 to 10^5.

It would not be unreasonable to expect 80% conversion within a matter of tens of seconds at typical concentrations of cationic initiator ($AlCl_3$ + HCl in this example).

13.4 Polymerizations with Particulate Stereoregulating Catalysts

13.4.1 Introduction

In Chapter 8, Ziegler-Natta and other stereoregulating catalysts are discussed. The majority of these are particulate (colloidal) rather than micromolecular and soluble. Consequently these polymerizations are heterogeneous at least in the sense that the propagation reaction takes place on a solid surface, not in homogeneous solution.

Some of these polymerizations are doubly heterogeneous: not only does the propagation (monomer added to polymer) take place at a solid surface, but also the polymer being produced crystallizes, forming yet another solid phase. The thermodynamic reason for this kind of "precipitation" polymerization is the same as for those cited in section 13.2. The continuous phase in most of these polymerizations is an alkane (cyclohexane, etc.) which is inert toward the catalyst, dissolves the monomer, and will dissolve the polymer if the polymer is not crystalline. In the Unipol and related processes for polyethylene polymerizations the continuous phase is a gas stream, consisting only of the monomer, and the dispersed phase consists of particles of silica-supported catalyst of a size range appropriate to fluidized bed operation.

13.4.2 Micromolecules, Macromolecules and Heterogeneous Catalysts

When reagent *micromolecules,* e.g. CO, O_2, react in the presence of heterogeneous catalysts like platinum, chemisorption at the interface occurs, leading to the production of a product micromolecule (e.g., CO_2) and its desorption. The micromolecules have dimensions comparable to the interatomic spacing of the catalyst surface, and no significant mechanical strain is imposed at the surface by reason of attachment of reagent or product molecules. Access of reagent micromolecules is limited only by the fraction of the catalyst surface already occupied. Just beyond the adsorbed layer, counter diffusion of reagent and product micromolecules readily occurs. (That the catalyst can be inactivated by "poisons" is irrelevant to the present discussion.)

Totally different conditions exist when *macromolecules,* e.g. poly(ethylene), poly(propylene), or poly(1,4-cis isoprene) are produced by heterogeneous catalysts, and it is important to understand the consequences of these conditions.

Fig. 13.5. schematically presents a physical scenario for three different cases: (a) The macromolecular chain as it is formed *does not crystallize,* and is solvated by the continuous phase; (b) the macromolecular chain does *crystallize* by folding upon itself as it is formed on the surface of a catalyst particle, despite the presence of the

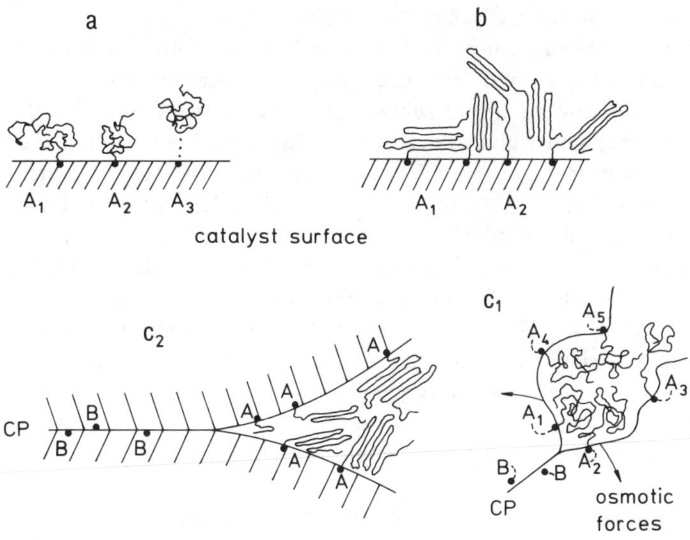

Fig. 13.5. General scheme — heterogeneous catalysis of polymerization. (a) Active sites A_1, A_2 produce random coiling chains (polymer fully soluble in solvent medium). Fluctuation has just detached a chain from site A_3.
(b) — Active sites A produce semi-crystalline chains (e.g. poly(ethylene)). Monomer diffuses through microporous polymer layer to active sites. Layer does not diffuse off surface. - In microporous catalyst, active surfaee is generated by growth of semi-crystalline (c_1) or amorphous chain (c_2). Monomer (gas or liquid) diffuses through microporous polymer to sites. Polymer expands, produces stress on cleavage plane CP. When CP splits, the sites B will become active sites A

continuous phase; (c) new catalyst surface is generated by *stress* imposed at the interface by the growing polymer.

Case (a): Suppose the macromolecule as it is formed is random coiling, thus noncrystallizing. The diameter of the coil will vary approximately with the square root of the molecular weight and would be measured in tens and then hundreds of Ångström units. This has two consequences:

(1) Random coils growing on adjacent sites of a catalyst surface soon begin to interfere because of lack of space available; (2) as the random coil becomes larger, it will experience increasing thermal fluctuations via the solvent which surrounds and permeates it. Thus at some stage of the chain growth, a fluctuation will be sufficient to detach the chain from the catalytic surface, for example site A_3 of Fig. 13.5. (a). The interference of the two chains shown growing on sites A_1 and A_2 might produce an even earlier detachment.

For a chain to become detached from the surface, the bond between the metal atom fixed in the surface and the chain would have to be terminated. A plausible mechanism would be hydride transfer from the *last unit* to be formed on the macromolecule to the catalyst site. According to the concepts developed in chapter 8, each new monomer unit to be added has to be inserted into the bond between the metal sites and the polymer; thus this "bond" is to be viewed as peculiarly active in its transition state (as new monomer is added), a state in which evidently the "mistake" of hydride transfer can occur even without added hydrogen to control molecular weight.

This case (a) applies to all the *elastomers* synthesized by heterogeneous catalysis, thus poly(isoprene), poly(butadiene), and ethylene propylene random copolymers, including EPD (ethylene-propylene-diene) rubber. This case is also relevant to the synthesis of poly(propylene) homopolymer to the extent that a fraction of the polymer produced is insufficiently isotactic to crystallize at the polymerization temperature.

Case (b): As shown in Fig. 13.5. (b) the macromolecule folds upon itself and neighboring macromolecules are producing a microporous "membrane" above the catalyst surface, through which the monomer must *diffuse* to arrive at the surface for insertion and which *obstructs* the release of polymer molecules being formed.

The consequences in respect to the kinetics are not clear. While it has been argued that crystallization of the polymer is the main reason for the rapid decrease in productivity (= moles monomer converted per gram catalyst per hour), other studies show that simple ageing of the catalyst in the continuum at the same temperature without monomer, and thus without polymerization, leads to nearly equal *decreases* in productivity.

Clearly the physical situation is different from case (a). The macromolecules being stabilized in a crystallized *"membrane" layer* are far *less* affected by *Brownian motion* than they would be if they existed in a random coiling state.

In this case, one may suppose that the "membranes" will become progressively buckled and cracked as polymerization proceeds and that like ice floes at the edge of a lake in winter, the "microflakes" will be sheared off.

The scenario of case (b) is appropriate to all polyethylene types (high density, low density) if the reaction temperature is less than about 100 °*C*. If the reaction temperature is greater than 120 °*C* and the continuous phase is an alkane like cyclohexane, case (a) instead is realized, i. e., the polymer cannot crystallize.

Case (c): New catalytic surface is generated in microporous particles of supported catalyst by expansion of entrapped polymer. In c_1 random coiling (non crystallizing) polymer is shown growing from active sites A. Entrapped in a pore smaller than the random coil, the polymer chains will develop osmotic pressure which will be communicated to the walls, since monomer and solvent will attempt to dilute the segments of the growing chains, but these cannot escape from the pore. The pressure, depending on polymer segment concentration can reach tens to hundreds of bar, eventually bursting the pore (as shown by arrows) and exposing new sites B along the plane of cleavage CP. In c_2 a crystallizing polymer is contemplated. Growth of the crystalline chains in the permeable "membrane" overlayer, between sites A_1 and A_2, A_3 and A_4 will lead to a splitting along a cleavage plane CP exposing new sites B.

The case shown in Fig. 13.5. (c_2) underlies the exploitation of silica based catalysts in gas-fluidized beds, discussed in 13.4.4.

13.4.3 Polymerizations by Heterogeneous Catalysts with Continuous Liquid Phase

When inert alkanes, e. g. hexane, are used as a continuum in which the Ziegler-Natta or metal oxide catalyst is suspended, and monomer is then introduced into this system, we have to distinguish three possibilities of phase properties which will considerably affect reactor design and operation:

Possibility (1): The polymer generated is non-crystallizing and dissolves completely in the continuous phase. The *only* other phase is the heterogeneous catalyst. In this case the viscosity will increase more or less in accord with the 2.5 power of polymer concentration and the 3.4 power of weight average molecular weight (see equation 11.1).

Depending therefore on what molecular weight is sought, polymer concentration may be practically limited to around 20 *vol.-%* or less, in order to keep mixing (and heat transfer) feasible by conventional equipment. The viscosity will be one to three orders of magnitude greater than that of the original continuous phase.

This case will apply to all elastomers, both dienes and random copolymers of ethylene and propylene. It will apply to *all* polymerization reactions by Ziegler-Natta or metal oxide catalyst if the reaction temperature exceeds about 120 °*C*, which ensures solubility of poly(ethylene) and of isotactic poly(propylene).

Possibility (2): The polymer generated is nearly completely crystallizing under the operating conditions (hexane as solvent, temperature below about 100 °*C*). In this case there are three phases: The liquid continuum, the crystalline polymer, and the heterogeneous catalyst. The latter two will exist as one material, i. e., the polymer consists of a permeable "membrane" of crystallized chains lying on top of, or within the pores of the catalyst.

Under these conditions we can expect the viscosity of this slurry to be very much lower than that of the solution (first possibility). For example, 20 *vol.-%* dispersed phases (polymer and catalyst) would scarcely double the viscosity of the monomer-hexane continuous phase.

This possibility is illustrated by the homopolymerization of ethylene, and its copolymerization with minor amounts of 1-pentene, 1-hexene, etc. (for the purpose of introducing short chain branching) in hexane or similar solvents, and at temperatures less than the crystalline melting point in that solvent, approximately 100 °*C* for ethylene homopolymer in hexane and 80 °*C* for the 1-hexene or 1-pentene copolymer. Should, through some transient episode, the reactor temperature exceed and then fall below the crystalline melting point, gross *agglomeration* of the polymer might be expected. Modern Ziegler Natta catalysts yield nearly 100% isotactic polypropylene which would fit this category.

Possibility (3): Some of the polymer crystallizes while another fraction does not. This is the case with *propylene* polymerization below about 100 °*C* on the older Ziegler-Natta catalysts. The more isotactic material is crystalline, the more heterotactic is soluble in hexane even at room temperature. Incomplete evidence suggests that either the poly(propylene) chain is rather highly isotactic (say $>80\%$) or relatively heterotactic, around 40% or more. There does not seem to be a continuous spectrum of order from "perfect" isotacticity to strictly atactic polymer (25% *iso*, 25% *syndio*, 50% *hetero* triads), as if the catalytic sites were of two kinds, one producing very well ordered chains, the other not.

As a consequence, the reaction system will lie between (1) and (2); the continuous phase will be less viscous than in (1), but more viscous than in (2).

Reactors and Reactor Conditions:

In view of these considerations, Ziegler-Natta polymerizations of ethylene or of propylene, and metal oxide polymerizations of ethylene are carried out at moderate temperatures of 60°–90 °*C* to ensure that the crystallizable polymer (polyethylene or isotactic polypropylene) does not go into solution. At these temperatures, with carrier liquids such as hexane, pressures of only a few bar are required to dissolve the monomer in the liquid.

Polymer properties are more or less independent of conversion in these heterogeneous polymerizations; fresh monomer can be fed indefinitely to the catalytic surface, provided it is not blocked by an impermeable layer of polymer; and the crystalline polymer floc can be separated from the liquid continuum by settling and decantation owing to density difference.

These facts in combination make it possible to carry out these heterogeneous polymerizations in a wide variety of reactors and under remarkably different conditions. *Batch reactors* are feasible: Monomer and catalyst are charged into an autoclave containing hexane at 80 °*C* (air and moisture contact avoided), and the batch is stirred gently to keep the crystalline floc suspended, until an arbitrary percent (perhaps 80%) of the monomer has been converted. This can occur in a matter of a few minutes. The batch is then cooled and dumped. Residual monomer and hexane are recovered, while the polymer floc, free of these, is contacted with alcohol which converts Ziegler-Natta catalysts to soluble alkyl esters.

With efficient catalysts, *continuous* ethylene polymerizations approximating plug flow can be operated with vertical towers in which catalyst and monomer are injected at the top along with some make-up carrier liquid, and floc is continuously withdrawn at the bottom, as a concentrated suspension in hexane. In this case the colloidal catalyst particle, by growing crystalline polymer on itself, becomes increasingly dense, and therefore has an increasing settling velocity. Then the larger particles migrate most rapidly to the bottom of the tower. Conversion occurs in a short time, measured in seconds or minutes.

13.4.4 Polymerization by Heterogeneous Catalysts in a Gas Fluidized Bed

When a gas is passed upwards through a bed of solid, not coherent, and uniform particles of appropriate diameter (~ 100 μm diameter) at increasing velocity, a *critical velocity* is reached at which the bed begins to *expand* and *roil* ("fluidized bed reactor"). Further increase in gas velocity serves to expand the bed somewhat more, but not until the gas velocity reaches nearly 100 fold the critical velocity for fluidization does the particulate material get blown out of the bed and carried away.

Because of the *excellent* heat and mass transfer they provide, fluidized beds have been widely exploited for reactions of gases with catalytic surfaces that have been deposited on (and in) particles like silica of size appropriate to fluidization. Fluidized beds operate nearly isothermally if properly designed. *Large gas recycle* ratios are employed because of the need for high gas velocities. This will be made more evident in the next section. In all cases other than polymerizations, both products as well as reagents are *gaseous* under the conditions of operation of the bed. It is impossible to conceive of any other realistic mode of operation: For example, if the reaction products were liquid, not gaseous, the particle surfaces, wet with this liquid, would stick upon collision and grow into aggregates which would then fall out of the fluidizing gas stream.

Polymerization through the use of a fluidized bed could be accomplished if four conditions are satisfied at the same time: (1) The monomers being polymerized are completely *gasified* (vaporized), not liquid. (2) Under operating conditions, polymer being formed is *not* "sticky". Thus it must be crystalline or glassy, not rubbery or fluid. (3) Polymerization must arise isothermally from activating sites on (or in) the suspended particles, the size of which is very important, for if polymerization ensued in the gas phase, through some agent mixed with gaseous monomer, very fine material would be generated that would escape the bed at once. (4) The polymer being formed on the active surface, (whatever the mechanism of activation is) must remain microporous to some degree, to allow access of monomer through the polymeric layer to the active sites, for if the polymer formed a dense impermeable membrane, the intended reaction would totally stop at a very early stage. These conditions are satisfied for polyethylene and isotactic polypropylene below their respective melting points.

For fluidized bed operation, Fig. 13.6., it is mandatory to use silica particles of the right size range (~ 100 μm) as the support for the active catalyst on which the polymerization will occur. Larger particles cannot be readily fluidzed in a gas stream, and particles smaller than ~ 10 μm are too easily blown out of the bed. In the Unipol process, ethylene without or with a few mol percent of co-monomer (1-pentene, 1-hexene, etc.) is polymerized on silica supported CrO_3 to form unbranched ("high density") or short branched ("low density") polyethylene respectively.

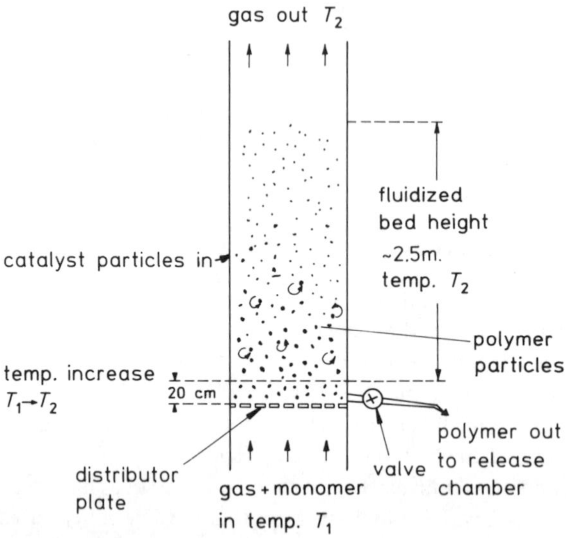

Fig. 13.6. "Fluidized bed" process for the polymerization of ethylene

The silica supported catalysts are so efficient (productivity = $\sim 10^4$ kg polymer produced/kg catalyst + support) that subsequent extraction is unnecessary. The presumed reason for this efficiency, in addition to the selection of the particular active agent, for example CrO_3, is that the porosity of the silica (pore size, pore volume fraction) allows ingress of monomer, polymerization in the pores, and fracture of the silica skeletal structure exposing new active catalyst.

Silica supported catalysts (Ziegler-Natta or metal oxide) are not limited to gas fluidized reactors. They are widely used in liquid continuous phase slurry polymerizations (section 13.4.3) of ethylene and of propylene when the conditions lead to insoluble crystalline polymer, thus below ~ 100 °C and with active catalyst that in the case of propylene yields predominantly isotactic, therefore insoluble polymer.

13.5 Interfacial Polymerization and Interfacial Reactions

13.5.1 Introduction

Certain polymerizations from monomer, and most reactions on polymers, are confined to a sharply-defined interfacial zone. Thus the systems are heterogeneous rather than homogeneous, and the observed rates of reaction depend, as one would expect, on the relation of the magnitude of chemical reaction rate to the mass transport rates of reagents and products to and from the reaction zone.

As with all heterogeneous reactions, one can construct innumerable scenarios from these variables.

13.5.2 Polymer Formation at an Interfacial Zone Between Two Immiscible Phases Each Containing a Reagent Monomer

Levenspiel (1962) has treated the cases of reagent A in one phase α reacting with reagent B in a second phase β, (α and β immiscible) at the interface or reaction zone between the two, producing *product R*, when A, B and R are micromolecules, and the phases containing A and B are of low viscosity. For example, when the reaction rate is *infinitely fast* concentration profiles of A, B, and R might be as illustrated in Fig. 13.7.

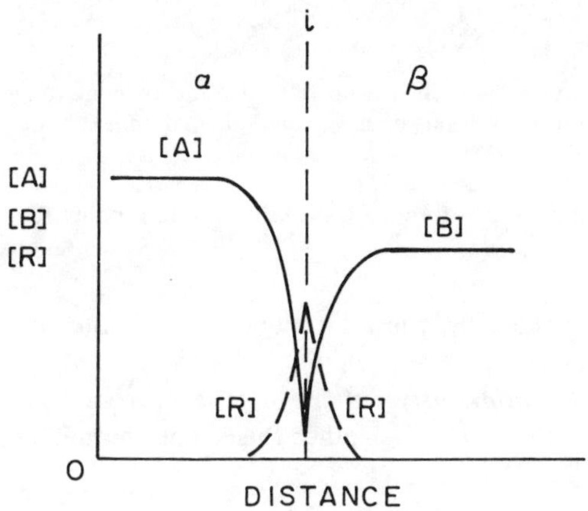

Fig. 13.7. For the hypothetical infinitely fast reaction A + B → R, molar concentrations [A] in phase α, [B] in phase β, and [R] in α and β as a function of distance from interface i

In polymerizations that approximate "infinite" rate, the fact that R is polymeric rather than micromolecular and has zero solubility in either phase α or phase β leads to profoundly different consequences.

Two such reactions are:

acid chloride + amine ⟶ amide

isocyanate + amine ⟶ disubstituted urea.

The reaction product of the diacid chloride and diamine is, for example, nylon 6–10. The reaction product of the second pair is a polyurea or poly(urethane-urea). The formation of R at the interface may form a substantially impermeable "membrane" and thereby stop the reaction or slow it to a very low value, depending on the degree of permeability of the "membrane" R to A and B.

In *microencapsulation* this self-limiting result is desired. For example phase α containing reagent A and a non-reactive compound X is dispersed as droplets in phase β. Immediately spherical shells of polymer R are formed on the drops of α. The amount of reagent A present in α may be so fixed that it is totally consumed while reagent B remains in excess in phase β (Fig. 13.8.). Thus X (a drug, a dye etc.) is *encapsulated* in

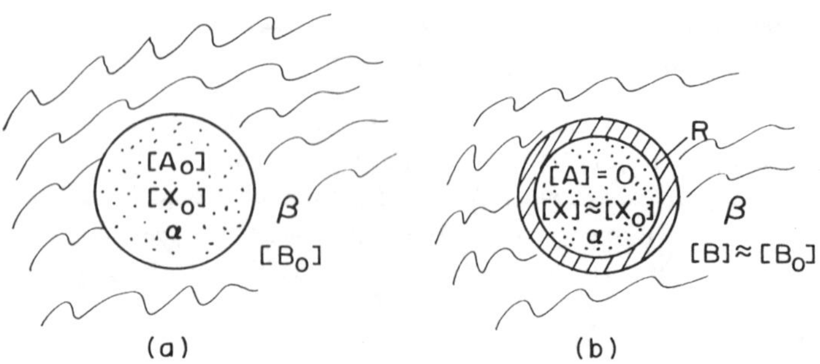

Fig. 13.8. Component A, initial concentration $[A_0]$, in a drop of phase α also containing component X, concentration $[X_0]$, exposed to continuous phase β containing component B of concentration $[B_0]$.
(a) Immediately after formation of drop.
(b) After all A has been consumed by reaction with B forming a shell of reaction product R

the shell R. the microcapsules are recovered by microfiltration for their intended application.

Membranes are formed (Fig. 13.9.) by *continuously* withdrawing the macromolecular "membrane" exposing phases α and β directly to each other, thereby promoting the reaction between A and B.

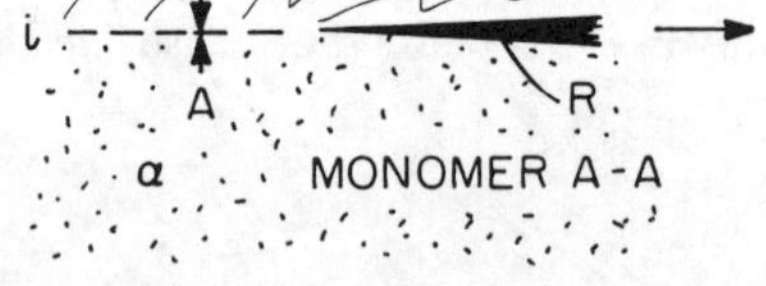

Fig. 13.9. Interfacial Polymerization: Monomer having active ends A in more dense phase α reacts with monomer having active ends B in less dense phase β to produce polymer R as a semicrystalline film. As R is withdrawn (→) fresh interface (– – –i) is exposed between α and β

In the reaction of a diamine A dissolved in a phase α like water, with a diacid dichloride B in a phase β such as dichloromethane, a competitive reaction is possible, viz.

$$B + 2H_2O \longrightarrow \text{carboxylic acid} + 2HCl$$

whereas the desired reaction is,

$$B + A \longrightarrow R + HCl \qquad (R = \text{nylon})$$

Also a second competitive reaction is possible:

$$A + H^+ \longrightarrow AH^+ \text{ (protonated amine)}$$

This can be avoided by including in phase α a neutralizing agent for the hydrogen chloride liberated by the reaction of B with A, since once the acid chloride has hydrolysed to carboxyl, the carboxyl end is totally unreactive, i. e., a dead end.

The reaction of diamines A in water α with diisocyanates B in a dry organic solvent β involves the competitive reaction,

$$B + H_2O \longrightarrow \text{amine,}$$

i. e., B becomes A.

From these interrelationships it is clear that a wide variety of reaction rates and product R characteristics are to be expected, depending on reagent concentrations, rate of withdrawal of the product membrane R, and the permeability of R to A and to B.

Unlike the conceptual scheme of Fig. 13.7. due to Levenspiel, the real situation in the examples cited is the following:

As the product R is created, it either has an A terminated end which will react with B unless A is protonated, or it has a B terminated end which will react with A unless water from phase α "kills" it. Thus if the competitive reactions are avoided very long polymeric chains can result in the product R.

In the case of nylon 6–10, crystallization by chain folding takes place as illustrated in Fig. 13.10., leaving an active function A on the end of a short sequence of units that

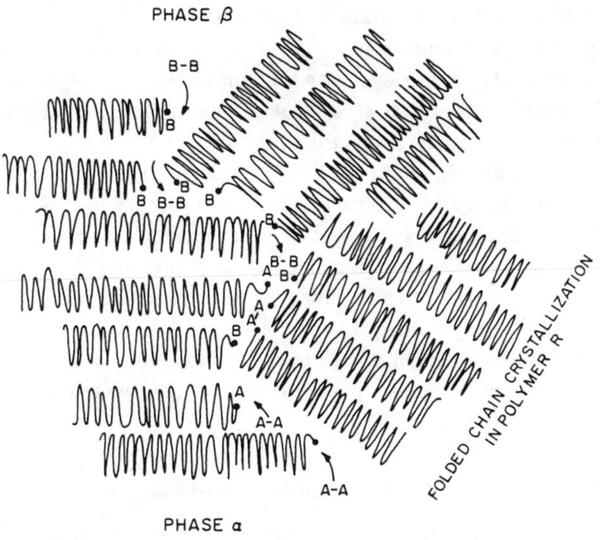

Fig. 13.10. Schematic detail showing diffusion limitations to the reaction of A with B, when one active end is fixed to an immobilized polymer chain

have not been incorporated by chain folding. The difunctional molecules B-B diffuse through the "pores" of the imperfectly crystallized nylon 6–10, thereby increasing chain length by one unit, to be followed by diffusion of A-A from the other side of the polymer layer. The process of addition must obviously cease when internal free volume (the space around chain end A in Fig. 13.10.) is exhausted, or when the overlapping polymer material around chain end A becomes so dense as to exclude B-B.

13.5.3 Reactions on Polymers at Interfacial Zones

13.5.3.1 Hydrolysis of Xanthate Cellulose

Fig. 13.11. depicts the formation of cellophane sheet (tubing) by extrusion of the viscous cellulose xanthate solution (cf. Section 9.3.3.3), shown as phase β, into a dilute sulfuric acid solution, phase α. Depending on the acid concentration and the concentration of cellulose xanthate in phase β, the "tightness" of phase δ: regenerated cellulose is variable. The higher the acid concentration, the more rapid the hydrolysis and phase separation of the insoluble cellulose in a thin layer, phase δ_1 adjacent to a reaction zone γ, in which the xanthate [—CH_2—O—CS_2Na] is hydrolysed to [—CH_2OH], beyond which is the unreacted xanthate solution, phase β.

Because the hydrolysis of xanthate to hydroxyl is very rapid, the thickness of zone γ remains small, as the thickness of phase δ increases and that of phase β decreases.

Because the rate of diffusion of H^+ from phase α into and through δ decreases drastically after first contact between pure phases α and β, from which δ is generated, the morphology (openness, effective porosity) of phase δ_1 varies from the tight skin at the interface with α to loosest inside (δ_2). The process described underlies the generation of asymmetric ultrafiltration membranes of cellophane.

As a counterexample, if very low concentrations of H^+ were used in phase α and if the xanthate solution were supported as a thin layer on a horizontal casting plate, a very gradual hydrolysis of xanthate and a slow phase separation of the cellulose would lead to more isotropic film without the characteristic tight skin.

The case shown in Fig. 13.11. is analogous to the case considered in section 13.5.2 in which the product R becomes a thin zone offering increasing resistance to reagents A and B which from it. In the present instance, while A might be interpreted at the free ion H^+, B is a xanthate group bound to a polymer chain and thus the diffusion coefficient $D_{A\alpha}$ of A through phase α is orders of magnitude greater than $D_{B\beta}$, the diffusion coefficient of xanthate through phase β.

As in the case of interfacial polymerization of nylon, the product generated at the interface is a semicrystalline polymer forming a partial barrier to the two reagents.

13.5.3.2 Hydrolysis of Poly(methyl acrylate) Presented as Cross-Linked Beads

To form ion-exchange beads, suspension polymerized methyl acrylate with divinyl benzene as cross-linking agent is suspended in solution such as 1 Normal NaOH. This results in the hydrolysis:

The phases present and the diffusing species are shown in the idealized Fig. 13.12. Phase α is the NaOH solution, phase β is the cross-linked poly(methyl acrylate) and phase γ is the reaction product, viz., a swollen network of acrylic acid sodium salt residues. Because the diffusion coefficient of Na^+ and OH^- is substantially zero in phase β, the reaction proceeds in a rather sharp interfacial zone. However, unlike the idealization of Fig. 13.12., the interfacial zone becomes highly irregular rather than remaining spherical, as shown in Fig. 13.13. The course of reaction on 0.7 *mm* diameter beads is shown over a period of 9 hours by sectioning the beads serially into equatorial discs after recovery at various times, followed by exposure to the dye Azure A which stains only the hydrolysed, hydrophilic polymer. The PMA core is transparent. Fig. 13.13. shows that the γ phase, initially a thin spherical shell, becomes progressively

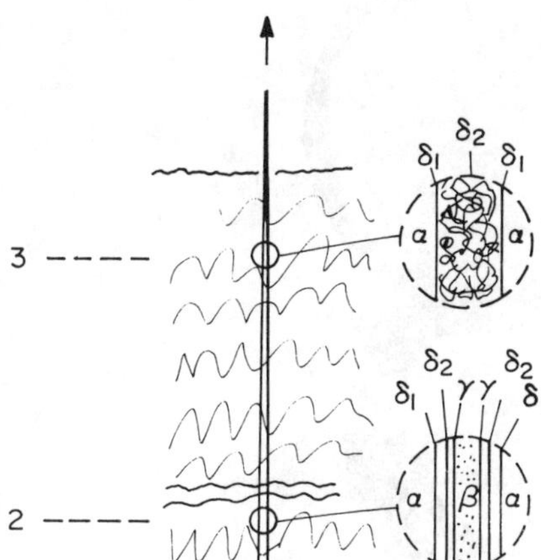

Fig. 13.11. Hydrolysis of cellulose xanthate by extrusion as a sheet into an acid bath. Phases: α, dilute sulfuric acid solution; β, cellulose xanthate solution; γ, reaction layer in which cellulose is in both forms; δ_1, regenerated cellulose "skin"; δ_2 more porous regenerated cellulose. Level 1, just above extrusion slot; level 2, part way to surface; level 3, just below surface, hydrolysis complete, no residual xanthate

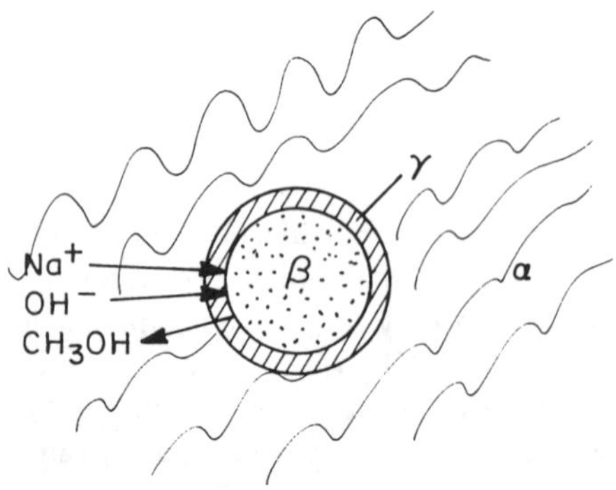

Fig. 13.12. Idealized shrinking core model of hydrolysis of crosslinked poly(methyl acrylate), PMA, phase β by exposure to sodium hydroxide solution, phase α, forming the spherical shell phase γ, cross-linked sodium salt of poly(acrylic acid), NaPAA

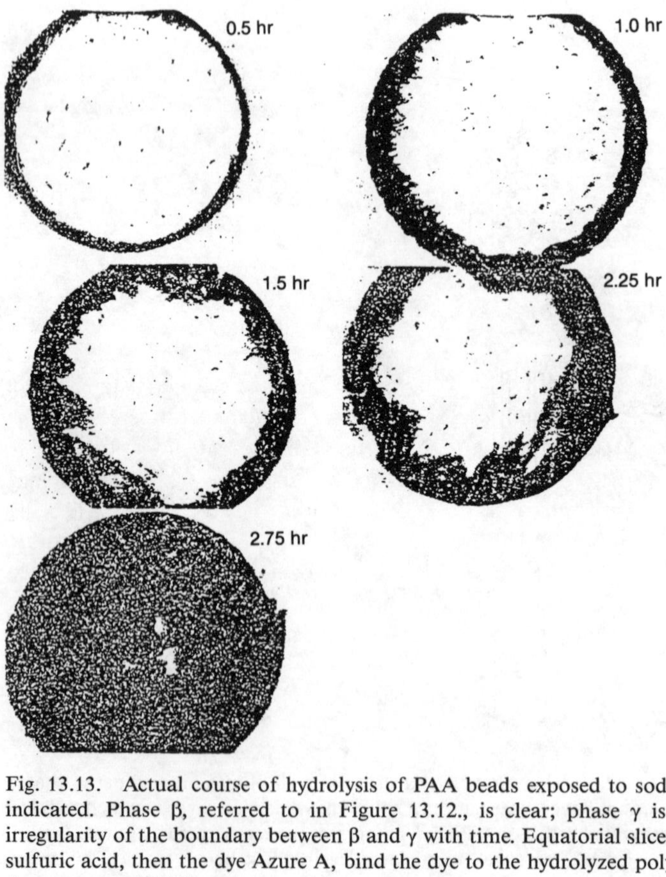

Fig. 13.13. Actual course of hydrolysis of PAA beads exposed to sodium hydroxide for time indicated. Phase β, referred to in Figure 13.12., is clear; phase γ is dark. Note increasing irregularity of the boundary between β and γ with time. Equatorial slices of beads, washed with sulfuric acid, then the dye Azure A, bind the dye to the hydrolyzed polymer (PAA), but not to the precursor (PMA)

more irregular. This is believed to result from the fissures created by the swelling stresses imposed between the unreacted core (phase β) and the *swollen* reacted phase γ, leading to a kind of "positive feedback" reaction. Where the fissures are created, the reagent (Na^+OH^-) can diffuse in more readily, causing local hydrolysis, local swelling thereafter, and thus further fracture.

Similar effects are noted in radiation grafting of monomer from a liquid phase α, for example N-vinyl pyrrolidone in water, to a polymeric phase β, for example a silicone rubber poly(dimethyl siloxane), forming an interfacial zone of hydrophilic poly(N-vinyl pyrrolidone) grafted on PDMS chains. The thickness of the interfacial zone becomes remarkably non-uniform as grafting proceeds under ionizing radiation (from an electron accelerator), evolving from an initially very thin uniform layer to a layer characterized by gross roughness. In this instance the "positive feedback" effect arises from the solubility of the aqueous monomer solution in the already reacted polymer, both locally swelling the rubber network and increasing the local content of monomer, so that the local rate of grafting is the greatest next to where it has already occurred to the greatest extent. Thus when a *uniformly transformed thin surface* is required on a polymer, rather than transformation throughout the bulk of the polymer as was the case with the ion exchange beads formed from PMA, intense short term exposure to the transforming reagent medium is mandatory. Examples are to be found in ion plasma gas treatments, and high dose rate ionizing radiation processes (U. V., electron irradiation).

Problem 13.1

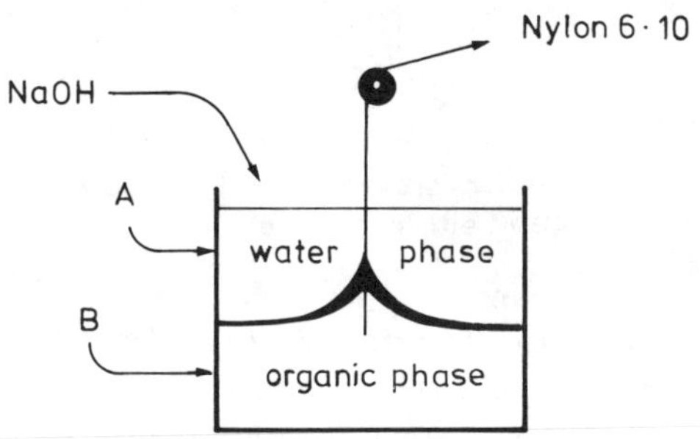

Nylon 6–10 is made by interfacial polycondensation of hexamethylene diamine (A) and sebacyl dichloride (B). The open unsteady state system is shown schematically in the Figure. 5 liters of the aqueous solution of A and 5 liters of the organic solution of B are used. The Nylon 6–10 is being withdrawn at the rate of 0.1 g/min. The two constituents A and B will be fed at appropriate rates to maintain bulk concentrations at 1 mol/l.

Sodium hydroxyde will be fed at a rate such as to titrate the HCl formed.

a) At what mass rate should these three be fed?

b) How long will it take for NaCl to reach a 2 molar concentration?

c) What else will probably have to be fed to the system?

Appendices A: Examples of Polymerization or Cross Linking Processes in Reactors which Impart Special Shape

A.1 Soft Contact Lenses from Poly(hydroxyethyl Methacrylate)

One way of generating *hydrophilic contact lenses* is based on the radical polymerization of hydroxyethyl methacrylate (HEMA) with ethylene glycol dimethacrylate (EGDMA) as crosslinking agent, accomplished in a centrifugal mold (Fig. A.1). The liquid monomer mixture receives its convex curvature by contact with the mold and its concave side by the action of gravity to generate a curve whose radii are set by angular velocity of the mold. Held at a temperature of around 60 °*C*, and containing a radical initiator appropriate to this level, the co-monomers undergo polymerization with simultaneous crosslinking. Since T_G is lower than the reaction temperature the conversion can be carried to nearly 100%. When the lens is stripped from the mold, it retains the shape imposed on the spinning liquid. The angular velocity of the mold

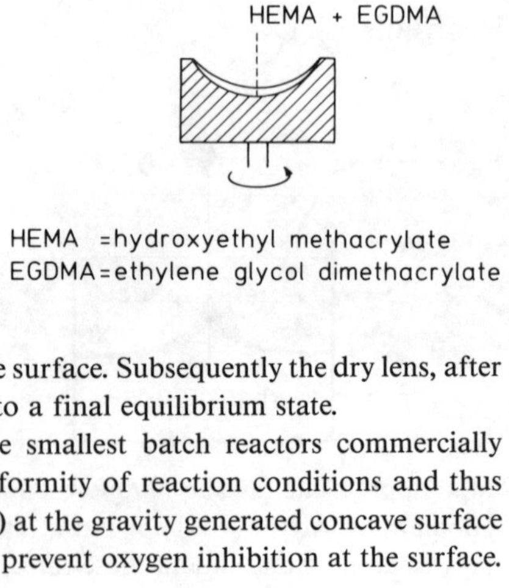

Fig. A.1. Centrifugal mold for poly(hydroxyethyl methacrylate) contact lenses (cross-section)

governs the radius of curvature of the concave surface. Subsequently the dry lens, after removal from the mold, is swollen in water to a final equilibrium state.

This case probably represents one of the smallest batch reactors commercially employed for polymerization. To insure uniformity of reaction conditions and thus isotropy of the lens, avoidance of oxygen (air) at the gravity generated concave surface is essential. Thus inert gas would be used to prevent oxygen inhibition at the surface.

A.2 Glass-Fiber Reinforced Polyester

Polyesters of low to medium molecular weights, synthesized by condensation of diols (ethylene gylcol, propylene glycol), saturated diacids (adipic, phthalic), and unsaturated diacids (maleic, fumaric) are dissolved (while hot) in styrene monomer containing free radical inhibitor to prevent onset of radical polymeriaztion. Volumetric ratio of polyester to styrene is approximately 1 : 2 and on a molar basis the styrene to (maleate + fumarate) ratio is about 4 : 1. Thus the "precursor" material is a syrup of relatively low viscosity, easily pourable, which contains a "prepolymer" — the unsaturated polyester, dissolved in the "solvent" styrene. The latter subsequently acts as co-monomer in copolymerization with maleate and fumarate units in the polyester chains. Since each polyester chain carries several unsaturated units, a network can readily be developed after initiation of radical polymerization.

The "batch reactor" in this case takes the form of a *pair of molds* (1) and (2), shown in Fig. A.2, defining a space of the order of 0.2 to 1 *cm,* into which unwoven or woven

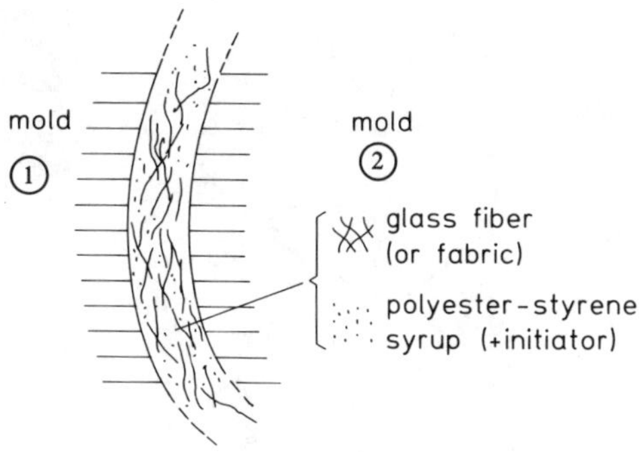

Fig. A.2. "Batch reactor" to manufacture glass-fiber reinforced polyesters

glass fiber material has been introduced. Just prior to use, the polyester-styrene syrup receives a free radical initiator, which has a half life of around 3 hours at room temperature. In this case the total "reactor" can have the substantial dimensions of a 20 meter yacht hull, in marked contrast to the 0.5 *ml* volume of the contact lens reactor of Fig. A.1. The "syrup" is introduced preferably from below so as to displace air, without entraining bubbles, and to wet the glass filaments. Conversion of styrene to a copolymer, crosslinked through maleate/fumarate units, takes place to an extent, even at room temperature, exceeding 96%. Since the diffusion coefficient of styrene through the network should abruptly decrease by orders of magnitude as the glass transition temperature of the monomer-polymer mixture approaches the reaction temperature, one would expect considerable percentages of styrene remaining in the final product. It appears that the very substantial surface area of glass filament per unit volume plus additional surface area due to pigments may provide alternative pathways for diffusion of monomeric styrene out of the article.

While this important process is treated as a homogeneous reaction — the assumption being that free radical copolymerization takes place in a single homogeneous phase — the fiberglass and the filler could be incorporated *covalently* into the reaction by pretreatment of the glass surface with a vinyl silane such as trichlorovinylsilane (TCVS), as in the following steps: (1) Coat glass surface with TCVS; (2) imbed in styrene-polyester and set off radical initiation.

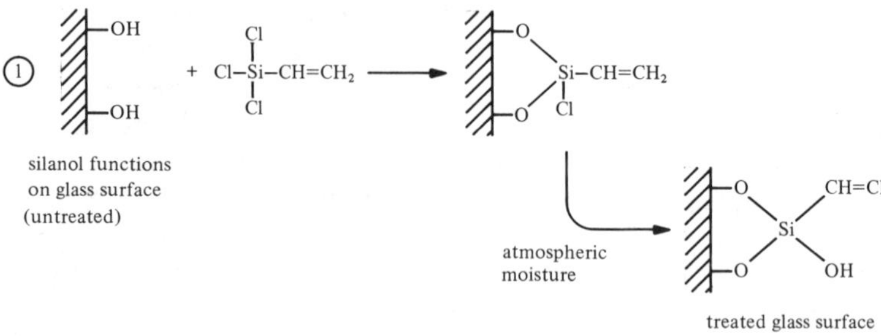

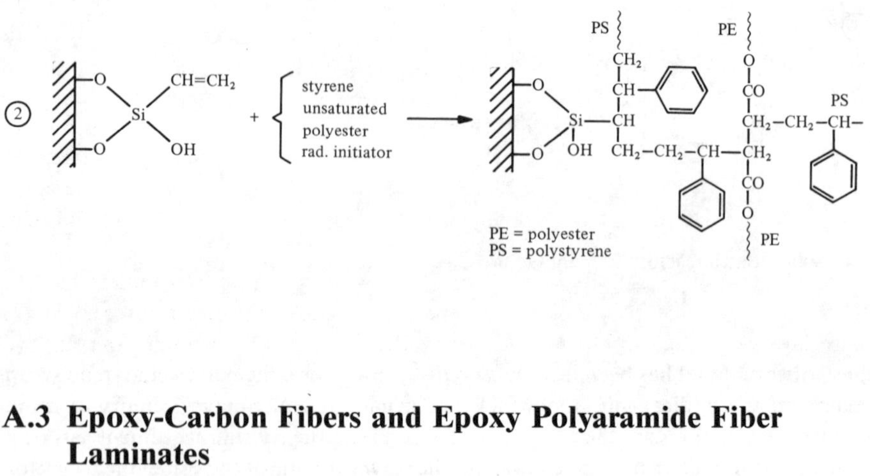

A.3 Epoxy-Carbon Fibers and Epoxy Polyaramide Fiber Laminates

Very large aircraft assemblies (cargo doors, wing panels, etc.) are made by the process shown in Fig. A.3. The starting material consists of sheets of "B-stage" epoxy resin containing reinforcing fibers (glass, carbon, or polyaramide). "B-stage" means that the epoxy-amine mixture, initially liquid, with which the fibers were impregnated (hence the word *"pre-preg"*), has been advanced to a degree of conversion slightly less than that leading to gelation. At room temperature the "resin" (the epoxy-amine) is glassy, thus not self-adherent. To obtain maximum mechanical strength the sheets are stacked so that the direction of their fibers lie at different angles, for example in a sequence: (1) 0°, (2) 45°, (3) −45°, (4) 90°.

The *batch reaction*, often taking as long as 18 *h*, is the final conversion of the epoxy resin to a very tightly cross-linked network. This is accomplished in *autoclaves* of sometimes gigantic proportions (large enough to contain a trailer truck).

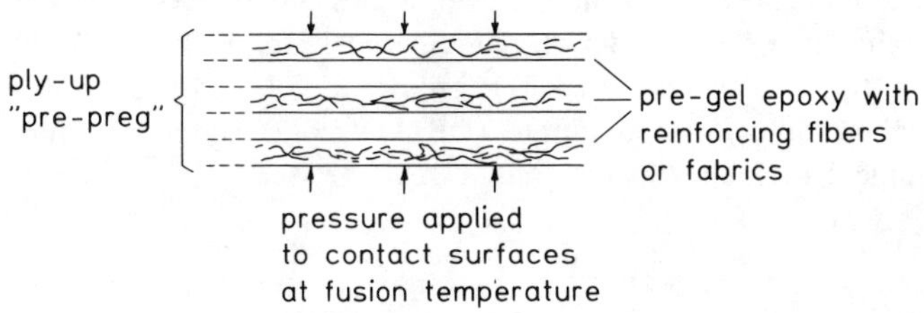

Fig. A.3. Procedure used to manufacture fiber reinforced epoxy resins

In the first part of the reaction, entrapped air is removed from between the sheets by gentle pressure (exerted for example in a vacuum bag)*) as the temperature is raised to a level so as to melt the resin without having it become too fluid (e.g., 80 °*C*); otherwise it would be expelled from the fiber layers. In this first stage, the sheets come into intimate contact following expulsion of air, and the B-stage resin — now above its glass transition temperature — begins again to react (oxirane coupling to amine) so that the gel stage is reached, and a continuous network is formed throughout the multilayered laminate. In some of the massive composites made by this technique, section thicknesses may be around 25 *cm*. Thus it is essential to design the heating schedule so as to avoid too rapid reaction which, in the case of thick sections, could lead to destructively high levels of temperature at the midplane due to rapid release of enthalpy of polymerization.

As the network formation proceeds, the glass transition of the resin continues to rise. Thus to insure nearly complete conversion, the final temperature reached in the autoclave may be 150 °C to 170 °*C*, just above the final T_G. Several hours at this temperature may be required, followed by slow cooling to avoid establishing temperature gradients which could lead to warping and non-uniform residual stress.

The final composite article removed from the autoclave will range in mass from several kilograms to several thousand kilograms and in dimension from one to tens of meters.

While we treat this as a homogeneous batch reaction, it could become heterogeneous, if the surface of the fiber is capable of reaction with the epoxy resin. For example, implantation of amine groups via ion plasmas onto the surface of carbon and polyamide fibers will make these reactive towards oxirane in the resin, leading to covalent coupling.

*) The sheets are placed inside a giant bag which is then evacuated so as to bring full atmospheric pressure to bear from the ambient hot air onto the layers.

A.4 "RIM" — Reaction Injection Molding of Polyurethanes

In order to produce large objects (e. g. motor vehical front assembly) at reasonable speed the technique sketched in Fig. A.4, is increasingly employed. In this case the "batch reactor" is the mold for the object (here shown as parts 3–4). Pre-inserted in this cavity (5) is reinforcing filamentary material, glass fiber for example. Two cyclically filling and emptying piston-cylinders are used: (1) contains a diol (and triol) with a urethane catalyst such as dibutyl tin dilaurate (DBT DL); (2) contains molten diisocyanate, for example 2,4-toluene diisocyanate (TDI).

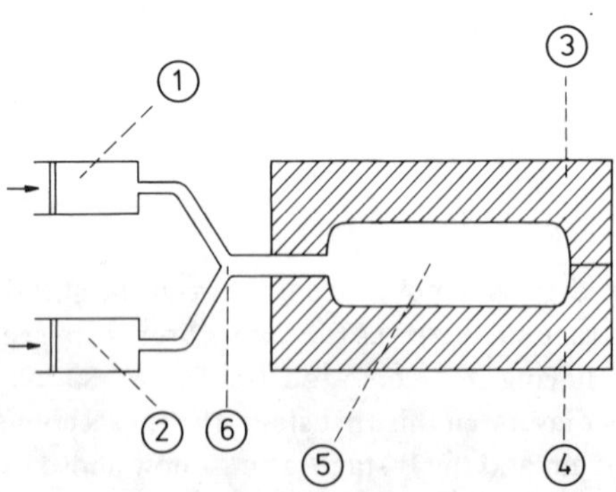

Fig. A.4. Reaction injection molding (RIM) device to produce reinforced polyurethanes.
1 — cylinder charged with diol and triol plus dibutyltin dilaurate
2 — cylinder charged with molten diisocyanate
3 — half of split mold
4 — half of split mold
5 — cavity to be filled with reinforcing fiber, then with mixed liquid from nozzle
6 — nozzle for mixing streams from 1 and 2

At the beginning of the cycle after each cyclinder has been separately filled, (in volumetric ratios appropriate to the stoichiometry; —NCO + —HO → —NHCOO—), the cyclinders are simultaneously discharged, their streams meeting and mixing at nozzle (6) just before the injection port of the mold 3–4, and enter the mold as a mixed and reacting stream.

Owing to the catalyst*⁾, the entire urethane formation is complete within a *few minutes*, and the mold is then split open to discharge the massive product. Significant temperature gradients can exist.

The product owes its fixed shape primarily to strong hydrogen bonding between the urethane groups formed in the hydroxyl-isocyanate reaction. Therefore, as the reaction proceeds and urethane groups are formed, a *network* is generated by association of urethanes to form junctions.

---

*⁾ For example dibutyltin dilaurate $(C_4H_9)_2\,Sn(OCOC_{11}H_{23})_2$.

In addition to this "virtual" or "physical" network formation, trifunctional precursors may be used (triols, or triols previously reacted with excess diisocyanate to form triisocyanates) to create true covalent junctions.

The separate reagent streams have viscosities of the order of a few centipoise, and thus mixing is readily achieved by the collision of the jets formed prior to the injection port.

This system is considered to be two phase, not homogeneous, because of the fact that the polyurethane chains simultaneously associate in a network while growing at their ends, out of "their" continuum. Shortly the network becomes the continuum phase, leaving the unreacted diol and diisocyanate dispersed therein to be consumed. Despite the significant temperature rise produced by the nearly adiabatic reaction of the components ($\sim 80\ °C$), the transition temperature for the relaxation of the associated urethane groups is considerably higher still. If this would be a homogeneous system like polymethyl methacrylate, the reaction would stop by immobilization of monomer. Not so here; rather as in vinyl chloride polymerization, monomer can freely diffuse to the reactable ends of already "frozen" macromolecules.

As in the case of epoxy-fiber laminate it is possible to perform this reaction in two phases, in yet another sense: Use of polyaramid or carbon fibers, that have previously been treated with an ammonia ion plasma and thus contain amine groups on the surface. The result would be *direct grafting* of the polyurethane via the diisocyanate component to the fiber with formation of diurea bonds.

A.5 Reaction Injection Molding of Nylon-6 via Anionic Polymerization of Caprolactam

The rapid anionic ring opening polymerization of caprolactam leads to massive objects by in situ generation of nylon-6.

The process*) consists of heating ε-caprolactam to $150\ °C$ and adding lithium hydride (initiator) and acyl lactam (promoter), just as the "shot" is injected into the mold cavity. Rapid ring-opening addition occurs, and within 2 to 4 minutes nearly complete conversion is achieved. The maximum temperature of polymerization, achieved adiabatically, is about $200\ °C$, thus *below* the melting temperature of pure nylon-6 ($221\ °C$), yet addition of monomer occurs readily, attesting to the two phase nature of this reaction, and the evolution of a continuous crystalline but microporous — and thus permeable — phase through which monomers can diffuse to add to the living chain ends.

Usually anionic polymerizations are carried out via homogeneous solution polymerization. This *solvent-free process* is a notable exception. As with reaction injection molding of polyurethanes, the nylon-6 moldings can be formed on polyaramid fabrics for reinforcement, i. e. the polyaramid filaments are placed in the mold before it is closed, and the very fluid monomer readily wets it before becoming transformed into the polyamide.

---

*) See chapter 7, activated monomer polymerization.

A.6 "Optical" Poly(methyl Methacrylate)

"Casting" of poly(methyl methacrylate) into optically clear articles of uniform refractive index such as rods, and sheets is accomplished by using the shapegiving mold as the batch reactor. Thus, the casting process requires that the product be removed in the shape imposed by the mold. The mold must be disassembled into parts, to release the contents, and since the temperature of the contents must be well below the glass transition (~100 °*C*), the walls must be cooled.

The cycle would typically consist of the following steps (see Fig. A.5):

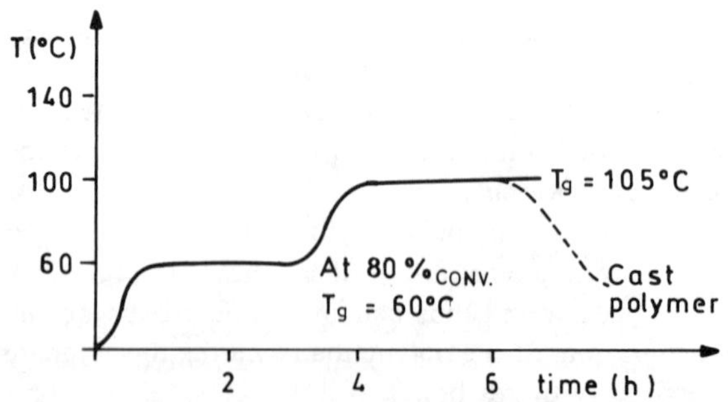

Fig. A.5. Temperature versus time diagram for poly(methyl methacrylate)

(1) Charge the assembled mold with a solution containing poly(methyl methacrylate) (~20 *vol.-%*) in methyl methacrylate monomer (~80 *vol.-%*) plus initiator (e.g. AIBN). (One could start with pure monomer, but the volumetric shrinkage would be greater).

(2) Using the heated walls of the mold, raise the temperature to ~60 °*C*; hold at this temperature for 2–6 *h* attaining about 80% conversion. The glass transition will increase to about 60 °*C* at 80% conversion and propagation will become diffusion limited.

(3) Raise the temperature to ~105 °*C* and hold for 2–4 *h* to carry conversion close to 100%.

(4) Drop the temperature to ambient at a rate which will keep the temperature gradient in the cast polymer at less than 10 $°C \cdot cm^{-1}$. When midplane temperature is below ~60 °*C* (well below glass transition), open mold and remove product.

Because of the relatively long reaction times and slow release of enthalpy of polymerization, thermal gradients in the material contained in the mold are insignificant, and thus all elements of the cast article undergo the *same* time-temperatue history, thus evolve into polymeric substance of virtually identical properties. In general, the thicker the section, the less must be the reaction rate (initiator concentration) and the longer the cycle, to avoid significant temperature gradients. Hence the refractive index is remarkably constant and *isotropic* throughout these "crystal clear" castings. It would be possible to produce castings in far shorter time using either more active initiation, higher temperature, or both. This would lead generally to lower average molecular weight and potentially to non uniform material properties, as a consequence of thermal gradients.

Appendix B: High Pressure Radical Polymerization of Ethylene

Prior to the development of Ziegler-Natta and metal oxide catalysts, polyethylene was made exclusively by *high pressure* radical polymeriation. This is still the *only* way by which ethylene can be copolymerized with vinyl acetate or ethyl acrylate to make special copolymers designated EVA or EEA.

The necessity for high pressure results from the fact that $k_p/k_t^{1/2}$ is far too small at ordinary pressures. For example at 130 °*C*, it is 0.05 at 1 *bar*, but 0.7 at 2500 *bar*. Increasing temperature to 200 °*C* results in a value of $k_p/k_t^{1/2}$ of about 3 at 2500 *bar*.

In view of the magnitude of pressure required, the most common form of reactor was heavy-walled tubular, and this was operated in nearly steady *continuous flow*.

The following example illustrates some of the unusual features of the process. Ethylene (critical temperature 10 °*C*) is compressed to 1300 *atm* and heated to 200 °C, under which conditions it has a density of 0.45 [g/cm^3].

The compressed gas is caused to flow through 235 meters of pipe, having an inside diameter *D* of 2 *cm*, the pipe being arranged in parallel 6 meter lengths connected by 180° semicircular couplings. An initiator like tert.-butyl hydroperoxide, which has a half life of about 40 *s* at 200 °*C* is injected into the ethylene as it enters the tubular reactor.

Mass flow rate in one example is 1 [$Kg \cdot s^{-1}$], which at density $\rho = 0.45$ [g/cm^3] corresponds to a volumetric flow rate of 2400 [$cm^3 \cdot s^{-1}$]. Since the pipe cross-section is 3.14 cm^2, the average velocity $\bar{V}$ is close to 1000 [$cm \cdot s^{-1}$]. If one takes the viscosity η to be about 0.005 *poise*, the Reynolds number $Re = \dfrac{D \cdot \bar{V} \rho}{\eta}$ is about 2×10^5.

Thus the flow in the pipe enters well into the *turbulent* regime.

Typically 15% conversion of monomer occurs during flow through the reactor, the unreacted ethylene being *recycled*.

For the case illustrated, the residence time, computed on volume of pipe divided by influent volumetric flow rate $\dot{V}$ [$cm^3 \cdot s^{-1}$] is about 30 *s*. The quoted rate of feed, 1 [$kg \cdot s^{-1}$], corresponds to a molar feed of 35.7 [$mol \cdot s^{-1}$]. A conversion of 15%, or 5.1 *mol* would produce, from enthalpy of polymerization (22,000 [$cal \cdot mol^{-1}$]) about 110,000 [$cal \cdot s^{-1}$] or 460 [$kJ \cdot s^{-1}$].

The need for *heat transfer* becomes obvious, since if the generated heat is stored adiabatically in the reaction mixture (taking heat capacity as about 0.5 [$cal \cdot g^{-1} \cdot K^{-1}$]), the temperature rise would be $\Delta T \cong (110{,}000)/(1{,}000)\ (0.5) \cong$ 220 °C.

The fluid entering the reactor, initially a homogeneous gas of density 0.45, must evidently become two phase as polymerization ensues. The polymer probably has density of about 0.75 [g/cm^3] at the temperature of the reaction (200 °C). As a consequence of conversion of 15% of the entering gas to polymer, the final volume fraction of polymer in the fluid leaving the pipe will be about (0.15/0.75)/[(0.15) (0.75) + (0.85) (0.45)] or about 0.09, and the gas phase 0.91. Thus the average velocity will not be greatly reduced.

A particularly interesting question arises as to the nature of the flow near the exit: Given 9 volumes of molten polymer in 91 volumes of gas, is the polymer an aerosol, convected by gas? Or is the gas entrapped as bubbles in a foam of molten polymer? Or is there separation with the polymer flowing along the lower surface of the pipe, with gas moving at much higher velocity above it?

Mention is made in the literature of the necessity of *periodically "blowing down"* the reactor in order to remove polymer which builds up in the lower surface of the pipe.

Polyethylene produced by the high pressure, radical initiated process is characterized with the following properties:

(1) Melting point in the range 105°– 112 °C (as compared to linear polyethylene, which is melting at 135 °C).

(2) Short branches of 2, 3, or 4 carbons, every 50 to 100 carbon atoms along the main chain, as revealed by nuclear magnetic resonance studies.

(3) A few long branches — the number widely variable. The reduction in melting point is the result of the imperfections introduced by the short branches. Essentially the melting point is comparable with that of a copolymer of ethylene with 1-butene, 1-pentene and 1-hexene.

Since no comonomers are in fact introduced, the short branches in the polymer chain must be the consequence of a kinetic event involving only polyethylene. This has been postulated as the *"back biting" reaction:*

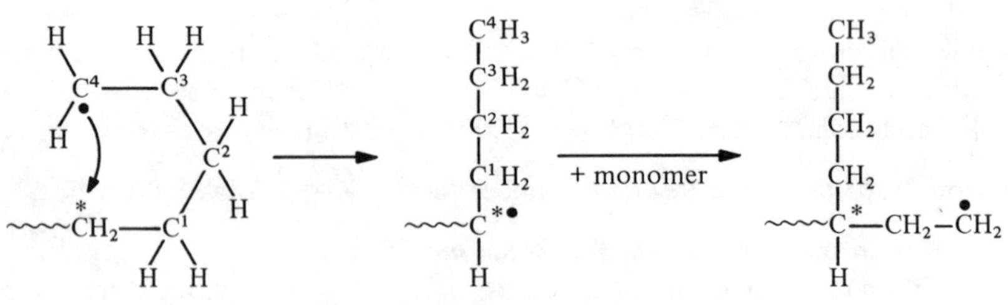

short branch formation in polyethylene

In the scheme shown, the growing chain, carrying a terminal radical on the carbon labelled 4, attacks the carbon labelled*, abstracting a hydrogen atom. This leads to a n-butyl side branch, and propagation continues from carbon*.

In order to explain the variation in short side branch frequency ρ_s with temperature and pressure, it was postulated that the rate of backbiting R_{bb} is first order in [$M^\bullet$], thus:

$$R_{bb} = k_{bb} [M^\bullet]$$

The rate of addition of monomer is the rate of propagation R_p

$$R_p = k_p [M^\bullet][M]$$

Defining $\rho_s \equiv R_{bb}/R_p$ gives:

$$\rho_s = \frac{k_{bb}}{k_p\,[M]}$$

Thus if k_{bb}/k_p increases with temperature, ρ_s will also increase, at constant monomer concentration [M]. If pressure is increased at constant temperature, [M] must increase and thus ρ_s must decrease.

Long branching density, ρ_1, the result of a growing chain radical attacking at an already finished and different polymer chain, is given by the expression

$$\rho_1 = -C_p \left[1 - \frac{\ln(1 - \pi)}{\pi} \right]$$

where

π = fractional conversion

ρ_1 = fraction of all units which carry a long branch

$C_p = k_{tr,p}/k_p$

These expressions should not be carried beyond qualitative interpretation, since the real system consists of two phases, i.e., the polymer melt at density of about 0.70 is in a gas (monomer) of density 0.45 [$g \cdot cm^3$]; in the molten polymer some gas is dissolved, presumably in proportion to pressure, but quantitatively unknown.

Solutions of the Problems

Problem 1.1

Assuming the fractions to be monodisperse (which means that for each of them $\overline{M}_w = \overline{M}_n$) the weight and number average molecular weights are given by:

$$\overline{M}_w = \sum_{i=1}^{5} w_i M_i = 199\,500$$

$$\overline{M}_n = \frac{\sum_i n_i M_i}{\sum_i n_i} = \frac{1}{\sum_{i=1}^{5} \frac{w_i}{M_i}} = 101\,000$$

Loss of some low molecular weight constituent during the fractionation process would affect the number average molecular weight far more than the weight average.

Problem 1.2

The probabilities of formation of isotactic, syndiotactic and heterotactic triads are governed, — in the Bernouillian case — by a single parameter, p_i, probability of formation of isotactic diads ($p_s = 1 - p_i$). They are given by:

$$i = p_i^2$$

$$s = (1 - p_i)^2$$

$$h = 2p_i\,(1 - p_i)$$

Fig. L 1 shows the variation of the *i, s, h* triad contents as a function of p_i. Any sample exhibiting proportions of *i, s, h* triads compatible with this scheme can be considered *Bernouillian*. It is strictly the case of the triad distribution a) and approximately that of the case b).

Problem 1.3

a) $$X_n = \frac{\sum N_i X_i}{\sum N_i} = 6.67$$

b) $$X_w = \frac{\sum m_i X_i}{\sum m_i} = 7.5$$

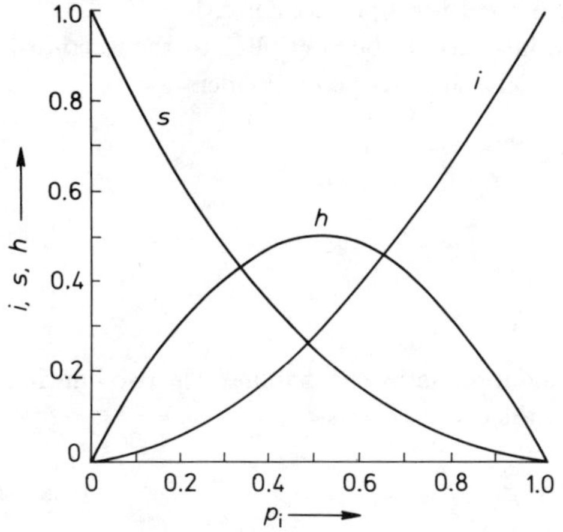

Fig. L1. Plots of *i, s, h* versus p_i (Bernouillian distribution of triads)

c) For the 10-bead chains it is impossible to have both ends red. Thus we have to look for the 5-bead chains, as for these if one end is red the other end must be red. Randomly, one half of the 5-bead chains must start with a red bead. Therefore the probability to select a 5-bead chain with a red bead at the end is:

$$p = 1/2 \text{ (number of 5-bead chains)/(total nb of chains)}$$

$$p = 1/3$$

If the probability of first selecting the end bead is included, then:

$$p = (1/3)(3/10) = 1/10 = 0.1$$

Problem 2.1

Bis(β-hydroxyethylterephthalate), (I)

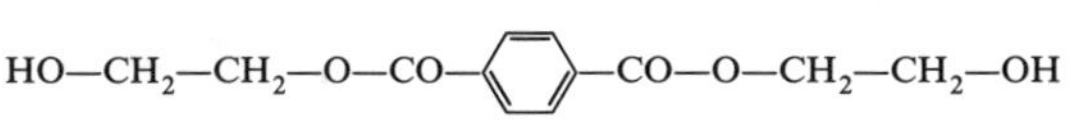

(I)

when heated under proper conditions undergoes selfcondensation according to:

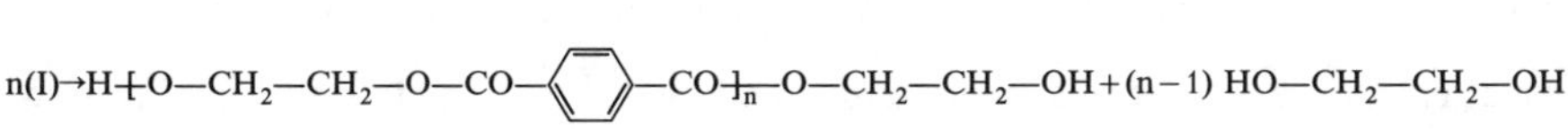

Once 98% of the total glycol has been released the degree of conversion π is 0.98. Thus:

$$\overline{P}_n = \frac{1}{1-\pi} = 50 \qquad \overline{P}_w = \frac{1+\pi}{1-\pi} = 99$$

and the polydispersity index is 1.98, as expected for a polycondensation.

To get the molecular weights the DP values have to be multiplied by the weight of the repeat unit (which is that of the base unit, since it is a selfcondensation).

$$\overline{M}_n = 9600 \qquad \overline{M}_w = 19000$$

Problem 2.2

a) Nylon-6,6 is made from adipic acid and hexamethylene diamine. The two limiting cases that are considered yield the same molecular weights:

Case I: Extent of reaction $\pi = 0.995 = p; \quad r = 1$

$$\overline{P}_n = \frac{1}{1 - \pi} = 200 \qquad \overline{P}_w = 399$$

Case II: Stoichiometric ratio $r = 0.99; \quad \pi = 1 = p$

$$\overline{P}_n = \frac{1 + r}{1 - r} = 199 \qquad \overline{P}_w = 399$$

b) To determine the value of i at which the weight fraction w_i reaches a maximum, one has to find the value of i at which the derivation of w_i with respect to i is zero:

$$w_i = (1 - p)^2 \, i p^{(i-1)}$$

$$\frac{d w_i}{di} = (1 - p)^2 \, [p^{(i-1)} + i \, p^{(i-1)} \ln p] = 0$$

p (probability for a given function to have reacted) being close to unity one can set: $q = 1 - p$ and $\ln p = \ln(1 - q) = -q + \ldots$ Thus one obtains the relationship searched for:

$$i = \frac{1}{1 - p}$$

which means that the maximum of the weight distribution function in the "most probable" case is located at the number average degree of polymeriaztion $\overline{P}_n$. [An alternate way to get this result is to calculate $\dfrac{d \ln w_i}{di}$ and to set it equal to zero]

c) Sum of mole fractions of molecules exhibiting degrees of polymerization 1 through 5: ($p = 0.995$)

$$\sum_1^5 x_i = (1 - p) \sum_1^5 p^{(i-1)} = 0.0247$$

Sum of the corresponding weight fractions

$$\sum_1^5 w_i = (1 - p)^2 \sum_1^5 ip^{(i-1)} = 3.7 \times 10^{-4}$$

Problem 2.3

a) The acid catalyzed polyesterification obeys a second order rate law:

$$-\frac{d[a]}{dt} = k'[a][b]$$

Since the stoichiometric ratio is unity this reduces to

$$-\frac{d[a]}{dt} = k'[a]^2$$

which upon integration yields:

$$\frac{1}{[a]} - \frac{1}{[a]_0} = k't$$

On the other hand the number average degree of polymerization is, by definition:

$$\bar{P}_n = \frac{N_0}{N} = \frac{[a]_0}{[a]}$$

It follows that:

$$\bar{P}_n = 1 + [a]_0 k' t$$

Knowing that $\bar{P}_n$ increases from 8 to 50 over a period of 300 minutes (18000 *s*), the apparent rate constant is

$$[a]_0 k' = \frac{42 [a]_0}{18000} = 2.3 \times 10^{-3} [a]_0$$

b) The equilibrium constant K defined as:

$$K = \frac{[\text{—COO—}]\ [H_2O]}{[\text{—COOH}]\ [\text{—OH}]}$$

is equal to 1 at 109 °*C*. Let us call Y the mole ratio of water to carboxylic functions. Then:

$$K = Y\frac{[\text{—COO—}]}{[\text{—OH}]} = \frac{[a]_0 p Y}{[a]_0(1 - p)} = 1$$

If we set:

$$\bar{P}_n = \frac{1}{1 - p} = 50$$

it follows that $p = 0.98 \doteq \pi$

Consequently:

$$Y = \frac{0.02}{0.98} = 2.04 \times 10^{-2} = \frac{1}{49}$$

The average mass of polyester per ester link (base unit) is: $= \dfrac{284}{2} = 142$, and there are in the average 49 links per macromolecule. Thus the weight ratio of water to polymer is:

$$\left(\frac{1}{49}\right)^2 \cdot \frac{18}{142} = 5.3 \times 10^{-5}\ [g/g]$$

(It is hard to attain such low water contents)

c) If 2% of the glycol is lost, but no adipic acid, the stoichiometric ratio r is 0.98 and the highest possible degree of polymerization (for $\pi = p = 1$) is given by:

$$\overline{P}_n = \frac{1+r}{1-r} = 99$$

Thus: $\overline{M}_n = 14000$

Problem 2.4

A crosslinked polyurethane is to be made from a diisocyanate and a polymeric tetrol. Let us call $\underline{a}$ the alcohol functions and $\underline{b}$ the isocyanate functions. All $\underline{a}$ functions belong to tetrafunctional ($f = 4$) base molecules thus $\rho = 1$. The stoichiometric ratio r, defined as [b]/[a] ranges from 1/4 to 4.

The general formula (3.27):

$$\alpha = \frac{\pi^2 \rho r}{1 - \pi^2 r(1-\rho)}$$

reduces to:

$$\alpha = r\pi^2$$

and the critical value of α at which gelation occurs is given by:

$$\alpha_c = \frac{1}{f-1} = \frac{1}{3}$$

To get gelation:

$$\alpha > \alpha_c; \quad \text{or} \quad r\pi_c^2 > \frac{1}{3}, \quad \text{or} \quad \pi_c > \frac{1}{\sqrt{3r}}$$

Thus one can plot the critical values of the conversion π at which gelation occurs as a function of the stoichiometric ratio r (Fig. L2). For the lowest value of r no gel can ever be obtained.

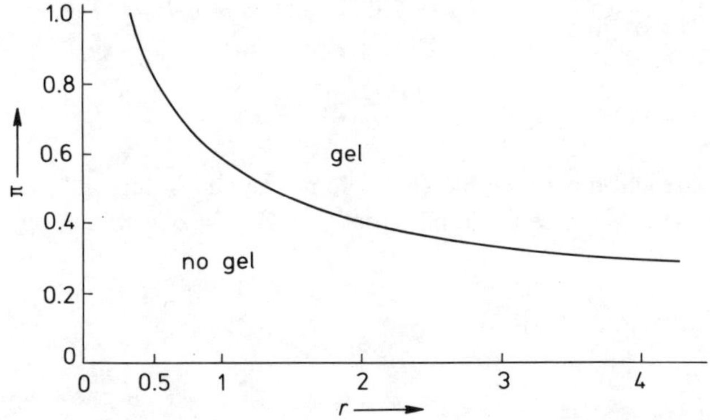

Fig. L2. Plot of π_c (critical conversion degree at which gel formation occurs) versus the stoichiometric ratio r

Problem 2.5

a) Assuming 1 *mol* of each reactant, initially:

$$HOOC(CH_2)_8COOH + H_2N(CH_2)_6NH_2 \longrightarrow \text{polyamide} + 2\,H_2O$$

202 g/mol, 116 g/mol, 282 g/repeat unit, 18 g/mol

Assuming weight loss is due exclusively to evaporation of water formed, at 99% conversion:

Weight loss = $0.99 \cdot 2 \cdot 18 = 35.64$ *g* and % weight loss = $35.64/318 = 11.2\%$.

The number average degree of polymerization is given by:

$$P_n = 1/(1 - \pi) = 100$$

and the corresponding molecular weight:

$$M_n = P_n \cdot (282/2) = 14100$$

b) Since the density of the melt is 0.9 *g/ml* and the mole weight of a repeat unit is 282 *g/mol*, the number of base units (or amide functions) formed at 100% conversion is:

$$[CO\text{—}NH] = 2 \cdot (0.9/282) \cdot 1000 = 6.38\ mol/l$$

At 99% conversion:

$$[CO\text{—}NH] = 6.38 \cdot 0.99 = 6.32\ mol/l$$

Consequently: $[\text{—NH}_2] = [\text{—COOH}] = 0.01 \cdot 6.38 = 0.0638\ mol/l$
Since:

$$\frac{[\text{CO—NH}] \cdot [\text{H}_2\text{O}]}{[\text{—NH}_2] \cdot [\text{—COOH}]} = 100, \qquad [\text{H}_2\text{O}] = 100\,\frac{(0.0638)^2}{6.32} = 0.064\ mol/l$$

c) Assuming that reverse reaction is negligible, that volume stays constant, and that the the density of the melt is 0.9 g/ml, which implies that $[\text{—NH}_2]_0 = 6.38\ mol/l$, one can write:

$$[\text{—NH}_2]_0 \frac{d\pi}{dt} = k\,[\text{—NH}_2]_0\,(1 - \pi)^2$$

or

$$\frac{d\pi}{(1 - \pi)^2} = k\,[\text{—NH}_2]_0\,dt$$

Integration leads to:

$$\frac{1}{1 - \pi_1} - \frac{1}{1 - \pi_0} = k\,[\text{—NH}_2]_0\,(t_1 - t_0)$$

since at $t = 0$ $\pi = 0$ and at $t = 6.2$ hr $\pi = 0.99$, the rate constant k is found to be:

$$k = 2.5\ l/mol \cdot hr$$

Problem 2.6

Since there is no loss or gain of water upon mixing the two polyamides, P_n does not change: $P_n = 100$; it can be expected that for each of them $P_w = 200$.

The repeat unit of nylon-11 (which is also its base unit) has a molecular weight of 183 g/mol, whereas for nylon 6–10 the two constituting base units have molar weights of 114 and 168 g/mol, respectively. The average molecular weight of the base unit for the mixture is thus given by:

$$\frac{100 \cdot 183 + 50\,(114 + 168)}{200} = 162\ g/\text{base unit}$$

Consequently the number and weight averages of the molecular weight of the melt are:

$$M_n = 16\,200 \qquad M_w = 32\,400\ g/mol$$

Problem 2.7

a) By definition:

$$r = \frac{4x + 2y}{2z} \qquad \rho = \frac{4x}{4x + 2y}$$

The branching probability is expressed as:

$$\alpha = \frac{\pi_a \pi_b \rho}{1 - \pi_a \pi_b (1 - \rho)} = \frac{\pi_a^2 \rho r}{1 - \pi_a^2 r(1 - \rho)}$$

(since $\pi_b = \pi_a \cdot r$).

The condition for network formation is:

$$\alpha_c = 1/(f - 1)$$

In the present case $f = 4$ and $\alpha_c = 1/3$.

Consequently:

$$\alpha = \frac{\pi_a \rho r}{1 - \pi_a^2 r(1 - \rho)} > \frac{1}{3}$$

If one sets $\pi_a = 1$, one gets the relationship between stoichiometric ratio and the proportion of tetrafunctional molecules necessary to attain gelation:

$$r + 2\rho r - 1 > 3$$

b) Using the same expressions as above, one can determine that for $r = 1$ and $\rho = 0.5$, gelation will occur when the conversion has attained the value: $\pi_a = 0.707$.

At this stage, the probability of occurence of elastic chains of various lengths is given by the expressions:

$$P_i = \pi_a [\pi_a^2 r (1 - \rho)]^j \pi_a r \rho$$

For	$j = 0$	$P_1 = 0.25$
	$j = 1$	$P_3 = 0.0625$
	$j = 2$	$P_5 = 0.0156$
	$j = 3$	$P_7 = 3.9 \cdot 10^{-3}$
	$j = 4$	$P_9 = 9.8 \cdot 10^{-4}$

(The sum of the P's is about 0.33, indicating that at the conversion chosen, when gelation has just occured, only one third of the chains formed are elastically effective.)

Problem 3.1

The probability p of chain growth in a free radical polymerization, in the presence of a transfer agent S, is expressed by:

$$p = \frac{k_p [M]}{k_p [M] + k_t [M^\bullet] + k_{tr} [S]}$$

If termination occurs solely by disproportionation it means that the distribution of molecular weights involves no macromolecules formed upon recombination. Thus

$$x_i = (1 - p)p^{(i-1)}$$

$$w_i = (1 - p)^2\, ip^{(i-1)}$$

are the expressions of the distribution functions by number (molar) and by weight. This is the most probable, or Schulz-Flory distribution. See Fig. 2.1 in chapter 2.

Problem 3.2

Bulk free radical polymerization of various monomers requires first to calculate the initial monomer concentrations $[M]_0$ knowing the density ρ and the molecular weight m_0 of the monomers:

$$[M]_0 = \frac{1\,000\,\rho}{m_0} \text{ in } [mole/l]$$

For methyl acrylate (MA).......... : $[M]_0 = 11.04$
methyl methacrylate (MMA) ... : $[M]_0 = 9.4$
styrene (St) .... : $[M]_0 = 8.75$
vinylacetate (VA) .... : $[M]_0 = 10.81$

a) The initial rate of polymerization is given by:

$$(R_p)_0 = -\frac{d[M]}{dt} = k_p \left(\frac{2k_d f}{k_t}\right)^{1/2} [M]_0[I]^{1/2}$$

The values found are in $[mole \cdot l^{-1} \cdot s^{-1}]$.

For MA.......: $(R_p)_0 = 5.01 \times 10^{-4}$
MMA.....: $(R_p)_0 = 8.6 \times 10^{-5}$
St ....: $(R_p)_0 = 1.5 \times 10^{-5}$
VA ....: $(R_p)_0 = 4.8 \times 10^{-4}$

b) The steady state radical concentration in the initial stage of the reaction can be obtained from:

$$2k_d f\,[I] = k_t\,[M^\bullet]^2$$

wherefrom one obtains:

$$[M^\bullet] = \left(\frac{2k_d f}{k_t}\right)^{1/2} [I]^{1/2}$$

The values found are in $[mole \cdot l^{-1}]$,

For MA.......: $[M^\bullet] = 2.2 \times 10^{-8}$
MMA.....: $[M^\bullet] = 1.3 \times 10^{-8}$
St ....: $[M^\bullet] = 1.2 \times 10^{-8}$
VA ....: $[M^\bullet] = 1.2 \times 10^{-8}$

c) Rate constants follow Arrhenius' law, and the value of k_T at a temperature T is related to its value k_{T_0} (known) at a temperature T_0 by:

$$\ln\frac{k_T}{k_{T_0}} = \frac{\Delta E}{R}\left(\frac{1}{T_0} - \frac{1}{T}\right)$$

For MMA at 40 °C the rate constants are:

$$k_d = 4.8 \times 10^{-7}\ [s^{-1}];\ k_p = 449\ [l \cdot mol^{-1} \cdot s^{-1}];$$
$$k_t = 4.4 \times 10^{-7}\ [l \cdot mol^{-1} \cdot s^{-1}]$$

wherefrom the initial rate of polymerization is found to be

$$(R_p)_0 = 1.4 \times 10^{-5}\ [mol \cdot l^{-1} \cdot s^{-1}],$$

roughly 6 times slower than at 60 °C.

For MMA at 80 °C the rate constants are:

$$k_d = 1.1 \times 10^{-4}\ [s^{-1}];\ k_p = 1051\ [l \cdot mol^{-1} \cdot s^{-1}];$$
$$k_t = 5.5 \times 10^{7}\ [l \cdot mol^{-1} \cdot s^{-1}]$$

and the initial rate constant is:

$$(R_p)_0 = 4.4 \times 10^{-4}\ [mol \cdot l^{-1} \cdot s^{-1}]$$

d) The kinetic chain length is given by:

$$\nu = \frac{k_p[M]}{k_t[M^{\bullet}]}$$

and the initial degree of polymerization is equal to ν or to 2ν depending upon whether termination occurs by disproportionation or by recombination. The values found are the following:

For MA.......	: $\nu = 5.8 \times 10^4$;	$\overline{M}_n = 10 \times 10^6$
MMA.....	: $\nu = 1.0 \times 10^4$;	$\overline{M}_n = 1 \times 10^6$
St	: $\nu = 1.8 \times 10^3$;	$\overline{M}_n = 3.8 \times 10^5$
VA	: $\nu = 5.7 \times 10^4$;	$\overline{M}_n = 4.9 \times 10^6$

e) The initial value of $\overline{M}_n$ for the MA polymerization has already been calculated $(\overline{M}_n)_0 = 10 \times 10^6$. The value of $\overline{M}_n$ for the polymer formed when already 95% of the monomer is converted is given by $\overline{M}_n = (1 - \pi)\ (\overline{M}_n)_0$, since the initiator concentration is kept constant: $(\overline{M}_n)_{95\%} = 5 \times 10^5$.

To calculate the temporal average $\langle \overline{M}_n \rangle$ one has to evaluate the time required to reach this conversion ($\pi = 0.95$), and to calculate the number of efficient radicals that have been formed during that time, [I] being kept constant.

Upon integration of the rate equation one gets:

$$\ln \frac{[\mathrm{M}]_0}{[\mathrm{M}]} = -\ln(1-\pi) = k_\mathrm{p}\left(\frac{2k_\mathrm{d}f}{k_\mathrm{t}}\right)^{1/2}[\mathrm{I}]^{1/2}t$$

wherefrom t can be calculated:

$$t = 6.6 \times 10^4\ [\mathrm{s}] = 18.3\ [\mathrm{h}]$$

The total number of efficient radicals results from:

$$\Sigma\,[\mathrm{M}^\bullet] = 2k_\mathrm{d}\,f[\mathrm{I}]\,t = 5.6 \times 10^{-4}\ \text{radicals}$$

The temporal average of $\langle \overline{M}_\mathrm{n} \rangle$ results:

$$\langle \overline{M}_\mathrm{n} \rangle = \frac{[\mathrm{M}]_0\,\pi}{\Sigma\,[\mathrm{M}^\bullet]}\,m_0 = 1.6 \times 10^6$$

Problem 3.3

a) The half-life time of AIBN is derived from the first order rate law of its homolytic decomposition:

$$[\mathrm{I}] = [\mathrm{I}]_0\,e^{-k_\mathrm{d}t}$$

$$t_{1/2} = \frac{\ln 2}{k_\mathrm{d}}$$

To evaluate k_d at 50° or 80 °C the same method is used as in the preceding problem. The results are:

- at 50 °$C \ldots k_\mathrm{d} = 0.21 \times 10^{-5}\ [s^{-1}]$; $t_{1/2} = 3.3 \times 10^5\ s = 91\ h$
- at 60 °$C \ldots k_\mathrm{d} = 0.85 \times 10^{-5}\ [s^{-1}]$; $t_{1/2} = 8.1 \times 10^4\ s = 22.6\ h$
- at 80 °$C \ldots k_\mathrm{d} = 6.3 \times 10^{-5}\ [s^{-1}]$; $t_{1/2} = 6.3 \times 10^3\ s = 1.8\ h$

b) Solution polymerization of MMA at 60 °C using AIBN as initiator:
The initial rate of polymerization is given by

$$-\frac{\mathrm{d}[\mathrm{M}]}{\mathrm{d}t} = R_\mathrm{p} = k_\mathrm{p}\left(\frac{2k_\mathrm{d}f}{k_\mathrm{t}}\right)^{1/2}[\mathrm{I}]_0^{1/2}\,[\mathrm{M}]_0$$

$R_\mathrm{p} = 8.22\ [mol \cdot l^{-1} \cdot s^{-1}]$.
The time necessary to convert 1/10 of the monomer into polymer ($\pi = 0.1$) requires integration of the above equation:

$$t = \frac{-\ln(1-\pi)}{k_p\left(\frac{2k_d f}{k_t}\right)^{1/2}[\mathrm{I}]^{1/2}}$$

$$t = 2560\ s = 43\ min$$

(In this integration [I] was considered as constant, which is justified since 43 *min* is small as compared with the half-life time of AIBN at 60 °*C*).

The conversion degree π attained after 10 hours of polymerization is obtained by integration of the rate equation:

- Assuming [I] to be constant:

$$-\ln(1-\pi) = k_p\left(\frac{2k_d f}{k_t}\right)^{1/2}[\mathrm{I}]^{1/2}\, t$$

- Taking into account the exponential decay of its concentration:

$$-\ln(1-\pi) = 2k_p\left(\frac{2f}{k_d k_t}\right)^{1/2}[\mathrm{I}]_0^{1/2}\,(1-\mathrm{e}^{-k_d t/2})$$

The values obtained are respectively (after 10 hours)

$$\pi = 77\% \text{ and } \pi = 75\%$$

The difference is still not very large, but increases with reaction time.

The rate of polymerization after 10 hours of reaction (taking account of the actual concentration of AIBN) is:

$$R_p = 1.62 \times 10^{-5}\ [mol \cdot l^{-1} \cdot s^{-1}]$$

(5 times less than at the beginning)

The molecular weight $\overline{M}_n$ of the poly(methylmethacrylate) formed initially is given by

$$\nu = \overline{P}_n = \frac{k_p[\mathrm{M}]}{(2k_d f k_t)^{1/2}[\mathrm{I}]^{1/2}}\,; \quad \overline{M}_n = 48\,300$$

provided termination occurs only by disproportionation.

If a transfer agent T is added at a concentration [T]:

$$\frac{1}{\overline{P}_n} = \frac{1}{(\overline{P}_n)_0} + C_T\frac{[\mathrm{T}]}{[\mathrm{M}]}\,; \quad \overline{M}_n = 23\,600$$

Problem 3.4

The rate of polymer formation is given by:

$$R_p = \frac{\mathrm{d}[\mathrm{P}]}{\mathrm{d}t} = k_p[\mathrm{M}^\bullet][\mathrm{M}] = k_p[\mathrm{M}^\bullet][\mathrm{M}]_0\,(1-\pi)$$

The rate of formation of branch units by transfer to the polymer is obtained from:

$$\frac{d[B]}{dt} = k_{trP}[M^\bullet][P] = k_{trP}[M^\bullet][M]_0\,\pi$$

From these two equations one obtains:

$$\frac{d[B]}{d[P]} = \frac{k_{trP}}{k_p} \cdot \frac{\pi}{1-\pi}$$

Since:

$$C_p = \frac{k_{trP}}{k_p} \text{ and } d[P] = [M]_0\, d\pi$$

One obtains:

$$[B] = [M]_0\, C_p \int \frac{\pi\, d\pi}{1-\pi} = [M]_0\, C_p\, [-\pi - \ln(1-\pi)]$$

Here are some values of $A = -\pi - \ln(1-\pi)$ as a function of the conversion degrees π attained:

π	0.1	0.3	0.5	0.7	0.9
A	0.0536	0.189	0.387	0.72	1.35

The branching degree ρ is thus given by:

$$\rho = \frac{[B]}{[M]_0\,\pi} = C_p \frac{A}{\pi} = C_p\left[-1 - \frac{\ln(1-\pi)}{\pi}\right]$$

The number average degree of polymerization is *not* affected by transfer onto polymer, because the number of macromolecules formed does not depend on transfer to polymer (whereas it is increased by all other transfer processes).

Problem 3.5

a) Knowing the initial values of R_i and of R_p, the two expressions below can be used to calculate the parameters:

$$k_d f = \frac{R_i}{2[I]} = \frac{1.6 \cdot 10^{-10}}{2 \cdot 0.04} = 0.2 \cdot 10^{-8}$$

$$\frac{k_p}{k_t^{1/2}} = \frac{R_p}{[M]\,R_i^{1/2}} = \frac{6.7 \cdot 10^{-7}}{2 \cdot (1.6 \cdot 10^{-10})^{1/2}} = 2.53 \cdot 10^{-2}$$

b) The kinetic chain length is given by:

$$\lambda = R_p/(R_i)^{1/2} = 4000$$

If termination occurs only by disproportionation, $M_n = 416000$.

c) If account is taken of the transfer constant to the monomer

$$\frac{1}{P_n} = \frac{1}{P_{n0}} + C_M$$

$$P_n = 2990 \qquad M_n = 310000$$

d) The mole ratio of transfer agent to monomer is given by:

$$\frac{[T]}{[M]} = \left(\frac{1}{P_n} - \frac{1}{P_{n0}}\right)\frac{1}{C_T} = 0.250$$

Consequently:

$$[T] = 0.5\ mol/l, \text{ or } 77\ g/l$$

e) The apparent overall activation energy of polymerization is given by:

$$E_R = E_P + E_d/2 - E_t/2 = 23.3\ Kcal/mol$$

Applying Arrhenius law to calculate the values of R_p and R_i at 100 °*C*, one gets:

$$R_i = 1.8 \cdot 10^{-9} \text{ and } R_p = 2.73 \cdot 10^{-5}$$

f) To calculate the conversion degree attained after 2.5 hours one can write:

$$-\frac{d[M]}{dt} = \frac{k_p}{k_t^{1/2}} R_i^{1/2} [M]$$

Integration leads to:

$$\ln\frac{[M]_0}{[M]} = 0.322 \cdot (1.8 \cdot 10^{-9})^{1/2} \cdot 3600 \cdot 5 = 0.246$$

$$[M]_0/[M] = 1.28; \quad [M] = 1.56; \quad \pi = 0.22\ (22\%)$$

Problem 3.6

a) The instantaneous polymerization degree of the polymer formed is given by:

$$P_{ni} = \frac{k_p[M^*][M]}{fk_d[I]} = \frac{k_p[M^*]}{fk_d[I]}[M]_0(1 - \pi)$$

(note that the factor 2 cancels since termination occurs by recombination).

At $\pi = 0$ $P_n = 1000$; consequently the variation of the instantaneous degree of polymerization versus conversion can be expressed as:

$$P_{ni} = 1000\,(1 - \pi)$$

The time averaged (better conversion averaged) degree of polymerization is defined as the ratio of the number of monomer units tied up to the number of macromolecules formed.

$$P_n = \frac{\int_0^{\pi} [M]_0 \, d\pi}{\int_0^{\pi} \frac{[M]_0}{P_{n\pi}} \, d\pi} = \frac{1\,000\,\pi}{-\ln(1-\pi)}$$

For $\pi = 0.90$ one gets: $P_n = 391$
For $\pi = 0.99$ one gets: $P_n = 215$

Assumptions made:

- Rate of radical generation = Rate of termination (Steady state assumption, implying that $[M^*] = \text{cst}$)
- No transfer reactions
- Process is at constant volume
- Rate constants do not vary with conversion

Problem 4.1

To calculate the variation of the instantaneous composition of the copolymer formed, as a function of that of the monomer mixture it originates from, the equation to be used is

$$F_a = \frac{r_a f_a^2 + f_a f_b}{r_a f_a^2 + r_b f_b^2 + 2 f_a f_b} \qquad F_b = 1 - F_a$$

which, in the case of an ideal copolymerization, reduces to:

$$F_a = \frac{r_a f_a}{r_a f_a + f_b}$$

The curves obtained in the specific cases are shown in Fig. L3 and L4.

Problem 4.2

a) The probability of alternation is given by:

$$P_{alt} = p_{ab} \cdot p_{ba} = \frac{k_{ab}[B]}{k_{aa}[A] + k_{ab}[B]} \cdot \frac{k_{ba}[A]}{k_{ba}[A] + k_{bb}[B]}$$

Introducing the reactivity ratios r_a and r_b

$$P_{alt} = \frac{[B]}{r_a[A] + [B]} \cdot \frac{[A]}{r_b[B] + [A]}$$

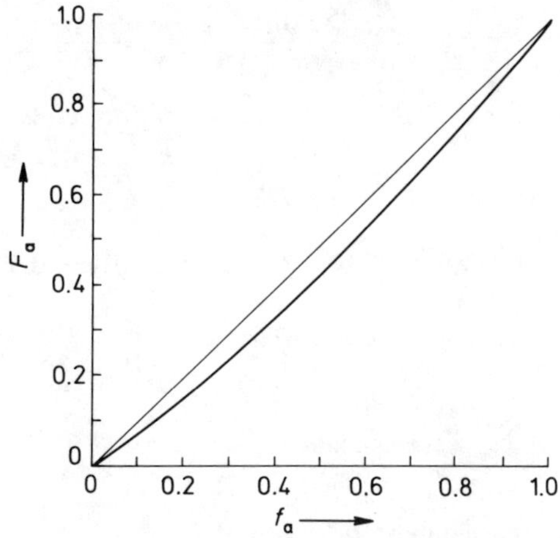

Fig. L3. Plot of the instantaneous copolymer composition: System styrene A- butadiene B

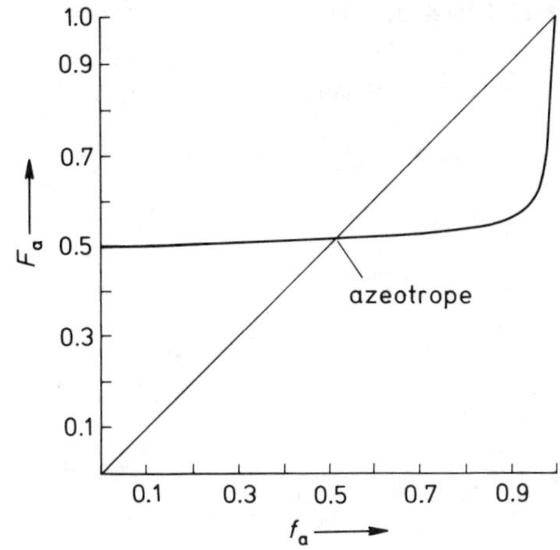

Fig. L4. Plot of the instantaneous composition of copolymer: System styrene A- maleic anhydride B

Calling:

$$x = [A]/[B]$$

$$P_{alt} = \frac{1}{(r_a x + 1)(r_b/x + 1)}$$

The probability of alternation is the highest for:

$$x^2 = \frac{r_b}{r_a}, \text{ or } \frac{[A]}{[B]} = \left(\frac{r_b}{r_a}\right)^{1/2}$$

In the system styrene-dimethylfumarate the ratio [A]/[B] for which the degree of alternation is the highest is 0.483.

b) The probability of diad formation can be expressed as:

$$p_{aa} = \frac{r_a[A]/[B]}{1 + r_a[A]/[B]}; \quad p_{ab} = \frac{1}{1 + r_a[A]/[B]} \ldots$$

The probability of triad formation results:

$$P_{aba} = F_a \cdot p_{ab} \cdot p_{ba} \ldots$$

The values to be found for $[A]/[B] = 0.483$ are:

$$P_{aaa} = P_{bbb} = 0.008$$

$$P_{aba} = P_{bab} = 0.381_5$$

$$P_{aab} = P_{bba} = 0.055 = P_{abb} = P_{baa}$$

Note that $\sum P = 1$.

c) The same method is used as above. The results are:

$$P_{aaa} = 0.085 \qquad P_{bbb} = 0.000$$

$$P_{aab} = 0.142 \qquad P_{baa} = 0.142$$

$$P_{bba} = 0.013 \qquad P_{abb} = 0.013$$

$$P_{aba} = 0.366 \qquad P_{bab} = 0.238$$

Here again $\sum P = 1$. Note that even for $x = 2$ the probability of alternation is high (0.604).

Problem 4.3

a) To produce a copolymer containing 20% of vinylacetate (A) and 80% of vinylchloride (B) the composition of the monomer mixture has to be calculated from:

$$\frac{d[A]}{d[B]} = \frac{1 + r_a[A]/[B]}{1 + r_b[B]/[A]}$$

Setting $x = [A]/[B]$ and solving the above equation with $d[A]/d[B] = 0.2/0.8 = 1/4$ the A content of the monomer mixture should be:

$$f_a = 0.33$$

b) Calling [M] the number of moles of monomers in the reactor (= [A] + [B]), [P] the number of polymer *units* formed (monomer units connected), and [J] the number of moles of monomer B added for make up during the process, one can set up the material balance: for vinyl acetate (none in the feed) one obtains:

$$0 = 0.33d[M] + 0.20d[P]$$

Thus:

$$\int_{M_0=100}^{0} d[M] = -0.61 \int_{0}^{100} d[P] = -61$$

Thus:

$$[M]_0 - [M] = 61$$

and $[M] = 39$

which is the number of moles of monomer remaining in the reactor when 100 moles of polymers *units* have been produced; this implies that 39 moles of vinylchloride had to be added during the process to maintain the composition of the monomer mixture constant at 33 *mol-%* vinylacetate and 67 *mol-%* vinylchloride.

One can check the total material balance:

$$\int d[J] = \int d[M] + \int d[P]$$

c) The triad distribution is calculated as in the previous problem. The values found are:

$$P_{aaa} = 0.002 \qquad P_{bbb} = 0.480$$

$$P_{aab} = 0.018 \qquad P_{baa} = 0.018$$

$$P_{abb} = 0.139 \qquad P_{bba} = 0.139$$

$$P_{aba} = 0.040 \qquad P_{bab} = 0.162$$

Problem 4.4

To get a copolymer containing 20 *mol-%* of styrene (A) and 80 *mol-%* of butadiene (B), the mole ratio of monomers can be evaluated from:

$$\frac{d[A]}{d[B]} = \frac{F_a}{F_b} = r_a \frac{[A]}{[B]}$$

since it is an "ideal" copolymerization. One obtains:

$$\frac{[A]}{[B]} = \frac{1}{3}$$

The volume occupied by 3 *moles* of butadiene is $3 \cdot m_0/\rho = 210$ *ml* and that of 1 *mole* of styrene is 116 *ml*, a total of 326 *ml*. Thus one liter of mixture contains 3.07 *moles* of styrene and 9.2 *moles* of butadiene.

Under the conditions given, the kinetic chain length is expressed as:

$$\nu = \frac{R_p}{R_i} = \frac{r_a[A]^2 + 2[A][B] + r_b[B]^2}{k_t^{*1/2}\left[\dfrac{r_a[A]}{k_{aa}} + \dfrac{r_b[B]}{k_{bb}}\right]} \cdot \frac{1}{R_i^{1/2}}$$

with: $k_t^* = F_a k_{taa} + F_b k_{tbb} = 29.5 \times 10^7$ $[l \cdot mol^{-1} \cdot s^{-1}]$

Knowing [A], [B], r_a, r_b and the homopropagation rate constants k_{aa} and k_{bb}, one obtains, for $\nu = 5000$,

$$R_i^{1/2} = 1.505 \times 10^{-5} \qquad R_i = 2.27 \times 10^{-10}$$

For AIBN at 60 °C
$k_d = 0.85 \times 10^{-5}$, and taking f as 0.5, one can write:

$$R_i = 2k_d f[I]$$

$[I] = 2.7 \times 10^{-5}$ $[mol/l]$ is the amount initiator necessary to fulfill the conditions.

Problem 4.5

Calling [P] the polymer concentration (in *mol/l* of mers) and [M] the remaining total monomer concentration, the material balance can be written as:

$$-d[P] = d[M]$$

The material balance on A is given by:

$$F_a\, d[P] = d([M]\, F_a)$$

resulting in:

$$\frac{d[M]}{[M]} = \frac{df_a}{F_a - f_a}$$

Inserting:

$$F_a = \frac{r_a f_a}{(r_a - 1)f_a + 1}$$

Integration of the above expression leads to:

$$\ln\frac{[M]_\pi}{[M]_0} = \frac{1}{(r_a - 1)} \ln \frac{(1 - f_a)_0}{(1 - f_a)_\pi}\frac{f_{a\pi}}{f_{a0}} + \ln\frac{(1 - f_a)_0}{(1 - f_a)_\pi}$$

where M_π is the number of moles of monomer (A + B) left at conversion π. Insert now: $f_{a\pi} = f_{a0}/10 = 0.05$ and find for the conversion:

$$\pi = 1 - ([M]_\pi/[M]_0) = 0.973$$

Applying the instantaneous copolymer composition equation yields:

$$F_a = 0.10 \text{ (for } f_a = 0.05)$$

Finally:

$$\langle F_a \rangle = \frac{[M]_0 f_{a0} - [M]_\pi f_{a\pi}}{[M]_0 - [M]_\pi} = 0.512$$

Problem 5.1

The equation to be used is:

$$f(i) = \frac{\nu^{i-1} e^{-\nu}}{(i-1)!}$$

knowing that ν is the average number of growth steps per macromolecule (i. e. $\nu = \overline{P_n} - 1$).

One can also operate by a recurrence procedure, since:

$$f(1) = e^{-\nu}; \quad f(2) = \nu e^{-\nu}; \quad f(3) = \frac{\nu^2 \cdot e^{-\nu}}{2}$$

Thus:

$$f(3) = \frac{\nu}{2} \cdot f(2)$$

$$f(4) = \frac{\nu}{3} \cdot f(3)$$

$$f(i) = \frac{\nu}{i-1} \cdot f(i-1)$$

it is easy to set up a computer program to carry out these calculations.

Problem 5.2

a) In the case of a first order rate law, integration leads to:

$$\ln \frac{[M]_0}{[M]} = -\ln(1 - \pi) = k_p [M^\bullet] t$$

If $k_p = 100$ $[l \cdot mol^{-1} \cdot s^{-1}]$, and $[M^\bullet] = 10^{-4}$ the time necessary to get to a conversion π of 90% is given by:

$$t = \frac{-\ln(1 - \pi)}{k_p[M^\bullet]} = 230\ s$$

regardless of the initial monomer concentration.

b) In the case of a second order rate law, the integration leads to

$$\frac{1}{[M]} - \frac{1}{[M]_0} = k_p[M^\bullet]\,t$$

or

$$\frac{1}{1 - \pi} = k_p[M]_0[M^\bullet]\,t$$

The obvious difference is that the rate of conversion depends here upon the initial monomer concentration. For $[M]_0 = 1\ [mol \cdot l^{-1}]$, $[M^\bullet]$ and k_p as above, the time necessary to reach 90% conversion is:

$$t = \frac{1}{(1 - \pi)\,k_p[M]_0[M^\bullet]} = 1\,000\ s$$

See Fig. L5 for the variation of conversion with time in both cases.

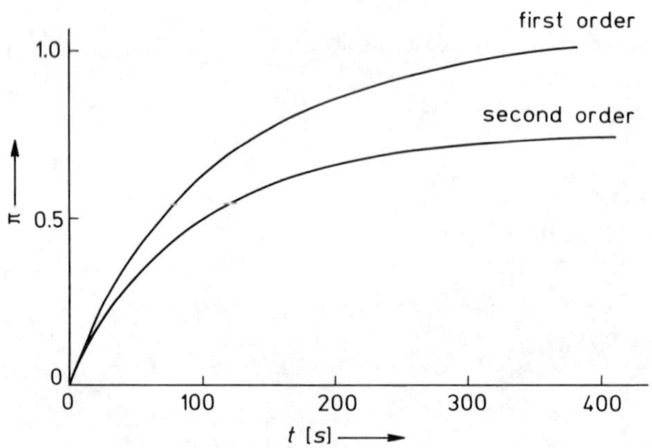

Fig. L5. Comparison between first and second order rate laws: Conversion versus time

Problem 5.3

a) Plot $\ln(1 - \pi)$ versus time. A straight line is obtained, which shows that the rate of polymerization is first order with respect to monomer concentration.

b) The slope of the line $(5 \cdot 10^{-3})$ is equal to $k_p[M^\ominus]$. Since $[M^\ominus]$ is equal to 10^{-3} $[mol \cdot l^{-1}]$ one obtains:

$$k_p = 5\,[l \cdot mol^{-1} \cdot s^{-1}]$$

c) After 10 *min* of reaction, π is equal to 0.95. Provided no transfer or termination is involved in the process, the number average molecular weight of the polymer formed is:

$$\overline{M}_n = m_0 \frac{[M]_0 \cdot \pi}{[M^{\ominus}]} = 98\,800$$

Problem 5.4

Plot the experimental values of k_p as a function of $\alpha = (K_D/[M^{\ominus}])^{1/2}$. The ordinate intercept yields $k_{p\pm} = 60$ $[l \cdot mol^{-1} \cdot s^{-1}]$ and the slope is $(k_{p-} - k_{p\pm})$, which is around 70000 $[l \cdot mol^{-1} \cdot s^{-1}]$:

$$k_p = k_{p\pm} + \alpha\,(k_{p-} - k_{p\pm})$$

If a homoionic salt is added, it is assumed that all the $K^{\oplus}$-ions originate from it. The dissociation equilibrium of $KB\phi_4$ yields its dissociation degree α_ϕ

$$\alpha_\phi = \left[\frac{K_\phi}{[KB\phi_4]}\right]^{1/2} = 0.11$$

Thus the potassium ion concentration is $[KB\phi_4] \cdot \alpha_\phi = 3 \times 10^{-3}$ $[mol \cdot l^{-1}]$ and then the dissociation degree α of the living sites is:

$$\alpha = \frac{[M^{\ominus}]}{[MK]} = \frac{K_D}{[K^{\oplus}]} = 2.6 \times 10^{-5}$$

The value of k_p that should be observed is thus:

$$k_p = k_{p\pm} + \alpha\,(k_{p-} - k_{p\pm}) = 62$$

Problem 5.5

In a racemic mixture the concentration of enantiomers is equal:

$$[R]_0 = [S]_0$$

When half of the monomers is consumed:

$$\frac{[R] + [S]}{[R]_0 + [S]_0} = 1 - \pi = \frac{1}{2}$$

Applying the equation

$$(1 - \pi)^{\rho-1} = \frac{1 + \alpha/\alpha_D}{(1 - \alpha/\alpha_D)^{\rho}}$$

and inserting

$$\alpha = \frac{1}{3} \alpha_D$$

yields to:

$$\rho = 0.37 = \frac{k_r}{k_s}$$

In other words $k_s/k_r = 2.71$, which means that the *S* enantiomer is more reactive than *R*.

Problem 5.6

x being the mole fraction of living ends that do not participate in coupling, and z being the mole ratio of difunctional terminator to the living ends, it is obvious that x must be equal to 0 if the number of living ends is equal or larger than the number of antagonist functions, i.e. if $z > 1/2$. The probability for an acid chloride function, picked at random, to find no partner to react is thus given by:

$$p = \frac{[\text{COCl}]_0 - [\text{LE}]_0}{[\text{COCl}]_0} = 1 - t$$

Consequently, the probability for an acid chloride function to react is given by:

$$t = [\text{LE}]_0/[\text{COCl}]_0 = 1/(2z)$$

since every acid chloride molecule carries two functions.

The probability for an acid chloride molecule to have reacted on both sides (yielding coupling) is thus equal to t^2 and the probability for such a molecule to have reacted only once (yielding functional polystyrene) is given by $2t(1 - t)$. The fraction of all living polymer molecules initially present which are functionalized is thus expressed as:

$$x = \frac{2t(1 - t)}{2t(1 - t) + 2t^2} = 1 - t = 1 - 1/(2z)$$

Examples: For $z = 1/2$ $t = 1$ $x = 0$

$z = 1$ $t = 1/2$ $x = 0.5$

$z = 2$ $t = 1/4$ $x = 0.75$

Problem 5.7

a) If there were no accidental protonic deactivation the number average degree of polymerization would be equal to 2000, since each growing site would experience 1000 growing steps.

The probability p for accidental deactivation is given by the mole ratio of water to living sites. It is equal to 0.1, i. e., one tenth of the sites are deactivated, and 9/10 remain active. Consequently the average number of growth steps per active site must be corrected as:

$$N = 1/(0.9 \cdot 10^{-3}) = 1\,111$$

Deactivation occurs at random. Thus, the probability that two sites remain active on a macromolecule is given by $(1 - p)^2$, and the probability that a macromolecule has one active site only is expressed as $2p\,(1 - p)$. The molar proportions of two-ended and one-ended macromolecules are thus:

$$(1 - p)^2/2 \quad \text{and} \quad 2p(1 - p)$$

respectively. In order to calculate the weight fractions of these species, account has to be taken of the fact that the average weight of a two-ended polymer molecule is twice that of a single ended one. Consequently the weight fraction of single ended macromolecules is given by:

$$w = \frac{2p\,(1 - p)}{2p\,(1 - p) + (1 - p)^2} = \frac{0.18}{0.18 + 0.81} = 0.178$$

b) The number average degree of polymerization of the polymer obtained can be expressed as follows, taking into account that the two-ended polymer molecules are twice as long as the single-ended:

$$P_{\mathrm{n}} = \frac{N2p\,(1 - p) + N\,(1 - p)^2}{2p\,(1 - p) + (1 - p)^2/2} = \frac{2N_0\,(1 + p)}{(1 - p)\,(1 + 3p)}$$

where N_0 is the average number of growth steps per active site if there were no deactivation at all.

Examples: If $p = 0$ $\quad P_{\mathrm{n}} = 2000 = 2N_0$

$p = 0.06$ $\quad P_{\mathrm{n}} = 1911 = 1.911\,N_0$

$p = 0.1$ $\quad P_{\mathrm{n}} = 1880 = 1.88\,N_0$

$p = 0.2$ $\quad P_{\mathrm{n}} = 1870 = 1.87\,N_0$

$p = 0.3$ $\quad P_{\mathrm{n}} = 1950 = 1.95\,N_0$

$p = 0.5$ $\quad P_{\mathrm{n}} = 2400 = 2.40\,N_0$

$p = 0.6$ $\quad P_{\mathrm{n}} = 2860 = 2.86\,N_0$

$p = 0.9$ $\quad P_{\mathrm{n}} = 10720 = 10.72\,N_0$

Problem 6.1

- The equilibrium reaction between BF_3 and H_2O can be written as:

$$BF_3 + H_2O \rightleftharpoons H^{\oplus}, BF_3^{\ominus}OH \qquad K = \frac{[H^{\oplus}, BF_3^{\ominus}OH]}{[BF_3][H_2O]}$$

- The initiation reaction (rate determining step):

$$H^{\oplus}, BF_3^{\ominus}OH + M \longrightarrow HM^{\oplus}, BF_3^{\ominus}OH \qquad R_i = k_i K[BF_3][H_2O][M]$$

- The propagation reaction:

$$HM_i^{\oplus}, BF_3^{\ominus}OH + M \longrightarrow HM_{i+1}^{\oplus}, BF_3^{\ominus}OH \qquad R_p = k_p[M^{\oplus}][M]$$

- The termination by ion-splitting:

$$HM_n^{\oplus}, BF_3^{\ominus}OH \longrightarrow HM_nOH + BF_3 \qquad R_t = k_t\,[M^{\oplus}]$$

Setting: $R_t = R_i$,
one obtains:

$$R_p = k_p \frac{k_i}{k_t} K[BF_3][H_2O][M]^2$$

Problem 6.2

In the absence of transfer reactions the degree of polymerization would be given by:

$$\overline{P_n} = \frac{k_p[M]}{k_t}$$

If transfer to the monomer occurs, the following equation has to be used:

$$\frac{1}{\overline{P_n}} = \frac{k_t}{k_p[M]} + C_M$$

First calculate the ratio k_t/k_p from the data at $-78\ °C$ where $\overline{M_n} = 100000$ ($\overline{P_n} = 1786$), knowing that $[M] = 1$ [*mol/l*]:

$$\frac{k_t}{k_p} = 4 \times 10^{-4}$$

Then apply the same equation to calculate $\overline{P_n}$ at $-50\ °C$ and at $-20\ °C$, knowing the corresponding values of the transfer constant:

At $-50\ °C$ $\overline{M_n} = 52800$; at $-20\ °C$ $\overline{M_n} = 22400$.

Problem 6.3

The bulk concentration of oxolane is $\frac{1000\,\rho}{m_0} = 12.3\ [mol \cdot l^{-1}]$. The amount of initiator necessary to yield a molecular weight of 10000 for 10% conversion is given by:

$$[I] = [M^{\oplus}] = \frac{m_0}{\bar{M}_n}[M]_0\,\pi = 8.9 \times 10^{-3}\ [mol \cdot l^{-1}]$$

The time necessary to get 10% conversion can then be calculated easily from the integrated rate law:

$$\ln \frac{[M]_0}{[M]} = -\ln(1 - \pi) = k_p [M^{\oplus}]\, t$$

$$t = \frac{-\ln(1 - \pi)}{k_p [M^{\oplus}]} = 5.9 \times 10^3\ s = 1.6\ h$$

The oxolane polymerization is considered here as irreversible.

Problem 6.4

This is a case of ideal copolymerization, since $r_a \cdot r_b = 1$. Consequently:

$$F_b = \frac{f_b}{r_a f_a + f_b} = \frac{1}{r_a (f_a/f_b) + 1}$$

If $F_{b0} = 0.05$ $\quad f_{a0}/f_{b0} = 7.6$ $\quad$ and $f_b = 0.116$

By trial and error one finds that at $\pi = 0.8\ f_b = 0.281$.
Thus at this conversion the average B-content of the copolymer is given by:

$$\langle F_b \rangle = \frac{f_{b0} - f_{b\pi}(1 - \pi)}{\pi} = 0.075$$

Problem 7.1

The equilibrium monomer concentration in a reversible polymerization is given by:

$$\ln [M]_e = \frac{\Delta H - T\Delta S}{RT} = \frac{\Delta H}{RT} - \frac{\Delta S}{R}$$

If the ΔH and ΔS values are expressed in $[cal \cdot mol^{-1}]$, and $[cal \cdot mol^{-1} \cdot K^{-1}]$ respectively, then $R = 2\ [cal \cdot mol^{-1} \cdot K^{-1}]$ the following results are obtained:

At: $\quad 0\ °C \quad [M]_e = 1.15\ [mol \cdot l^{-1}]$

$20\ °C \quad [M]_e = 2.3\ [mole \cdot l^{-1}]$

$40\ °C \quad [M]_e = 4.3\ [mole \cdot l^{-1}]$

Problem 7.2

First calculate the bulk molar concentration of oxolane

$$[\mathrm{M}]_0 = \frac{1\,000\,\rho}{m_0} = 12.3\ [mol \cdot l^{-1}]$$

a) The rate equation to be used is

$$R_\mathrm{p} = -\frac{\mathrm{d}\,[\mathrm{M}]}{\mathrm{d}t} = k_\mathrm{p}\,[\mathrm{M}^\oplus]([\mathrm{M}]_0 - [\mathrm{M}]_\mathrm{e})$$

and the initial rate of polymerization is:

$$R_\mathrm{p} = 9.1 \times 10^{-4}\ [mol \cdot l^{-1} \cdot s^{-1}]$$

b) The conversion attained after 15 minutes of reaction is obtained upon integration of the above equation:

$$\ln \frac{[\mathrm{M}]_0 - [\mathrm{M}]_\mathrm{e}}{[\mathrm{M}] - [\mathrm{M}]_\mathrm{e}} = k_\mathrm{p}\,[\mathrm{M}^\oplus]\,t$$

The remaining monomer concentration is $[\mathrm{M}] = 11.5\ [mol \cdot l^{-1}]$ which implies that $\pi = 1 - ([\mathrm{M}]/[\mathrm{M}]_0) = 0.06 = 6\%$.

c) The molecular weight of the polymer formed at that time is given by:

$$\overline{M}_\mathrm{n} = m_0 \cdot \frac{[\mathrm{M}]_0\,\pi}{[\mathrm{M}^\oplus]} = 17\,700$$

e) The time necessary to reach a conversion degree within 1% of the equilibrium monomer concentration is calculated using the same expression, in which one sets:

$$\frac{[\mathrm{M}]_\mathrm{e}}{[\mathrm{M}]} = 0.99 \quad \text{or} \quad \frac{[\mathrm{M}]}{[\mathrm{M}]_\mathrm{e}} = 1.01$$

One obtains:

$$t = 6.75 \times 10^4\ \ s = 18.7\ h$$

f) The molecular weight of the polymer formed is obtained as in c):

$$\overline{M}_\mathrm{n} = 240\,000$$

d) and f) At low conversions, due to the absence of transfer and termination, the molecular weight distribution is sharp (*Poisson-like*). At conversion close to the limiting value, due to reversibility, the polydispersity broadens and eventually reaches the *"most probable"* distribution.

Problem 8.1

In the usual copolymerization notation the triad probabilities are given by the expressions:

1) $P_{aaa} = F_a\, p_{aa}^2$ 2) $P_{bbb} = F_b\, p_{bb}^2$

3) $P_{aab} = F_a\, p_{aa}\, p_{ab}$ 4) $P_{bba} = F_b\, p_{bb}\, p_{ba}$

5) $P_{abb} = F_a\, p_{ab}\, p_{bb}$ 6) $P_{baa} = F_b\, p_{ba}\, p_{aa}$

7) $P_{aba} = F_a\, p_{ab}\, p_{ba}$ 8) $P_{bab} = F_b\, p_{ba}\, p_{ab}$

In the notation of steric configuration *m* and *r*, p_{ll} and p_{dd}, the analogues of p_{aa} and p_{bb} are identical, and equal to $k_m/(k_m + k_r)$.

Likewise, p_{ld} and p_{dl} the analogues of p_{ab} and of p_{ba} are identical and equal to $k_r/(k_m + k_r)$.

Consequently, the probability of formation of isotactic triads can be compared with the sum of (1) and (2):

$$i = (F_a + F_b)p_{aa}^2 = \left(\frac{k_m}{k_m + k_r}\right)^2$$

Similarly for the syndiotactic triads (eq. (7) and (8)):

$$s = (F_a + F_b)\,(p_{ab})^2 = \left(\frac{k_r}{k_m + k_r}\right)^2$$

while the heterotactic triads include all the conformations p_{lld}, p_{dll}, p_{ddl}, p_{ldd}, as expressed by:

$$h = 2(F_a + F_b)\,p_{aa}\,p_{ab} = 2\,\frac{k_m k_r}{(k_r + k_m)^2}$$

If $k_m = k_r$ then $i = \frac{1}{4}$; $s = \frac{1}{4}$; $h = \frac{1}{2}$

If $k_m = 3k_r$ then $i = \frac{9}{16}$; $s = \frac{1}{16}$; $h = \frac{6}{16}$

Problem 9.1

a) Upon hydrolysis every branch of structure (1) will become disconnected when poly(vinyl acetate) is hydrolyzed.

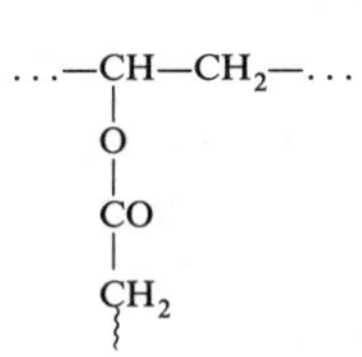

(1)

Only branches of the structure (2) will remain attached to the backbone.

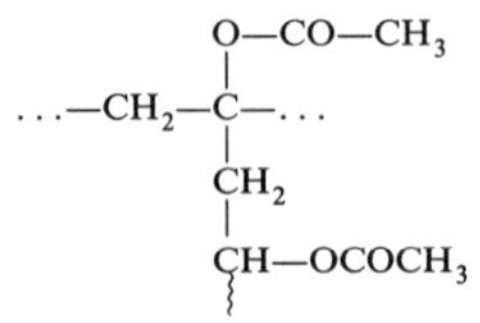

(2)

Suppose we consider 1000 initial macromolecules of DP = 2000.

After hydrolysis, but before oxidation, the total number of macromolecules is equal to 4000, as the average DP has been reduced by a factor 4.

Thus the number of hydrolyzable branches is equal to 3000, and the branching degree, i.e., the proportion of branches per vinylacetate unit in the initial polymer is:

$$\rho = \frac{3 \times 10^3}{2 \times 10^6} = 1.5 \times 10^{-3}$$

Periodate treatment breaks selectively the head-to-head units in poly(vinylalcohol)

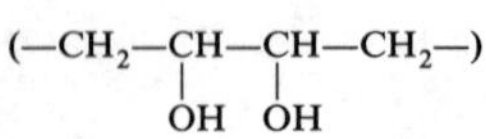

The number of such head-to-head units results from the decrease in polymerization degree in poly(vinylalcohol) upon this treatment:

$$\frac{2 \times 10^6}{166} - \frac{2 \times 10^6}{500} = 8048$$

and the proportion of head-to-head units in the initial polyvinylalcohol is: $8048/2 \times 10^6 = 4 \times 10^{-3}$.

b) Let us call A the acetate functions and O the hydroxyl functions in a partly hydrolyzed poly(vinyl acetate). If one parameter suffices to characterize the probability of hydrolysis of a given A function (let it be f) the triad probabilities should be:

$$\text{AAA} = (1 - f)^3$$

$$\text{AAO} = \text{AOA} = \text{OAA} = f(1 - f)^2$$

$$\text{AOO} = \text{OOA} = \text{OAO} = f^2\,(1 - f)$$

$$\text{OOO} = f^3$$

If the hydrolysis were autoretarded or autoaccelerated 2 or more parameters would be necessary to describe the triad distribution. In the former case the AOA and AAA triads would be favored with respect to OOO or OOA triads. In the latter case the substitution tends to become sequential, thus favoring the OOO triads.

Problem 9.2

a) The weight of the initial repeat unit of cellulose is: $w_i = 162$ *g/mol.*
The weight of the repeat unit of cellulose triacetate is: $w_f = 288$.
Thus the percentual weight increase upon chemical transformation is given by:

$$\Delta M = 100 \cdot (288 - 162)/162 = 77.8\%$$

b) Similarly for PVC, the weight of a repeat unit is 62.4 *g/mol* and that of a dehydrohalogenated unit is 26 *g/mol.* Consequently the weight decrease upon total dehydrochlorination is:

$$\Delta M = 100 \cdot (62.4 - 26)/62.4 = 58.3\%$$

c) The molecular weight of a polystyrene repeat unit is 104 *g/mol*, and that of a chloromethylated styrene unit is 152.4 *g/mol* and that of a unit fitted with a methyl-trimethylammonium chloride substituent is 211.4 *g/mol.*
Since the first reaction is carried out to 50% only, and the second involves solely the chloromethylated units the percentual weight increase is given by:

$$\Delta M = 100 \cdot [(104 + 211.4)/2 - 104]/104 = 51.5\%$$

(If the product is washed with NaOH to remove the Cl^- ions the final weight of the modified units is 193 and the percentual weight increase is 42.8%.)
d) The hydrolysis of poly(vinyl acetate) to poly(vinyl alcohol) involves similarly a percentual weight decrease of:

$$\Delta M = 100 \cdot (86 - 44)/86 = 48.8\%$$

e) Acetalisation of poly(vinyl alcohol) is an intramolecular cyclization process, the conversion of which is limited to 86.5%. Consequently the percentual weight increase upon acetalization is given by:

$$\Delta M = 100 \cdot [(0.865 \cdot 71 + 0.135 \cdot 44) - 44]/44 = 53.1\%$$

f) Hydrolysis of Poly(t-butyl acrylate) implies a weight decrease of:

$$\Delta M = 100 \cdot (128 - 72)/128 = 43.7\%$$

Problem 9.3

If butyl rubber containing 3% of isoprene units is crosslinked via all its isoprene units, the ratio ρ is equal to 0.03, and the condition for gelation to occur is expressed as:

$$\varepsilon = \rho\,(y_w - 1) > f - 1$$

where: ε is the average number of crosslinks to be formed from one chain to another and y_w is the weight average number of units in the chains prior to crosslinking (= M_w/m_0, m_0 being the molecular weight of the mer). f is the functionality of the branch points, here 3.

Thus: $y_w = 2/0.03 + 1 = 67.3$

Problem 9.4

The copolymer of methyl acrylate and vinyl acetate is obtained at constant feed ratio: $f_a/f_b = 3$. In other words: $f_a = 0.75$ and $f_b = 0.25$.
Applying the copolymerization equation one gets:

$$F_a = 0.966 \quad \text{and} \quad F_b = 0.0356$$

Since only 3.6 out of 100 units are B's we can suspect that they are fairly isolated along the chain. Indeed, the BB diad probability is:

$$p_{bb} = 0.001 \text{ (or 1 out of 1000 diads)}$$

and the BBB triad probability is nearly zero.
The cyclization reactions can only occur when A and B are adjacent in the chain. Since the BBB triad probability can be neglected, every B unit can be considered to be adjacent to at least one A unit, and is able to take part in a cyclization reaction. Since lactone formation involves one A and one B unit, in a sequence of 100 units 3.56 units of B and 3.56 A units will be involved, whereby the number of remaining A units is decreased by 3.56. Consequently the molar proportion of lactone rings formed is given by:

$$0.0356/0.9644 = 0.0369 \quad \text{(or 3.69\%)}$$

Problem 12.1

- Consider first the single phase polymerization:

$$R_p = k_p[\text{M}][\text{M}^\bullet]$$

$$[\text{M}]_0 = \frac{1}{3}[\text{M}]_{\text{undiluted}} = \frac{14.8}{3} = 4.93\ [mol \cdot l]$$

After 25% conversion have been attained:

$$[\text{M}] = 0.75\ [\text{M}]_0 = 3.7\ [mol \cdot l]$$

The rate of radical formation is $1 \times 10^{12} \left[\dfrac{radicals}{ml \cdot s}\right]$, which means that

$$[M^\bullet] = \left(\frac{1}{k_t}\frac{d[R^\bullet]}{dt}\right)^{1/2} = 2.1 \times 10^{-9}\ [mol \cdot l^{-1}]$$

Thus:

$$R_p = 7.73 \times 10^{-7}\ [mol \cdot l^{-1} \cdot s^{-1}]$$

In other words 4.62×10^{14} monomer molecules polymerize per *ml* an per *second.*

The degree of polymerization can be expressed as:

$$\overline{P}_n = \frac{R_p}{R_i/2} = 924 \quad (\overline{M}_n = 96000)$$

(If no solvent were used, the [M] value would be 3 times larger, increasing thus R_p and $\overline{M}_n$ by a factor 3)

- Consider next the emulsion polymerization.

The number n of particles per *ml* is given and the monomer concentration inside the particles is [M] = 7 [*mol/l*].

$$R_p = -n\frac{dm}{dt} = n \cdot \frac{1}{2} k_p[M] = 3.5 \times 10^{16}$$

R_p is expressed in monomer molecules polymerized per *ml* and per *second.* This is two orders of magnitude faster than in single phase polymerization.

The degree of polymerization is given by

$$\overline{P}_n = k_p[M]\,\Delta t$$

where

$$\Delta t = \frac{n}{(d[R^\bullet]/dt)} = 100\ s$$

Thus the $\overline{P}_n$ is also two orders of magnitude higher than in the single phase polymerization:

$$\overline{P}_n = 7 \times 10^4 \qquad (\overline{M}_n = 8.1 \times 10^6)$$

Problem 12.2

- Ideal case A: No transfer; 15 *s* of propagation, 15 *s* no propagation:

$$\overline{P}_n = 7500$$

(one macromolecule is formed during the 30 *s* period considered, in each particle):

$$-\frac{dm}{dt} = k_p[M] \cdot \frac{1}{2} = 250$$

is the number of monomers tied up per second in the particle.

- Case B

$$\overline{P_n} = \frac{\text{number of monomers tied up}/s}{\text{number of macromolecules formed}} = 2\,500$$

$$-\frac{dm}{dt} = k_p[M] \cdot \frac{1}{2} = 250 \text{ (as above)}$$

- Case C

The first radical entering grows for 20 *s*; the number of monomers tied into the chain is 10000. The second radical grows for 5 *s* before recombining with the first one.

Thus the resulting macromolecule has a $\overline{P_n}$ of 12500. The average rate of monomer consumption is:

$$-\frac{dm}{dt} = k_p[M] \cdot \frac{20 + 5}{30} = 416 \textit{ monomers/s}$$

Problem 12.3

From problem 12.1 it is known that $[M] = 7$; $k_p = 100$, $\frac{d[R^\bullet]}{dt} = 1 \times 10^{12}$ [*radicals* $\cdot$ $ml^{-1} \cdot s^{-1}$]; $n = 1 \times 10^{14}$ [particles $\cdot$ ml^{-1}]. At the time cited a single particle has a diameter $d = 1\,000$ Å.

- Case I

The volume growth ΔV during a "green light" period is:

$$\Delta V = 2\frac{dV}{dt} \cdot \Delta t = -2v_u q \cdot \frac{dm}{dt} \Delta t = v_u q k_p [M] \Delta t$$

The volume of a monomer unit v_u is given by:

$$v_u = \frac{m_0}{d} \frac{1}{N_a} = 112 \times 10^{-24} \; ml$$

and q is the reciprocal of the volume fraction of monomer: $q = 2$.

Thus:

$$\frac{\Delta V}{V_{\text{particle}}} = \frac{v_u \cdot q k_p [M] \Delta t}{\frac{\pi}{6} d^3} = 0.0303$$

- Case II: post-droplet state $d = 1000$ Å.

The total monomer present in a particle is:

$$\frac{V_{\text{particle}}}{q v_u} = 4.6 \times 10^6 \textit{ molecules}$$

The monomer polymerized during $\Delta t = 100\ s$ is given by:

$$k_p[M]\ \Delta t = 7 \times 10^4$$

The fraction of monomer consumed during a "green light" period is thus: 0.0152.

Problem 12.4

Calling butadiene A and styrene B, if the copolymer composition is:

$$F_a = \frac{r_a f_a}{(r_a - 1) f_a + 1} = 0.80 \text{ it means that } f_a = 0.75$$

The diad probabilities are then:

$$p_{aa} = 0.8 \qquad p_{ab} = 0.2$$
$$p_{bb} = 0.2 \qquad p_{ba} = 0.8$$

A sequence of 100 units (mers) in the chain contains 80 A and 20 B units.
Out of the 80 A units 64 have added A, and 16 have added B; out of the 20 B units 4 have added B, and 16 have added A.
The time necessary for these 100 units to be incorporated is:

$$t_{100} = 64\ (k_{aa}[A])^{-1} + 16\ (k_{ab}[B])^{-1} + 4\ (k_{bb}[B])^{-1} + 16\ (k_{ba}[A])^{-1}$$

The monomer concentrations in the particles (undiluted) can be calculated from the mole ratio of the constituents, knowing their densities, since:

$$3[A] \cdot 56/0.6 + [B] \cdot 104/0.9 = 1000\ (ml)$$

with:

$$[A]/[B] = 3 \quad \text{since } f_a = 0.75$$

consequently:

$$[A] = 7.77\ mol/l \quad \text{and} \quad [B] = 2.59\ mol/l$$

The dilution by polymer, at a conversion of 50%, implies that the concentrations are half the above values:

$$[A] = 1.29\ mol/l \quad \text{and} \quad [B] = 3.89\ mol/l$$

Thus,

$$t_{100} = 0.364\ s$$

The green light period is given by the ratio of the number of particles N to the rate of generation of radicals: It amounts to 10 *s*.

The number of monomer molecules incorporated during one green light period is also the number average degree of polymerization. It is thus given by:

$$P_n = 10/(t_{100}/100) = 2745 \text{ units (mers)}$$

Consequently the number average molecular weight of the polymer formed at that stage is:

$$M_n = 2745\,(0.8 \cdot 54 + 0.2 \cdot 104) = 175\,700$$

Problem 12.5

a) During the green light period 274 monomer units are adding per second to the radical in the particle. Another 274 molecules of monomer must diffuse out to keep the thermodynamic balance monomer/polymer in the particle. This makes a total of 548 molecules. But the red light period is as long as the green light period. Thus, the time averaged flux of monomer is:

274 monomer molecules/s

b) If the mean density of the polymer is 0.8 *g/ml* and the volume fraction of the polymer in the particle is 0.5, the volume of polymer per particle is given by:

$$(1/2)\,(\pi/6)\,D^3 = 0.262 \cdot 10^{-15}\ ml$$

The volume of a single polymer molecule involving 200 units of average molecular weight 64 is expressed as:

$$2000 \cdot 64/(0.8 \cdot 0.6 \cdot 10^{24}) = 2.67 \cdot 10^{-19}\ ml$$

Consequently the number of macromolecules per particle is given by the ratio of the two above figures: It amounts to 981 molecules/particle.

c) The particle volume being $0.523 \cdot 10^{-15}\ ml^3$, when it amounted to 1/3 of this value, it was: $0.175 \cdot 10^{-15}\ ml^3$, and the corresponding diameter 693 Å. The surface area of a sphere is given by $\pi\,D^2$, and its volume by $(\pi/6)\,D^3$. Assuming that each soap molecule occupies an area of 25 Å^2 at the interface when the particles were 1/3 of their present size, the area occupied by each of them at the present stage is:

$$(25) \cdot (1000/693)^2 = 52\,\text{Å}^2$$

Problem 13.1

a) The molecular weight of the repeat unit is 282 *g/mol.*
Consequently the rate of nylon withdrawing is:

$$0.1/282 = 3.55 \cdot 10^{-4} \, mol/min$$

The reactants must be fed to sustain reaction for no accumulation or decrease:

$$\text{Dichloride } (M = 239) : 3.55 \cdot 10^{-4} \cdot 239 = 0.085 \, g/min$$

$$\text{Diamine } (M = 116) : 3.55 \cdot 10^{-4} \cdot 116 = 0.041 \, g/min$$

NaOH must be added to completely neutralize HCl generated:

$$2 \cdot 3.55 \cdot 10^{-4} \cdot 39.9 = 0.028 \, g/min$$

b) The time necessary for NaCl to attain a 2 molar concentration, i. e. to have formed $2 \cdot 5 = 10$ moles of NaCl is given by:

$$t = 10/(2 \cdot 3.55 \cdot 10^{-4}) = 14\,100 \, min$$

$$t = 235 \text{ hours}$$

c) It will be necessary to replace the solvents which have been removed by the withdrawn nylon.

Bibliography

General References and References to Chapter 1

"Encyclopedia of Polymer Science and Technology" Second Edition John Wiley & Sons New York 1987

"Comprehensive Polymer Science" G. Allen, J. C. Bevington Eds. (The individual volumes are quoted in the corresponding chapters) Pergamon Press Oxford 1989

"Comprehensive Chemical Kinetics" C. H. Bamford, C. F. Tipper Eds. (Volume 14 is devoted to free radical polymerizations and volume 15 to non-radical polymerizations) Elsevier Amsterdam 1976

"Developments in Polymerization" R. N. Haward Ed. Applied Science Publishers 1979/1980

P. J. Flory "Principles of Polymer Chemistry" Cornell University Press Ithaca N. Y. 1953

H. J. Elias "Macromolecules" Hüthig & Wepf Verlag Basel 1975

G. Odian "Principles of Polymerization" McGraw Hill New York 1970

F. W. Billmeyer "Textbook of Polymer Science" John Wiley Interscience New York 1971

P. C. Hiemenz "Polymer Chemistry: Basic Concepts" Marcel Dekker New York 1984

H. R. Allcock, F. W. Lampe "Contemporary Polymer Chemistry" Prentice Hall Inc. Englewood Cliffs, N. J. 1981

F. A. Bovey, F. H. Winslow "Macromolecules: An Introduction to Polymer Science" Academic Press New York 1979

G. Champetier Ed. "Chimie Macromoléculaire" Hermann Paris 1972

"Polymerization Processes" C. E. Schildknecht, I. Skeist Eds. Interscience Publ. New York 1977

H. Sawada "Thermodynamics of Polymerization" Marcel Dekker New York 1976

"The Stereochemistry of Macromolecules" A. D. Ketley Ed. Marcel Dekker New York 1967

References to Chapter 2

"Comprehensive Polymer Science" Volume 5 (see above) D. H. Solomon "Step Growth Polymerization" Marcel Dekker New York 1972

References to Chapter 3

"Comprehensive Polymer Science" Volume 3 (see above) C. H. Bamford "Free Radical Polymerizations" Elsevier Scientific Publishing Company Amsterdam 1976

J. C. Bevington, J. R. Ebdon "Developments in Polymer Science" Vol 2 R. N. Haward Ed. Applied Science Publishers London 1979

References to Chapter 4

"Comprehensive Polymer Science" Volume 3 (see above)

"Copolymerization" G. E. Ham Ed. Interscience J. Wiley & Sons New York 1964

J. M. G. Cowie "Alternating Copolymers" Plenum New York 1985

References to Chapter 5

"Comprehensive Polymer Science" Volume 3 (see above)

M. Szwarc "Carbanions, Living Polymers & Electron Transfer Processes" Interscience New York 1968

M. Morton "Anionic Polymerization: Principle and Practice" Academic Press New York 1983

"Recent Advances in Anionic Polymerization" T. Hogen-Esch, J. Smid Eds. Elsevier Amsterdam 1984

"Anionic Polymerization: Kinetics, Mechanisms and Synthesis" J. E. McGrath Ed. ACS Symposium Series 166 (1981)

M. van Beylen, S. Bywater, G. Smets, M. Szwarc, D. J. Worsfold, Advances in Polymer Science 86 87 (1988)

"Ring Opening Polymerizations: Kinetics, Mechanisms and Synthesis" ACS Symposium Series 286 (1985)

"Ring Opening Polymerizations" K. J. Ivin, T. Saegusa Eds. Elsevier 1984

References to Chapter 6

"Comprehensive Polymer Science" Volume 3 (see above)

"The Chemistry of Cationic Polymerization" P. H. Plesch Ed. Pergamon Press Oxford 1963

J. P. Kennedy "Cationic Polymerization of Olefines: A Critical Inventory" Interscience Publishers 1975

J. P. Kennedy and E. Maréchal "Carbocationic Polymerizations" Interscience Publishers 1982

"Cationic Ring Opening Polymerization" S. Penczek et al. Advances in Polym. Science 37 (1980) and 68/69 (1985)

"Polymerization of Heterocycles (Ring Opening)" S. Penczek Ed. Pergamon Press Oxford 1976

References to Chapter 7

- Living Cationic Polymerizations

T. Higashimura, Sawamoto Macromolecules **17** 165 (1984), **18** 611 (1985)

Faust, J. P. Kennedy J. Polymer Science **A25** 1847 (1987)

- Group Transfer Polymerization

D. Y. Sogah et al. Macromolecules **20** 1473 (1987)

References to Chapter 8

"Comprehensive Polymer Science" Volume 4 (see above)

"Preparation and Properties of Stereoregular Polymers" R. W. Lenz, F. Ciardelli Eds. Reidel Publ. Cy. 1978

W. Kaminsky, H. Sinn "Transition Metals and Organometallics as Catalysts for Olefin Polymerization" Springer Verlag Berlin 1988

R. P. Quirk "Transition Metal Catalyzed Polymerizations" Cambridge University Press New York 1988
J. Boor Jr. "Ziegler-Natta Catalysis and Polymerization" Academic Press New York 1979
R. Schrock "Living Ring-Opening Metathesis Polymerizations" Advances in Polymer Science (1989)

References to Chapter 9

"Comprehensive Polymer Science" Volume 6 (see above)
"Degradation and Stabilization of Polymers" G. Geuskens Ed. Applied Science Publishers London 1975

References to Chapter 10

"Comprehensive Polymer Science" Volume 6 (See above)
R. J. Ceresa "Block and Graft Copolymerization" John Wiley & Sons New York 1973
Noshay and McGrath "Block copolymers" Academic Press 1977
"Thermoplastic Elastomers" N. Legge, G. Holden, H. Schröder Eds. Hansen Publishers 1987

References to Chapter 12

"Comprehensive Polymer Science" Volumes 4 and 7 (see above)
Developments in Polymer Science R. N. Haward Ed. Volume 2 Applied Science Publ. 1980
R. M. Fitch "Polymer Colloids" Plenum Press New York 1971
D. C. Blackley "Emulsion Polymerization: Theory and Practice" Applied Science Publ. London 1980

References to Chapter 13

"Comprehensive Polymer Science" Volumes 4 and 7 (see above)
"Transition Metal Catalyzed Polymerizations: Ziegler-Natta and Metathesis Polymerizations" R. P. Quirk Ed. Cambridge University Press 1988
S. Floyd, T. Heiskanen, W. H. Ray "Solid Catalyzed Olefin Polymerizations" Chemical Engin. Progress 84 56 1988
"Chemical Reaction Engineering" O. Levenspiel Ed. John Wiley & Sons 1962

Subject index

Petar R. Dvornic, Robert W. Lenz

High Temperature Siloxane Elastomers

1990, 328 pp., 83 figs., 55 tabs.,
hard cover, DM 138,—,
SFr. 118,—, US$ 85,—
ISBN 3-85739-118-9

The authors of this book have attemped to present here a coherent picture of the field of high temperature siloxane elastomers, these having attracted considerable research attention during the last two or three decades. The interest in these materials developed. In the main, from high demands for such elastomers for applications particulary in the aerospace, defense and computer industries.

Chapter I gives a general introduction in the field and briefly describes some of the characteristics of the siloxane bond which are of importance in the designing of high performance siloxane elastomers. After a concise discussion of the polyxsiloxanes and their properties in Chapter II, chapter III – V (organised in four sections) are devoted consecutively to (1) monomer synthesis, (2) polymer preparation, (3) polymer characterization and properies and (4) vulcanization and properties of the vulcanizates.

This book will provide research scientists and advanced graduate students with a single source of reference for information on the chemistry and property of the high temperature siloxane polymers. It covers all literature published up to 1987. As a result, the next produces a comprehensive coverage of essentially all of the most important developments in this field. There may exist some proprietory of classified materials which have not as yet been made published.

Hüthig & Wepf

Hüthig & Wepf
Scientific Publishers
Eisengasse 5
CH-4001 Basel

Hüthig & Wepf
Publishing Ltd.
212, North Barry Av.
Mamaroneck, N. Y.
10543-2826

Hüthig & Wepf Verlag
Postfach 10 28 69
6900 Heidelberg 1

Hüthig & Wepf

Esen Bekturov, E. A. Bakauova

Synthetic Water-Soluble Polymers in Solution

1986, 241 pp., hard cover,
DM 108,—
ISBN 3-85739-110-3

Water-soluble polymers play a very important rôle in nature. Polymer science and technology at the present time becomes more and more devoted to the investigation of water-soluble polymers, as can be seen from the topics of many national and international meetings on the subject.
Water-soluble polymers are used for precipitation of waste waters, for raising the yield of ore, separation of metals, diminution and stabilisation of hydrodynamic resistance, secondary and tertiary oil recovery, etc. In medicine and biology, water-soluble polymers are widely applied as blood plasma substitutes, for stabilization and purification of enzymes, an for programmed release of pharmaceuticals.

The book contains an overview of the physico-chemical properties of solutions of synthetic water soluble polymers and copolymers, polyacids, polybases, and non-ionic, hydrophilic macromolecular substances. The latest scientific work of recent years is dealt with, especially concerning molecular parameters and conformational changes of polymers.

Hüthig & Wepf
Scientific Publishers
Eisengasse 5
CH-4001 Basel

Hüthig & Wepf
Publishing Ltd.
212, North Barry Av.
Mamaroneck, N. Y.
10543-2826

Hüthig & Wepf Verlag
Postfach 10 28 69
6900 Heidelberg 1

Die Makromolekulare Chemie

Hüthig & Wepf

The journal's recent history began in 1939, when Prof. Hermann Staudinger assumed the editorship of the wellknown German review "Journal für Praktische Chemie" (journal for practical chemistry) and made it into the first publication dedicated mainly to macromolecular chemistry. After several changes, it was given its present title in 1947. Prof. Hermann Staudinger, Nobel Prize laureate in chemistry, was one of the founders of the science of macromolecular chemistry.

The titles of the journal and its sister publications have become an internationally recognized byword for one of the foremost sources of information on polymer science. Due to their international appeal, they are now edited almost exclusively in the English language.

Die Makromolekulare Chemie/ Macromolecular Chemistry and Physics
Founded by Hermann Staudinger continued by Werner Kern
Editor-in-chief:
Hartwig Höcker (Aachen)
ISSN 0025-116X

"Die Makromolekulare Chemie/Macromolecular Chemistry and Physics" contains contributions dealing with basic research in the field of chemistry, physical chemistry, and physics of macromolecules, including biopolymers.

The subscription includes 12 monthly issues plus 12 issues of **"Rapid Communication"**, a separate journal which offers highly important and most recent developments in the field of macromolecular science, published at very short notice.

Die Angewandte Makromolekulare Chemie/Applied Macromolecular Chemistry and Physics"
Editor-in-chief:
Dierich Braun (Darmstadt)
ISSN 0003-3146

"Die Angewandte Makromolekulare Chemie/Applied Macromolecular Chemistry and Physics" deals with macromolecular substances and polymers of practical interest. It includes fundamental and applied research on synthesis, modification, characterzation, analysis, structure, and properties of these substances. It covers all types of chemical reactions leading to or based on macromolecules including experiments, kinetics, mechanisms and technical processes. Other topics are chemical and physical properties of synthetic, semisynthetic, and naturally occurring polymers as well as the fundamentals of practical application.

Thus this publication is of great interest to chemists and physicsts in the area of functional polymers, engineering plastics, and additives to polymers, lacquers and coatings.

"Die Angewandte Makromolekulare Chemie/Applied Macromolecular Chemistry and Physics" is issued 10 times a year, each volume containing about 200 pages.

Die Makromolekulare Chemie/ Macromolekulare Symposia
Editor-in-chief:
Hartwig Höcker (Aachen)
ISSN 0258-0322

"Die Makromolekulare Chemie/Macromolecular Symposia" was first issued as a magazine in 1986 and is published approx. 10 times a year. It contains presentations in the field of macromolecular chemistry made at selected international meetings, especially those of the IUPAC. Papers and articles carried here will not be found in the same form in any other publication.

Hüthig & Wepf Scientific Publishers
Eisengasse 5
CH-4001 Basel

Hüthig & Wepf Publishing Ltd.
212, North Barry Av.
Mamaroneck, N. Y.
10543-2826

Hüthig & Wepf Verlag
Postfach 10 28 69
6900 Heidelberg 1

Hüthig & Wepf

P. Sigwalt, N. Spassky, H. Sekiguchi (Ed.)

Ring-Opening Polymerization

Lectures presented at the IUPAC Fifth International Symposium on Ring-Opening Polymerization, 1986 Blois, France

1987, 330 pp., hard cover,
DM 117,—
ISBN 3-85739-206-1

This book cantains the main lectures given at the 5th International Symposium on Ring-Opening Polymerization, held in Blois, France, June 1986 by 14 leading scientists from nine countries.
This edition is a milestone marking the present state of development of the chemistry of Ring-Opening Polymerization and thus a useful tool for all scientists interested in this field. In addition, it is a very helpful source of information because it contains 640 literature references related to this field.
The following four main topics have been treated during this symposium:

1. Polymerization of heteroycles and cycloolefins: Kinetics and mechanismus
2. New polymers: Tailor-made polymers an copolymers
3. Industrial polymers: Biomedical polymers and pharmaceutical materials: Syntheses and applications
4. Structure and physical properties oft the polymers.

Hüthig & Wepf
Scientific Publishers
Eisengasse 5
CH-4001 Basel

Hüthig & Wepf
Publishing Ltd.
212, North Barry Av.
Mamaroneck, N. Y.
10543-2826

Hüthig & Wepf Verlag
Postfach 10 28 69
6900 Heidelberg 1

UNIVERSITY OF WARWICK LIBRARY

3 DAY LOAN

Fines are due on all overdue items.
Telephone renewals: 024 7652 4438

DATE OF RETURN

1/5/08		

UW 9809461 0